# Creep Effect and Prediction Method of Dynamic Disaster of Surrounding Rock

Zhijie Wen · Jian Tao · Zhenqi Song · Yujun Zuo

# Creep Effect and Prediction Method of Dynamic Disaster of Surrounding Rock

Springer

Zhijie Wen
Mining College
Guizhou University
Guiyang, Guizhou, China

Jian Tao
Mining College
Guizhou University
Guiyang, Guizhou, China

Zhenqi Song
College of Energy and Mining Engineering
Shandong University of Science
and Technology
Qingdao, Shandong, China

Yujun Zuo
Mining College
Guizhou University
Guiyang, China

ISBN 978-981-96-7737-5    ISBN 978-981-96-7738-2 (eBook)
https://doi.org/10.1007/978-981-96-7738-2

This work was supported by National Natural Science Foundation of China (51974174), Shandong Province Outstanding Youth Fund (ZR2019YQ26) and Shandong Province College Youth Talent Introduction and Cultivation Support Program. Guizhou Science and Technology Program (Grant No.: LH[2024]-026).

© The Editor(s) (if applicable) and The Author(s) 2025. This book is an open access publication.

**Open Access** This book is licensed under the terms of the Creative Commons Attribution-NonCommercial-NoDerivatives 4.0 International License (http://creativecommons.org/licenses/by-nc-nd/4.0/), which permits any noncommercial use, sharing, distribution and reproduction in any medium or format, as long as you give appropriate credit to the original author(s) and the source, provide a link to the Creative Commons license and indicate if you modified the licensed material. You do not have permission under this license to share adapted material derived from this book or parts of it.

The images or other third party material in this book are included in the book's Creative Commons license, unless indicated otherwise in a credit line to the material. If material is not included in the book's Creative Commons license and your intended use is not permitted by statutory regulation or exceeds the permitted use, you will need to obtain permission directly from the copyright holder.

This work is subject to copyright. All commercial rights are reserved by the author(s), whether the whole or part of the material is concerned, specifically the rights of translation, reprinting, reuse of illustrations, recitation, broadcasting, reproduction on microfilms or in any other physical way, and transmission or information storage and retrieval, electronic adaptation, computer software, or by similar or dissimilar methodology now known or hereafter developed. Regarding these commercial rights a non-exclusive license has been granted to the publisher.

The use of general descriptive names, registered names, trademarks, service marks, etc. in this publication does not imply, even in the absence of a specific statement, that such names are exempt from the relevant protective laws and regulations and therefore free for general use.

The publisher, the authors and the editors are safe to assume that the advice and information in this book are believed to be true and accurate at the date of publication. Neither the publisher nor the authors or the editors give a warranty, expressed or implied, with respect to the material contained herein or for any errors or omissions that may have been made. The publisher remains neutral with regard to jurisdictional claims in published maps and institutional affiliations.

This Springer imprint is published by the registered company Springer Nature Singapore Pte Ltd.
The registered company address is: 152 Beach Road, #21-01/04 Gateway East, Singapore 189721, Singapore

If disposing of this product, please recycle the paper.

# Preface

Coal is an important material basis for the survival and development of human society, and is closely related to the national economy, social livelihood and national strategic competitiveness. At present, the world energy pattern has been profoundly adjusted, China's economic development has entered a new normal, and the supply-side structural reform has been continuously and deeply promoted. In recent years, the coal industry has accelerated the elimination of backward production capacity, orderly withdrawn from excess capacity, and gradually realized transformation and upgrading. Influenced by the national supply-side structural reform, the development of the coal industry is on a slowing down trend. Although the proportion of coal energy production and consumption has declined, coal still plays a pivotal role in the energy consumption structure. China's coal-rich, oil-poor and gas-poor energy endowment and the characteristics of the current level of national economic development have determined that the coal-dominated energy consumption structure will exist for a long time.

However, the reality of China's coal resources is particularly prominent, mainly in the following aspects: ① China's total proven coal resources account for 11% of the world's total, of which about $2.95 \times 10^{12}$ t belongs to the deep coal resources, accounting for 53% of the total coal resources. ② With the continuous mining of coal resources, shallow and deep coal resources have been mined out, and the resource reserves are getting smaller and smaller. ③ According to the statistics, the depth of coal mining is increasing by 8~12 m per year, especially in the eastern region, which has become the key area of deep mining. However, due to the complexity and specificity of the physical and mechanical environment of the deep strata, many new technical difficulties have appeared in the process of mining practice, such as the specificity of mine pressure manifestation, impact ground pressure, coal and gas protrusion, and large deformation of surrounding rocks. Therefore, it is urgent and necessary to meet and solve the new challenges and problems faced by deep coal resources mining. Realizing the safe and efficient mining of deep coal resources is related to the national economy and people's livelihood, and it is also related to the sustainable development of the coal industry.

The special engineering geological characteristics of deep coal rock body and its "three highs and one disturbance" geomechanical environment determine that the new problems faced by deep coal resource mining are significantly different from those faced by shallow coal resource mining. The development of new mining techniques, such as long-wall large-height working face rapid advancement complete set technology, leaving narrow and small coal pillars along the air excavation, and no coal pillar cutting top and leaving the roadway, etc., makes the popularization and application of large-scale mining possible. Consequently, the creep and drastic change of the peripheral rock in the large-scale quarry aggravate the creep instability of the deep coal rock body. The first cause of the dynamic instability of coal-rock bodies induced by deep coal resource mining, such as the large-area coming under pressure on the roof plate, the creep and large deformation of the surrounding rocks, the impact ground pressure, the coal and gas protrusion, and the rapid rheology of coal-rock bodies, is the geomechanical endowment environment of the deep coal-rock bodies. Reviewing the development history of coal industry, academician Zhenqi Song put forward the theory of "internal and external stress field" of mining field, practical mine pressure and rock layer control theory, and transferring rock beam theory, which summarized and revealed the model of mining field overburden structure evolution centered on overburden movement, and explored the evolution of mining field overburden structure centered on overburden movement. It summarizes and reveals the evolution model of overburden rock structure "centered on overburden rock movement", explores the movement law of overburden rock layer in the mining field, clarifies the distribution and manifestation of mine pressure and support pressure of surrounding rock, and realizes scientific prevention and control of coal-rock power disaster. Academician Minggao Qian put forward the theory of masonry beams and the theory of key layers, fully grasping the movement characteristics of key layers and laying a theoretical foundation for predicting the movement of overburden rock. It is based on the research results of many experts and scholars, especially the engineering experience of production line engineers and technicians. The authors elaborate on the relevant results on the theme of mining coal rock creep instability and prediction methods. Based on the basic concept of creep instability of coal rock body in quarry, this book puts forward the academic viewpoints and engineering practice of creep instability of large-scale peripheral rock structure in quarry, and reveals the development law of coal rock body in quarry from creep instability to drastic change and destruction. It is worth noting that, because the creep instability of coal rock bodies in quarries has different characteristics at different stages of development, the mechanism and control of creep instability of coal rock bodies in this book need to be explored in depth in future research.

This book focuses on the different processes, forms, characterization and prediction methods of the spatial-temporal effect, creep effect, decompression effect, dynamic impact effect and crushing effect of the creep instability to the drastic damage of the quarry coal rock body. The book is divided into 10 chapters, firstly, it introduces the basic mechanism and concept of coal rock body destabilization in quarry; secondly, it focuses on the principles and prediction methods of the creep destabilization of the roadway surrounding rock, large area roof, overlying rock layer of the mining area, and the roof of the synthesis mining face; secondly, it elaborates on the mechanism and prediction methods for the typical disasters such as impact pressure and coal and gas protrusion, etc., induced by the excavation of the deep coal rock body; and lastly, it elaborates on the monitoring and prediction of the creep destabilization of the coal rock body. Finally, the monitoring and prediction of coal rock body creep and instability are explained.

In this book the process of, writing first of all, special thanks to Prof. Deyun Zou; Shandong University of Science and Technology, School of Mining and Safety Engineering, academician Zhenqi Song, Prof. Yujing Jiang for improving the book research content to give pertinent advice; Shandong Energy Linyi Mining Group and other large coal enterprises of the relevant leaders and engineers and technicians for the book related to the enrichment of the information to give strong support; but also by the National Natural Science Foundation of China [52274130], Taishan Scholars Advantageous Characteristics of Disciplines Talents Team Support Program, Taishan Scholars Project of Shandong Province, Natural Science Foundation of Shandong Province (ZR2018MEE001), Qingdao Source Innovation Program (18-2-2-68-jch), Shandong Province University Scientific Research Program Project (Science and Technology) Key Project (J18K2010), and State Key Laboratory Open Fund (SHGF-18-13-30). We would like to express our sincere thanks to them for their hard guidance and help.

This book references and quotes the research results of many experts and scholars, formally their hard work in the early stages of this book, the smooth release of this book, thanks to their contribution to the development of science and technology for the coal industry wisdom and effort. Due to the limited level of the authors, and due to the complexity of the causes and mechanisms of creep destabilization of the coal body in the quarry, this book involves many concepts, ideas and understanding of the need for continuous improvement and development. Readers are kindly requested to criticize and correct this book.

Guiyang, China  Zhijie Wen
March 2025

**Competing Interests** The authors have no competing interests to declare that are relevant to the content of this manuscript.

# Contents

1 **An Introduction to the Principles and Prediction of Creep Stability of Coal Rock Bodies** .................................. 1
   1.1 Overview ................................................. 1
       1.1.1 Creep and Stability Issues ......................... 4
       1.1.2 Creep Destabilization and Classification of Coal Rock Bodies ...................................... 13
   1.2 Characteristics of Creep Destabilization of the Rock Layers Surrounding the Roadway .......................... 15
       1.2.1 Tunneling Construction ............................ 16
       1.2.2 Production Service Lanes ......................... 18
   1.3 Creep Stability Characteristics of Large Roof Slabs ........... 20
   1.4 Characteristics of Creep Destabilization of the Overlying Rock Layers in the Extraction Zone ........................ 21
   1.5 Characteristics of Creep Destabilization of Impacted Ground-Pressured Coal Rock Bodies ....................... 23
   1.6 Characteristics of Creep Destabilization in Gas-Abrupt Coal Seams ............................................... 23

2 **Principles and Prediction of Creep in the Perimeter Rock of the Quarry Roadway** ...................................... 27
   2.1 Overview ................................................. 27
   2.2 Creep and Phenomena of the Perimeter Rock of the Roadway ......................................... 29
       2.2.1 Mechanisms and Types of Perimeter Rock Damage in Deep Quarries ......................... 29
       2.2.2 Enclosed Rock Deformation and Damage Patterns and Basic Mechanisms ........................... 34
       2.2.3 Factors Affecting Perimeter Rock Stability ........... 37
       2.2.4 Physical Characteristics of Surrounding Rock and Control Program—Integration of Support Structure ......................................... 44

|  |  |  |  |
|---|---|---|---|
| | 2.3 | Mechanics of Anchor Mesh Cable Supported Structural Body ................................................. | 45 |
| | | 2.3.1 Anchorage Strength of Anchor Cable ............... | 46 |
| | | 2.3.2 Mechanical Effect of the Anchor Mesh Cable Structure Body on the Surrounding Rock ............ | 48 |
| | 2.4 | Monitoring Rationale and Methodology ..................... | 54 |
| | | 2.4.1 Alternative Information ........................... | 56 |
| | | 2.4.2 Evaluation of the Application of Monitoring Methods .......................................... | 57 |
| | | 2.4.3 Coal Pillar Support Pressure Monitoring ........... | 60 |
| | 2.5 | Application and Outlook for New Monitoring Technologies .... | 60 |
| | | 2.5.1 Acoustic Emission Monitoring Technology ........... | 60 |
| | | 2.5.2 Acoustic and Electromagnetic Wave Monitoring Technology ....................................... | 61 |
| | | 2.5.3 Acoustic Monitoring Applications ................... | 62 |
| **3** | **Principle and Prediction of Large Area Roof Plate Coming to Pressure in Working Face** ..................................... | | **67** |
| | 3.1 | Overview ................................................ | 67 |
| | 3.2 | Principles and Phenomena of Large-Area Roof Coming Under Pressure ........................................... | 68 |
| | | 3.2.1 Large Roof Coming Under Pressure and Dynamic Characterization ................................. | 68 |
| | | 3.2.2 Large-Area Top Plate to Pressure Principle ........... | 69 |
| | 3.3 | Monitoring Principles and Methods ......................... | 73 |
| | | 3.3.1 Monitoring Modeling ............................. | 73 |
| | | 3.3.2 Discriminative Method for Geological Survey Data .... | 74 |
| | | 3.3.3 Instrumented Real-Time Continuous Monitoring Methods .......................................... | 76 |
| | 3.4 | Techniques Faced by Large Roof Slabs Coming Under Pressure in the Working Face Problems ..................... | 80 |
| | 3.5 | Principles and Research Directions of Large-Area Roof Coming Under Pressure .................................... | 81 |
| **4** | **Principles of Creep in the Overlying Rock Layer of the Mining Zone and the Prediction of Ore Tremor** ......................... | | **85** |
| | 4.1 | Overview ................................................ | 85 |
| | 4.2 | Principles of Creep in the Overlying Rock Layers of a Large Mining Area .................................... | 87 |
| | | 4.2.1 Overlying Rock Layers and Their Movement Patterns in the Mining Area ....................... | 87 |
| | 4.3 | Mine Quakes Triggered by the Movement of Overlying Rock Layers in the Extraction Zone ....................... | 88 |
| | | 4.3.1 Mineral Earthquakes and Their Characterization in Old Mining Areas ............................... | 89 |

Contents                                                                          xiii

|  |  | 4.3.2 | Working Face Mechanism of Occurrence of Mining Earthquake in the Mining Area and Its Characterization .................................. | 90 |
| --- | --- | --- | --- | --- |
|  | 4.4 | Predictive Forecasting Methods and Case Studies ............. | | 93 |
|  |  | 4.4.1 | Acoustic Emission Monitoring and Fundamentals ..... | 93 |
|  |  | 4.4.2 | Ground Acoustic (or Geophonic) Monitoring Methods .......................................... | 95 |
|  | 4.5 | Prospect of Prediction Methods for Subsidence of Overlying Rock Layers in the Mining Area ................ | | 95 |
| 5 | **Principles and Prediction of Impact Ground Pressure in Coal Rock Bodies** .................................................. | | | 99 |
|  | 5.1 | Overview ................................................. | | 99 |
|  | 5.2 | Current Status of Research and Development on Impact Ground Pressure ......................................... | | 104 |
|  |  | 5.2.1 | Review of and Theoretical Development of the Understanding Impact Ground Pressure ........ | 107 |
|  |  | 5.2.2 | Development of Research Impact Ground Pressure Mechanisms ........................... | 107 |
|  |  | 5.2.3 | Review and Summary of Impact Ground Pressure Theory and Research Results ...................... | 109 |
|  | 5.3 | Twenty-first Century Impact Ground Pressure Gestation Mechanism and Creep Characterization ..................... | | 111 |
|  |  | 5.3.1 | Characterization of the Action of Effect the Shock ..... | 112 |
|  |  | 5.3.2 | Mechanism of Impact Ground Pressure and Impact Effect Equation ..................................... | 114 |
|  |  | 5.3.3 | Coal Rock Body Energy Storage Relationship with Impact Ground Pressure ...................... | 120 |
|  |  | 5.3.4 | Impact Pressure Triggers and Energy Conversion ...... | 122 |
|  | 5.4 | Predictive Forecasting Principles and Methods ................ | | 126 |
|  |  | 5.4.1 | Impact Propensity of Coal Rock Bodies .............. | 127 |
|  |  | 5.4.2 | Drill Chip Method ................................ | 128 |
|  |  | 5.4.3 | Shock Ground Pressure Acoustic Emission (Micro Seismic) Monitoring Method ...................... | 130 |
|  |  | 5.4.4 | Mineral Pressure Monitoring Method ................ | 133 |
|  |  | 5.4.5 | Extraction Stress Monitoring Method ................ | 134 |
|  |  | 5.4.6 | Electromagnetic Radiation Monitoring Method ....... | 135 |
|  | 5.5 | Outlook for Impact Pressure Prediction Research ............. | | 135 |
| 6 | **Principles and Prediction of Coal Seam Gas Outcrop Methods** .... | | | 137 |
|  | 6.1 | Overview ................................................. | | 137 |
|  | 6.2 | Principles and Phenomena of Coal Seam Gas Prolapse ......... | | 139 |
|  |  | 6.2.1 | Gas Highlighting Basic Factors and Energy ........... | 139 |
|  |  | 6.2.2 | Principles of Energy Transfer and Destruction in Coal Rock Bodies .............................. | 141 |

|  |  | 6.2.3 | Principles of Deformation(Elasticity)potential Measurement for Coal Rock Bodies ................ | 149 |
|---|---|---|---|---|
|  |  | 6.2.4 | Types of Coal and Gas Outcrops .................... | 150 |
|  |  | 6.2.5 | Prolapse of Coal and Gas at the Face ............... | 151 |
|  | 6.3 | Principles and Methods of Monitoring and Forecasting ........ | | 153 |
|  |  | 6.3.1 | Commonly Used Coal and Gas Outcrop Prediction Methods ......................................... | 153 |
|  | 6.4 | Principles for Predicting and Forecasting Impact Pressure and Salient Uniformity .................................... | | 154 |
|  |  | 6.4.1 | Impact Ground Pressure, Prominent Characterization Appearances ..................... | 154 |
|  |  | 6.4.2 | Unified the Mechanism of Impact Pressure and Protrusion Occurrence Understanding of ......... | 156 |
|  |  | 6.4.3 | Elastic Potential Energy in a Coal Rock Body Equation for the Impact Effect of .................... | 158 |
|  |  | 6.4.4 | Uniform Prediction and Forecasting Methods for Impact Pressure and Protrusion .................. | 159 |
|  |  | 6.4.5 | Analysis of the Direction of Development of Predictive Forecasting .......................... | 164 |
| 7 | **Principles of the Roof Slab Coming Under Pressure and Creeping Instability in the Integrated Mining Face** ........... | | | **167** |
|  | 7.1 | Overview .............................................. | | 167 |
|  | 7.2 | Recognition of the Subsidence of the Overlying Rock Formation and the Pattern of Pressure Coming from Its Top Plate ............................................... | | 168 |
|  |  | 7.2.1 | Basic Characteristics of Subsidence Movement of the Overlying Rock Layer in the Mining Area ...... | 169 |
|  |  | 7.2.2 | Academic Ideas and Current Developments ........... | 171 |
|  | 7.3 | Mechanics and Characterization of Subsidence Motion in Overlying Rock Formations ............................ | | 172 |
|  |  | 7.3.1 | Principles of Overlying Rock Movement and Changes in Subsidence Patterns and Phenology .... | 173 |
|  |  | 7.3.2 | The Movement of the Overlying Rock Formation and Interaction Between the Working Resistance of the Support ..................................... | 178 |
|  |  | 7.3.3 | Integrated Mining Face Roof Plate to Pressure Law and Hollow Area Subsidence Collapse Characterization Appearance ...................... | 189 |
|  | 7.4 | Characteristics and Technical Management of Shallow Buried Mining Roof to Pressure Manifestation ............... | | 190 |
|  | 7.5 | Influence of Filling of the Mining Airspace on the Pattern of Coming Pressure on the Roof Plate ...................... | | 192 |

|     | 7.6 | Determination of the Strength of the Top Plate to Pressure and the Selection of the Synthesized Mining Bracket .......... | 192 |
|     |     | 7.6.1 Strength of the Top Plate Coming Under Pressure and the Strength of the Support of the Mining Support ........................................... | 193 |
|     |     | 7.6.2 Initial Support and Working Resistance .............. | 199 |
| 8   | **Coal Rock Body Energy Principal Analysis Method** .............. | | 201 |
|     | 8.1 | Energy Balance Guidelines ............................... | 202 |
|     |     | 8.1.1 Single-Degree-of-Freedom Structural Systems ........ | 202 |
|     |     | 8.1.2 Multi-degree-of-Freedom Structural Systems ......... | 204 |
|     |     | 8.1.3 Combined Structural Systems ...................... | 206 |
|     | 8.2 | Determination of Critical Load by the Energy Method ......... | 214 |
|     | 8.3 | Nonlinear Stability Theory ............................... | 215 |
|     |     | 8.3.1 Symmetric Stable Branching Point Problems ......... | 216 |
|     |     | 8.3.2 Symmetric Unstable Branching Point Problems ....... | 220 |
|     |     | 8.3.3 Jumping Phenomena in Flat Arches ................. | 226 |
|     | 8.4 | Coal Rock Body Energy Principles ........................ | 229 |
| 9   | **Quarry Mine Pressure Monitoring Technology and Applications** ............................................. | | 237 |
|     | 9.1 | Research Methodology for Mine Pressure Monitoring Summary ............................................... | 237 |
|     | 9.2 | Purpose of Mine Pressure Monitoring in Quarries and Current Status of Technology Development .............. | 238 |
|     |     | 9.2.1 Purpose and Role of Mine Pressure Monitoring Studies ........................................... | 238 |
|     |     | 9.2.2 Current Status of Technology Development .......... | 241 |
|     | 9.3 | Information on Mine Pressure Observation Data for the Two Roadways in the Quarry ...................... | 243 |
|     |     | 9.3.1 Arrangement of Mine Pressure Monitoring Zones in the Two Runways of the Quarry .................. | 243 |
|     |     | 9.3.2 Dynamic Monitoring Curve Application for Cave Surrounding Rock ................................ | 244 |
|     | 9.4 | Integrated Mining Support Pressure Monitoring Methods ...... | 246 |
|     |     | 9.4.1 Bracket Pressure Monitoring Measurement Area Arrangement ..................................... | 246 |
|     |     | 9.4.2 Methods of Analyzing Pressure Curves for Integrated Mining Supports ..................... | 248 |
|     | 9.5 | Monitoring Equipment Performance Classification and Applications ........................................ | 249 |
|     |     | 9.5.1 Lane Displacement and Deformation Monitoring Instrumentation and Applications ................... | 249 |
|     |     | 9.5.2 Lane Roof Surrounding Rock Pressure Monitoring Instrumentation and Application .................... | 257 |

|  |  | 9.5.3 | Bracket Pressure (Operating Resistance) Monitoring Instruments and Applications | 261 |
|---|---|---|---|---|
|  |  | 9.5.4 | Real-Time Online Monitoring Systems and Applications | 265 |
|  | 9.6 | | Coal Rock Mass Catastrophe Prediction Methods | 275 |
|  | 9.7 | | Creep Monitoring Equipment Development Directions for Coal Rock Bodies | 280 |

**10 Principles and Applications of Acoustic Monitoring of Coal Rock Bodies** ... 283
- 10.1 Introduction to Coal Rock Acoustic Monitoring Technology ... 283
- 10.2 Laws of Sound Wave Propagation in Coal Rock Bodies ... 284
  - 10.2.1 Acoustic Waves in Coal Rock Bodies ... 284
  - 10.2.2 Reflection, Refraction, and Waveform Transformation of Acoustic Waves in Coal Rock Bodies ... 286
  - 10.2.3 Propagation of Sound Waves in Relation to Coal Rock Body Structure ... 288
  - 10.2.4 Acoustic Wave Bypassing and Scattering in Coal Rock Bodies ... 291
  - 10.2.5 Correlation of Coal Rock Body Traits with Acoustic Velocity and Applications ... 292
  - 10.2.6 Acoustic Frequency and Received Signal Frequency in Coal Rock Bodies ... 296
  - 10.2.7 Acoustic Emission Phenomena and the Kaiser Effect ... 297
  - 10.2.8 Acoustic Emission Signal Detection and Processing ... 299
  - 10.2.9 Relationships Between Coal Rock Acoustic Emission and Rock Stress ... 300
- 10.3 Creep Destabilization Signals of Coal Rock Mass and Monitoring Principles ... 302
  - 10.3.1 Acoustic Emission Principles and Monitoring of Coal Rock Bodies ... 303
  - 10.3.2 Acoustic Emission Monitoring System ... 310
- 10.4 Acoustic Detection Technology of Rock Anomaly Defects and Loose Zones ... 318
  - 10.4.1 Electromagnetic Radiation Monitoring Techniques ... 318
  - 10.4.2 Ultrasound Detection Technology ... 319
  - 10.4.3 Technology for Ground-Penetrating Radar Monitoring Applications ... 323
  - 10.4.4 Principles of Elastic Wave CT Technology ... 328
  - 10.4.5 Acoustic Monitoring Technology Use Case Profile ... 330

**References** ... 335

# Chapter 1
# An Introduction to the Principles and Prediction of Creep Stability of Coal Rock Bodies

## 1.1 Overview

What is the mine pressure? For a long time in the past, we used to call the pressure of surrounding rock due to deformation and falling rock on the support as mine pressure (mine pressure for short), which is the concept of traditional "narrow mine pressure", is incomplete, and is only a small part of mine pressure. With the continuous deepening and development of mining rock mechanics research, people have gradually realized that, between the perimeter rock roof and anchor network cable support, roof and stent is not the relationship between the load and support configuration, but a closely linked interaction, mutual influence and common bearing structural relationships (such as anchor network cable support, trellis beam support, comprehensive mining support, monolithic hydraulic pillars and other support structures development and application). In contrast, the comprehensive mining support and anchor net cable support have improved the stress and displacement state of the surrounding rock extremely well, and the sufficient support force and anchoring force have given full play to the coupling effect between the surrounding rock and the supporting structures, so that they can actively and proactively maintain the stability of the quarry and the roadway. Fundamentally speaking, geostress is the fundamental source of all underground engineering, including mine quarry, roadway mine pressure manifestation. Geopressure is the original geopressure that exists in the coal rock seams, also known as the original rock stress. In the absence of mining engineering disturbance, the rock body is in the original rock stress state. The underground space effect formed by the extraction of mine tunnel or quarry breaks the equilibrium state of the original rock stress. With the passage of time, the ground stress will be reduced, the elastic energy is released, which causes the coal rock body to move to the free surface (or space) and then deformation and damage, causing the original rock stress redistribution. This change process reflects the time and space effect of the surrounding rock deformation. Excessive displacement and stress concentration of the surrounding rock leads to local or overall instability and destruction of the rock body, which is the process and mechanism of mine pressure formation. Therefore, mine pressure is the mechanical effect of the coal rock body due to mining disturbance. It is related to the stress state of the coal rock body, the structure of the coal rock body, the physical and mechanical properties of the coal rock body, the engineering geological conditions, the time and space (also known as space–time effect) effect and other factors.

"Mine pressure manifestation can be manifested in various forms, such as deformation of coal rock bodies, micro or macro damage, movement and deformation of coal rock bodies, surface subsidence, decompression and expansion deformation of coal rock bodies, roadway undercutting, ganging, roofing, sectional shrinkage, damage to braces, collapse of quarry and so on. The form and degree of mine pressure are controlled and determined by the factors mentioned above. Mine pressure research content and engineering technology, including the formation of bulk mine pressure, deformation of the roof surrounding rock pressure, impact pressure and gas protrusion and other powerful impact effects and coal rock body creep effect and the formation of the expansion of the mine pressure.

1. Loose body ore pressure also known as loose ore pressure

The distinctive feature of this kind of coal rock body is the organization of loose structure, small volume porosity, disturbed by the local surrounding rock is very easy to produce broken, resulting in a large number of fragments collapsed to form a loose coal rock body on the support body pressure, which is the traditional "narrow sense of mining pressure" concept can be adapted to the only object. In this extreme case, the coupling between the support and the surrounding rock is lost because the support has been completely detached from the surrounding rock. Mineral pressure control must avoid this by preventing, as far as possible, damage to the surrounding rock from progressing to the point where it becomes loose and collapses. In many cases, but also to take the anchor reinforcement and other techniques to make the loose surrounding rock solidification or net control, to achieve again alone or with the support together with the role of common bearing.

2. Distortionary pressure

The pressure caused by the displacement of surrounding rock due to mining is the most basic form of mine pressure. Deformation mine pressure can be divided into elastic deformation mine pressure, plastic deformation mine pressure, elastic–plastic deformation mine pressure, rheological deformation mine pressure and so on according to the nature of the coal rock body and the type of deformation. In the case of good strong coal rock body, the displacement and deformation of the surrounding rock develops to a certain degree and stops, and the surrounding rock itself may be able to maintain stability without support. However, in most cases, the surrounding rock must be supported to prevent damage caused by excessive deformation. At this time, the deformation of the mine pressure characteristics and support methods and support structure are closely related. In the surrounding rock and support for the integration of conditions, the surrounding rock and support constitute a common carrier, they are interdependent, with mutual constraints, common deformation. If

## 1.1 Overview

the support measures are not taken in time or the support method is not appropriate, it cannot effectively improve the surrounding rock stress distribution state, inhibit the deformation of the surrounding rock, and then the surrounding rock will be damaged, collapse, and form the "loose mine pressure".

3. Pressure bumping and gas herniation

It is a kinetic phenomenon of coal rock body (protrusion is dominated by gas dynamics), which is a form of sudden release of a large amount of elastic deformation energy gathered in the coal rock body under certain incentives. When impact pressure or gas protrusion occurs, accompanied by a loud noise, the coal rock body in the form of particles and blocks and dust (or dust and gas) burst out, rushing to the roadway or quarry with great speed. Impact ground pressure and gas protrusion are mainly related to two factors. One factor is the structural nature of the coal rock body, with high strain energy (including gas energy) stored in the coal rock body's inherent conditions. Generally speaking, a hard and intact coal rock body is easy to store high-strain energy. Another factor is the external environment that generates high strain energy, such as the place where the ground stress is high and the surrounding rock stress is concentrated. With the increasing mining depth and ground stress, impact ground pressure or gas protrusion is more likely to occur in the deeper part of the stratum. Since the structural nature of coal rock body and ground stress are natural attributes of mining engineering, the main way to prevent and control impact ground pressure is to avoid unreasonably high stress concentration (i.e., to reduce the concentration of elastic energy) and all kinds of external factors inducing impact.

4. Expansion of the mineral pressure

The compressive stress damage phenomenon produced by the expansion and deformation of the coal rock body, the coal rock body expansion mineral pressure mentioned here is different from the principle of decompression expansion displacement deformation of the coal rock body. Expansion of the formation of the causal factors can be divided into two aspects: ① the composition of the rock contains swelling minerals and water-related, showing obvious hydrophilicity, water absorption, sludging and softening, mainly in the expansion of the rock body; ② the hardness coefficient of rock mass is very small ($f = 0.1 \sim 0.5$), due to the mineral lithology related to the mining space of the peripheral rock roof by the small mining disturbance and the spatial effect of creep, the formation of decompression and expansion of the resulting decompression and deformation of the coal rock body. After the creep effect induced by small mining disturbance and space effect, decompression and expansion are formed and continuous fall occurs. The size of the expansion pressure is related to the strength of its expansion characteristics, that is, it is directly proportional to the amount of expansion minerals. Expansion of the pressure of another important condition is the role of water or humid air; expansion of the rock body can only be expanded in contact with water. Therefore, one of the effective ways to prevent and control the expansion of mineral pressure is to weaken or control the influence of water on the expansion of the rock body.

## 1.1.1 Creep and Stability Issues

### 1.1.1.1 Creep Problems

The surface deformation (and deep displacement) and compressive stress of the coal rock body in the quarry are greatly influenced by the time factor, i.e., the role of spatial effects. Under certain test conditions, when the load (or stress) is kept constant, the deformation (or displacement) of the coal rock body with the development of time is called the creep phenomenon, which is a concept given based on the test results of laboratory specimens of very small-scale rock bodies. At present we consider this to be a relatively narrow concept. The creep phenomenon of large-scale engineering coal rock body should be understood as the period (or stage) of extremely slow development and change of stress and strain, such as deformation of roadway perimeter rock, roof to pressure, deep displacement and other phenomena of the initial changes can be understood as creep phenomenon.

According to classical mechanics, Hooke's law is the law of stress and deformation that we pointed out. Another type of creep phenomenon refers to the softening and expansion of a rock mass due to damping by water, which is collectively called soft rock. When the strain is constant, the process by which the stress in the rock mass decreases over time is called relaxation. This phenomenon, which occurs inside the coal rock body at this time, can be interpreted as the elastic energy stored in the coal rock body is converted into thermal energy, which gradually disappears with the passage of time. Therefore, our understanding of the complete destruction process of coal rock body can be understood as: early creep → late instability destruction. At present, the creep (or relaxation) properties of coal rock bodies are usually described by the equation of state with stress and strain versus time derivatives. A variable number of spring elements used for coal rock body properties, $K$, and piston elements, $\eta$, are used to describe the coal rock body creep model. A typical creep curve of a coal rock body is shown in Fig. 1.1. The test procedure of the creep curve is as follows: the rock specimen is kept under a constant pressure for a long period of time, then the deformation of the rock body can be observed to increase with time, and at the beginning of the pressurization, there is an instantaneous strain $U$. The figure by the is represented segment, which includes the elastic and plastic strains that are mainly characterized by elasticity. Because the action time of this segment is very short, so it can be considered time-independent. Creep development into the *OA AB* section (i.e., I region), the strain continues to increase, but the strain rate is decreasing, the curve is downward concave curve, this stage of creep is called the first stage of creep or transient creep. Between (i.e., region II), the strain develops at a constant rate, which is called the second stage of creep or constant creep, compared with this stage of the longest time. *B* and *C* After point (i.e. region III), the strain develops at an accelerated rate and the curve curves upward concavely, and this stage of creep is called the third stage of creep or accelerated creep, i.e., the drastic change stage. When the strain reaches a certain value $D$, then the leap (or drastic) instability damage or broken. The engineering physics meaning of the rock creep curve shows that the destruction

## 1.1 Overview

**Fig. 1.1** Principal curve of creep development in coal rock bodies

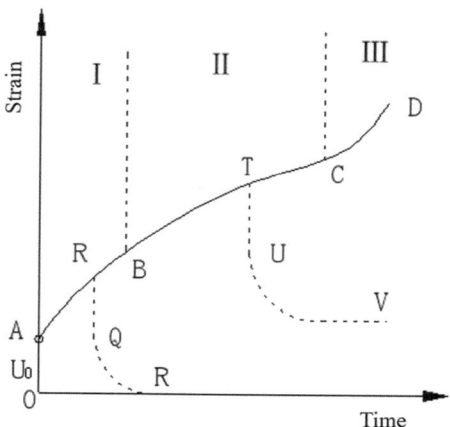

of coal rock body there is a creep development of different processes or stages, the combination of the process, stage tendency and range of the development of the degree and characterization of the different, the destruction is only the final result. The characterization of the different stages represents the tendency of the change of the coal rock body's properties and the temporary state.

If at the first stage of unloading (before point B), the recovery of rock mass deformation will gradually return to the original size along the route of the *PQR* curve, so the creep at this stage is elastic creep. The *PQ* segment represents the immediate recovery of deformation (fully elastic characteristics), while *the QR* segment represents a gradual return to its original state without permanent deformation. This elastic deformation after loading and immediately all occur, but with the growth of time gradually occur, so after unloading the deformation is not immediately all disappear, but gradually disappear the characteristics of hysteretic elasticity. If the second stage is unloaded, the strain will be restored along the curve TUV route, in which the TU section is immediately restored and the UV section is gradually restored, but it cannot be restored to the original size, and the rock mass retains a part of the permanent deformation. The two recovery curves of PQR and TUV show that the mechanical state of coal rock mass can control the deformation and deterioration of the surrounding rock through active support. As shown in Fig. 1.1, the deformation of each stage of the curve can be expressed by the curve drawn by the typical function equation. The creep curve equation is obtained by the least squares regression analysis using the experimental results, such as the power function equation and the exponential function equation.

Coal rock bodies are formed by crystal particles of different compositions, properties, sizes and shapes through a long process of sedimentary cementation, which in turn produces mechanical properties such as cohesion or strength. Therefore, the mechanism of creep formation is the change of fissure produced by the coal rock body after the impact action of stress or elastic potential energy. If this impact continues for a certain period of time, the cracks within the coal rock body will gradually open, propagate and converge. When a crack is formed, the stress (elastic energy) in the highly concentrated stress region will be transferred to the low stress intensity region (i.e., low-energy region) that is easy to transfer the release of this elastic energy, and this change in stress causes the internal expansion and displacement of the rock body to the deformation of the mining space. In other words, the creep criterion of quarry coal rock (stress or elastic potential energy or form) is: internal decompression and expansion, resulting in cracks and fissures to the surface deformation, and the formation of perimeter loosening ring is exactly this principle.

The study and understanding of the creep properties (or more broadly the rheological properties) of coal rock bodies is very important for the prediction and control of coal rock body catastrophes in mining sites. The creep process of coal rock body has unpredictability, which occurs inside the coal rock body structure and its artificial structure inside the stress effect or elastic energy impact effect and affect the stability of its inside to outside. Such as mining perimeter rock anchor network cable support structure, with the extension of time and performance changes, can make the anchor network cable structure of the anchoring force decline or functional failure; newly excavated bare road perimeter rock loosening circle formation there is also a time effect problem; coal rock body internal or deep tectonic changes and fault movement is a creep process; there is a tendency to impact or gas protruding coal seam of the rock door uncovering process, with the excavation progressively close to the coal seam With the excavation progressively approaching the coal seam, a very high elastic potential energy is gradually formed in the coal seam to transfer and accumulate in the direction of excavation, which is also a creep development process.

### 1.1.1.2 Stabilization Issues

Engineering is often encountered in the pressurized rods, such as lifting screw studs, crank linkage mechanism in the connecting rod, machine tool screw, piston compressor piston rod, diffraction frame structure of the pressure rod, this type of pressurized rods are often due to the stability of the problem of loss of load-bearing capacity, the knowledge of this aspect of the mechanics of materials textbooks can be seen. In fact, in civil engineering, water conservancy, mining, machinery, petrochemical, aerospace, biomedical and other projects there are stability problems. For example, the water tower, rigid frame, roof alley gangs, shell roofs, stents, aircraft and other structures, in many projects if not properly considered will also be due to instability and loss of load-bearing capacity, this kind of engineering accidents in many fields. With the development of engineering technology, the stability problem has attracted more and more attention, has become an important branch of mechanics, in a wide range of applications in the process of continuous development, and has become the theoretical basis of many engineering technologies.

The relative distance between any two points of mass within an object should change when the external force on the object changes, and this class of objects is

## 1.1 Overview

known as deformable solids or simply variable objects. For variable objects, two basic assumptions are usually introduced:

(1) Assumption of continuity: the assumption that an object is dense, filled with medium everywhere in its entire volume, and has no voids. The elasticity and deformability (or elasticity) of coal rock bodies should be higher than those possessed by rocks; the tensile strength of coal rock bodies should be greater than that of rocks.
(2) Homogeneity assumption: the mechanical properties are considered to be identical at all points. The cohesion (or bonding) and strength of the same rock layer are the same.

For the continuous uniform deformable body, the external force and deformation are related by the constitutive relation. If the loading path described by the constitutive relation is the same as the unloading path, the above object is called a continuous and uniform elastomer. If these paths are described by a linear relationship (generalized Hooke 's law), then such an elastomer is called a linear elastomer. In addition, if the mechanical properties of the object in all directions are the same, such an elastic body is called an isotropic body. The object of mining engineering research is mainly a continuous and uniform elastic body. Both elastic body and linear elastic body should consider the problem of elastic stability or balance.

The problems that can be solved by using the basic equations of elastic mechanics in mining engineering are quite limited. When dealing with structural problems in engineering, some assumptions have to be made to simplify the problem. It should be pointed out that the choice of simplified assumptions depends largely on the relative dimensions of structural elements in three-dimensional space. Based on the spatial coordinates of structural elements (mechanical structural elements), structural elements in engineering can be attributed to the following four categories:

(1) The dimensions in all three directions have the same magnitude (e.g., sphere, short or medium-length cylinder).
(2) The dimensions in one direction are much larger than the dimensions in the other two directions that have the same magnitude (e.g., rods, columns, beams).
(3) The dimensions in one direction are much smaller than the dimensions in the other two directions having the same magnitude (e.g., thin plates, thin shells).
(4) The dimensions in one direction are much larger than the dimensions in the other two directions that have different magnitudes (e.g., thin-walled beams of section).

Structural elements of category do not have stabilization problems, all other structural elements have stabilization problems.

If the particle is subjected to a certain force at any position in space, and the magnitude and direction of the force are determined by the position of the particle, the force is called the field force. If the work done by the field force acting on the particle is related to the starting and ending position of the particle, and has nothing to do with the motion path of the particle, the field force acting on the particle is called force or conservative force. A mechanical system with constant constraints under the

**Fig. 1.2** Schematic representation of the three basic equilibrium forms

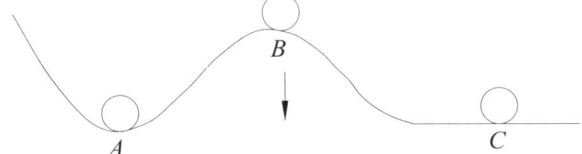

action of conservative force is called a conservative system. The so-called constant constraint means that the mathematical expression of the constraint equation, that is, the constraint condition, does not include time t. The pressure and support in mining engineering, the transfer and release of elastic energy in coal and rock mass have the characteristics of conservative force and unsteady constrained mechanical system. In the conservative system, the force can be expressed by the function of system potential energy, and the stability problem in the conservative system is concerned in the engineering. Taking the ball shown in Fig. 1.2 as an example, the three basic equilibrium forms and concepts of the object are revealed.

A ball weighing g is at rest at the disjoint points A, B, and C of a surface which has zero curvature along the direction normal to the figure. Points A, B, and C on the surface with zero slope represent static equilibrium positions, and the equilibrium characteristics of these points are essentially different: at point A, if the ball is slightly perturbed (small displacement, small velocity), it will swing back and forth around the equilibrium position A, and finally remain at rest at point A. This kind of equilibrium is known as a stable equilibrium; at point B, if the ball is slightly perturbed, it will tend to move away from the equilibrium position B, and it cannot come back to its original equilibrium position. At point B, if the ball is slightly perturbed, it will tend to leave the equilibrium position B, and cannot return to the original equilibrium position, this equilibrium is called as unstable equilibrium; at point C, if the ball is slightly perturbed, it will tend to remain in the perturbed position, this equilibrium is called with the equilibrium or neutral balance. It is worth noting that the three forms of equilibrium mentioned above: stable, unstable, and entourage equilibrium are all for small scales, i.e., the above definitions depend on tiny perturbations. If the perturbation is not tiny, it may happen that the sphere is unstable for a small range but stable for a large range, as at point B in Fig. 1.3a, or that the sphere is stable for a small range but unstable for a large range, as at point A in Fig. 1.3b.

The equilibrium and stability problems shown in Figs. 1.2 and 1.3 are shown as the equilibrium and stability problems of elastic systems. The equilibrium state of elastic systems in engineering can be summarized into three forms: stable equilibrium, unstable equilibrium and contingent equilibrium. If the elastic system can return to its original equilibrium position or has a tendency to return to its original equilibrium position after slightly deviating from its equilibrium position, the original equilibrium state is called a stable equilibrium state. If it continues to deviate, it is called an unstable equilibrium state. At this time, the elastic system loses stability, referred to as instability or buckling. The accompanying equilibrium state is usually an intermediate state of transition from stable equilibrium to unstable equilibrium.

# 1.1 Overview

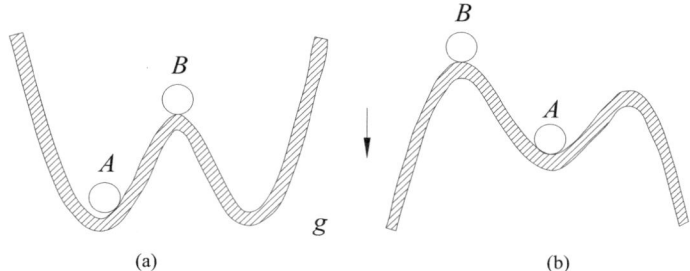

**Fig. 1.3** Schematic diagram of the principle of equilibrium of an elastic system

The equilibrium stability of an elastic system depends on the loading mode, geometric characteristics of the structure, constraints and other factors, and its forms of instability can be categorized into the following four types.

(1) Branch point destabilization. By the mechanics of materials can be known, for the axial compression of the ideal equal straight rod (see Fig. 1.4a), when the load P1 is less than the critical value of Pcr, the compression of the rod to impose a small interference to make it bend. Once the interference is withdrawn, the rod necessarily restores the original straight line equilibrium form, i.e., the straight-line form is the only equilibrium form before the instability, at which time the straight-line equilibrium form is stable. With the rod midpoint deflection f as the horizontal coordinates, axial pressure P as the vertical coordinates, then any point P1 on the vertical axis (f = 0) represents a straight-line equilibrium state, we call the OA line for the original equilibrium path, as shown in Fig. 1.4b. When the load exceeds Pcr to reach P2, then the lever can be a straight line state can also bear the bending state, but the straight line state is unstable, in other words, if the lever is subjected to minor interference, it cannot restore the straight line state and will continue to bend to the point B in Fig. 1.4, at this time the value of the deflection is f2, the curve AB is called the second equilibrium path. The intersection points A of the original equilibrium path and the second equilibrium path is called the branch point, and the load corresponding to this point is called the critical load Pcr, and the state corresponding to this point is called the critical state. The equilibrium state corresponding to before the critical state is called the pre-buckling equilibrium state, while the equilibrium state after exceeding the critical state is called the post-buckling equilibrium state. In Fig. 1.4, since the second equilibrium path has a horizontal tangent at the branch point A, the deflection of the compression rod cannot be determined in a neighborhood of first-order infinitesimals, which means that the same load can correspond to an arbitrary form of tiny bending equilibrium, which is the case of as-you-go equilibrium.

If the coordinate parameters are transformed so that the shortening of the compression rod $\Delta$ is the horizontal coordinate, the corresponding graph as shown by the curve OAC in Fig. 1.5 corresponds to the branch-point instability of the column in Figure (b) under the action of the pressure P. The curve in Fig. 1.5 $OA'C'$ shows the corresponding graph of Figure (a) for the branch point instability of a rectangular thin

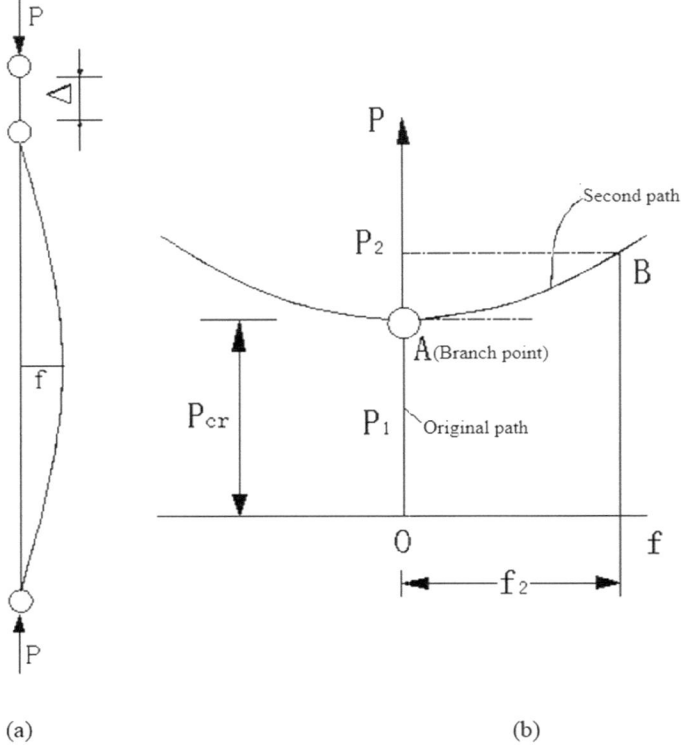

**Fig. 1.4** Branch point destabilization and follow-on equilibrium

plate under in-plane pressure Nx, where $\Delta$ represents the shortening of the plate in the load direction.

In summary, branch point instability is characterized by the presence of another adjacent equilibrium near the original stable equilibrium, and the stability of the two different equilibria is about to shift at the branch point.

(2) Extreme point instability. Some of the elastic system in the project in the form of balance in the event of instability does not occur branching phenomenon, that is, there is no branching point, its stability is characterized by the load and deformation of the relationship between the characteristic curve with the so-called extreme point, such as the so-called point A shown in Fig. 1.6. After the load reaches the maximum load value corresponding to the extreme point, the deformation increases rapidly and the load decreases, indicating that the bearing capacity of the elastic system decreases rapidly, such as the instability of the coal pillar and the roof plate.

(3) Jump destabilization. The behavior of jumping instability phenomenon is a sudden change event, the equilibrium system jumps from one equilibrium form to another new equilibrium form, the displacement of the new stable equilibrium form is usually larger than that of the previous unstable equilibrium, such as Fig. 1.7a A →

## 1.1 Overview

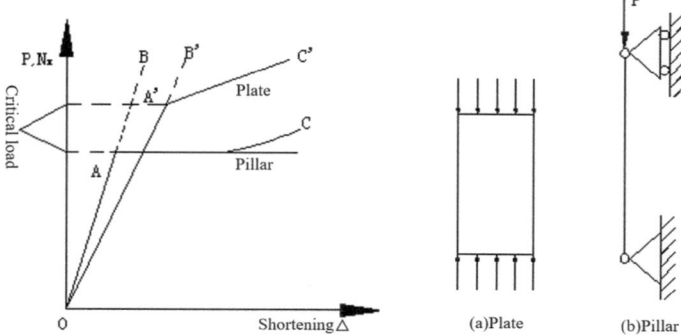

**Fig. 1.5** Schematic diagram of branch point destabilization of a plate body

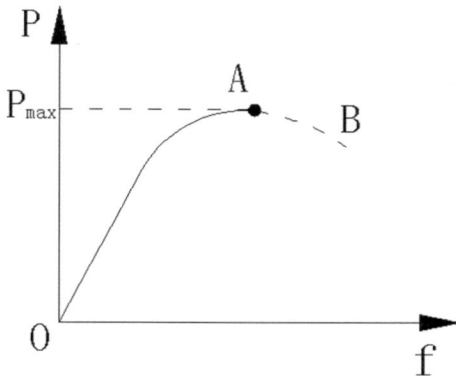

**Fig. 1.6** Extreme points of support loads and the principle of instability

B → C change process of the displacement trajectory. Typical examples of this type of instability are the subsidence deformation of the top rock layer, and the jumping of some kind of flat arch or transfer rock beam under the action of transversely distributed pressure, such as the flat arch beam shown in Fig. 1.7b.

(4) Finite perturbation buckling. In practice, some elastic structures, due to the significant reduction in stiffness after buckling, in order to maintain the equilibrium of the buckled state, accordingly, the load must be greatly reduced. For example, the buckling of a thin cylindrical shell under axial pressure and the buckling of a whole thin spherical shell under uniform external pressure belong to this type, as shown in Fig. 1.8. In Fig. 1.8a, $\Delta x$ represents the axial load per unit length; in Fig. 1.8b, q represents the uniform external pressure; V0 is the original volume of the sphere, and $\Delta V$ is the amount of change in volume during loading. For the above structure, a finite perturbation during the quasi-static loading process can then allow the structure to transition from an unflexed equilibrium form to another non-neighboring flexed equilibrium state before the load reaches the classical flexure load, and thus this form of flexure is named finite perturbation flexure.

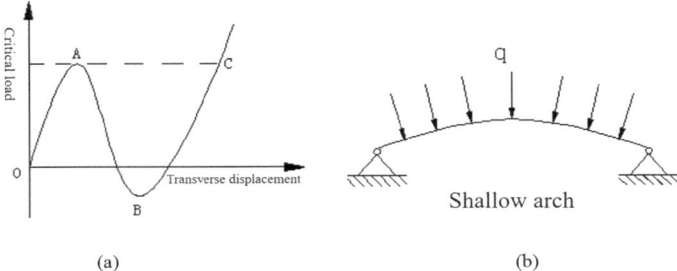

**Fig. 1.7** Principle and phenomenon of jump destabilization

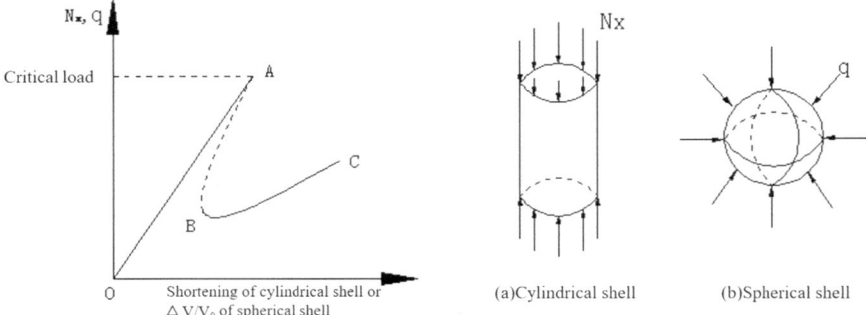

**Fig. 1.8** Buckling equilibrium principle for column-shell type elastic structures

There are three common discriminative criteria for evaluating the stability of equilibrium states in engineering, namely, the hydrostatic criterion, the energy criterion, and the kinetic criterion.

(1) The hydrostatic criterion, also known as the microturbulence criterion, the main point is that, assuming that there is an equilibrium state with infinitesimal difference near the branch point, the difference between it and the original equilibrium state can be regarded as a microturbulence, listing the equilibrium differential equations of the microturbulence, and the problem is reduced to the eigenvalues of the differential equations, then the conditions for the system destabilization can be obtained.

(2) The energy criterion, the gist of which is that a mechanical system consisting of an elastic system and an external load is in stable equilibrium if the total potential energy of the system is minimized with respect to all adjacent states.

(3) Dynamics criterion, The main point of the dynamic criterion is that in the generalized coordinate space of finite degree of freedom, the equilibrium state of a system whose position is described by coordinates ( i = 1,2,…, n) is $q_i = 0$, and the speed of the system changing with time is $\dot{q}_i$. If the system deviates from its equilibrium position, but the initial values $q_i^0$ and $\dot{q}_i^0$ can always be found, so that

## 1.1 Overview

in the later motion, $|q_i|$ and $|\dot{q}_i|$, do not exceed some pre-determined boundaries, the system can be considered to be in a stable equilibrium state.

In addition, the discrimination on stability can be studied on a small scale, i.e., in a tiny region adjacent to the original equilibrium state only. However, in order to reveal the instability phenomenon more deeply and comprehensively, it is sometimes necessary to explore it on a large scale. The former is based on the small-deflection linear theory, while the latter involves the large-deflection nonlinear theory. Although the former carries certain limitations, the important conclusions of many research topics can often be obtained from the linear theory, so the linear theory is still an important foundation of the stability theory.

The importance of loads in practical engineering with respect to the stabilization of elastic systems is a key term. Instability and damage in mining and civil engineering are almost invariably load-related. Therefore, a brief description of buckling loads and critical loads is given, and in principle there is a difference between these two loads. The former is the load corresponding to the buckling of an elastic system during loading; the latter is the load obtained from a mathematical model of the elastic system under a given load. Since the latter is based on the elastic–plastic theory eigen structural equations and is usually used as an eigenvalue for a particular eigenvalue problem, determined by the eigenequation, this load is often referred to as the critical load or critical value. Since there is only one form of buckling observed in the test, i.e., the form of buckling corresponding to the smallest eigenvalue, it is the smallest eigenvalue that is the critical load. In analyzing and studying the instability damage of specific engineering objects, we call it the minimum critical load or minimum critical value.

### 1.1.2 Creep Destabilization and Classification of Coal Rock Bodies

Compared with the coal rock body (layer) affected by mining disturbance, the original coal rock body maintains its own inherent physical and mechanical properties (such as compressive strength, shear strength, toughness and stiffness such as) the stress field the stress field, longitudinal and transversal of the rock body and the during the formation period of the formation, and cohesion potential energy of strain and elasticity (quantity geologic is in a relatively balanced and stable state. Quarry coal rock body in bearing force or extrusion or by elastic strain energy impact moment its performance and behavior will change, this change is sometimes violent sometimes slow, but as long as it is a sustained impact, and ultimately from the creep stage into the destabilization damage. Although the mechanical elements and mechanical structure that trigger the creep destabilization of the coal rock body in the quarry are different (the coal rock body and the structural mode of the three elements of the force), but in the pre-stabilization of the coal rock body can be regarded as a slow change to a drastic change in the course of the pre-stabilization of the coal rock

**Fig. 1.9** The development of creep in coal rock bodies

body, i.e., a creep to the drastic change (destabilization) of the evolution, as shown in Fig. 1.9

When creep occurs in the coal rock body, the wave energy impact of the elastic stress wave triggers the coal rock body to produce physical information phenomena, such as acoustic emission or micro seismic phenomena, fissure penetration or propagation phenomena in the surrounding rock, the phenomenon of elevated dust and gas concentration in the space, the phenomenon of temperature increase caused by the impact friction between the coal mass points (or particles or crystals), electromagnetic radiation (release) and the phenomenon of electrical conductivity change. The intensity of the various physical phenomena released from the coal rock mass is related to the inherent physical and mechanical properties of the coal rock mass. These phenomena and changes in their characterizing parameters can be monitored in situ and in the laboratory by means of corresponding monitoring instruments. Therefore, on the basis of this principle it can be stated:

(1) Due to the different mechanical models composed of the three elements of coal rock body mechanics, the degree of manifestation and characterization of coal rock body creep and instability will be different, which will also produce different disaster damage phenomena.
(2) The same mechanical three elements of the mechanical model in the creep instability process tends to have many changes, so there is the same creep instability process has different modes of monitoring equipment and monitoring methods.

The selection of monitoring mode and method should take into account the mechanical elements and the intensity of the phenomenon in the stage of inducing creep and instability of coal rock body, and the basic principle of selection is to take the physical change phenomenon that is easy to implement and effective, real-time is good, the signal is strong, and easy to identify and utilize as the target of monitoring, and this principle is the prerequisite for effective monitoring.

There are many phenomena in the process of creep instability induced by the mechanical model of coal and rock mass under the action of different mechanical elements. Therefore, when considering the monitoring and control methods, we must first choose an effective prediction and prediction monitoring method. A different mechanical model plays a key role in the occurrence of disasters such as large-scale collapse of stope and roadway roof, rock burst and gas outburst. There is a different process stage of inducing creep instability of coal and rock mass; there is a different law of evolution; they all have a different phenomenon characterization and signal release process. But they all have or may have one or more of the same catastrophic information signals.

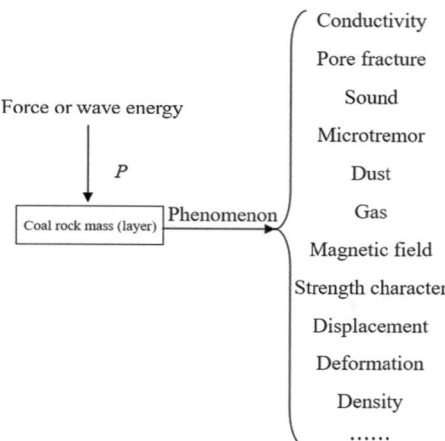

**Fig. 1.10** Physical signals generated or released by coal rock bodies

In general, from the monitoring point of view, the formation process of coal and rock disasters has a unified mechanical performance characterization and creep development law, and produces a unified information signal. Different signals have different emphasis on information or dominant signals, as shown in Fig. 1.10.

Therefore, when studying the mechanisms and control methods for creep and instability disasters in coal rock masses (stress fields or energy), attention should be given to analyzing and understanding the micro-scale variations and phenomena that occur during the creep and instability process of coal rock masses (layers). By utilizing micro-scale informational signals, one can predict large-scale catastrophic instability signals in coal rock masses (layers). It is crucial to establish a spatiotemporal framework that reflects and predicts the mechanical structural model and elemental composition of the coal rock mass (layer) in the working face, which can trigger creep and catastrophic changes in the rock layer. The selection, observation, and monitoring analysis of the dominant (informational signal) elements in the creep instability process becomes the technical key for the prevention and control of instability disasters. Constructing the spatiotemporal structure of the spatial mechanical model for the working face is fundamental for investigating creep instability theory and guiding engineering practice.

## 1.2 Characteristics of Creep Destabilization of the Rock Layers Surrounding the Roadway

Based on the technical considerations of both roadway construction and production service stages, the creep instability of the roof and surrounding rock layers (masses) in the working face roadway is divided into two phases: the roadway excavation construction phase and the production service phase, for discussion and analysis.

## *1.2.1 Tunneling Construction*

The roadway excavation activity disrupts the structural characteristics and stress equilibrium of the original rock mass, leading to changes in the stress field of the surrounding rock. This causes a redistribution of stress, forming a new stress field, which is referred to as the secondary stress effect, as shown in Fig. 1.11. During the process of stress redistribution, the surrounding rock, initially under the original rock stress field, undergoes decompression and expansion. In some regions, the stress distribution becomes more concentrated, with some areas transitioning from compressive to tensile stress, while others experience compression or tension. The coal rock mass (layer) exhibits a general tendency of expansion, leading to displacement and cracking in spatial directions. The surface of the rock wall may deform, leading to fracturing or instability, phenomena that result from the natural stress decompression equilibrium effect. Therefore, the decompression equilibrium effect can be classified into two scenarios: (1) the decompression equilibrium effect of the surrounding rock formed by the natural distribution of in-situ stress at the excavation face (head) during roadway excavation; and (2) the decompression equilibrium effect of the surrounding rock induced by mining disturbances during the service period of the roadway. Regardless of the conditions, the decompression equilibrium effect leads to the expansion and loosening of the surrounding rock, with the longitudinal (depth-wise) compressive stress increasing to a maximum value. It is important to note that the author does not support the common engineering practice of using techniques such as blasting for loosening to relieve pressure, which is believed to transfer and increase stress at deeper levels. Even if stress increases, it is temporary. In such cases, the final state or natural equilibrium of the stress change eventually aligns with the original rock stress balance, with only the thickness of the loosening zone changing. As shown in Fig. 1.12, the decompression expansion displacement and other changes of the surrounding rock in the newly excavated roadway occur under uncontrolled conditions and are primarily induced by stress field changes (ignoring the influence of mining vibrations), ultimately forming a loosening zone. The thickness of the loosening zone is directly related to the difference in stress before and after decompression. According to geological structure theory, the geological environment of mining engineering is formed by the deposition of several layers of strata, exhibiting clear layered structural characteristics, which is typical of the layered coal rock mass.

As shown in Fig. 1.13, using the N1 rock layer as an example, the Earth has undergone a series of dramatic changes over millions of years, with geological formations evolving in response to the long developmental process. As a result, sedimentary rock layers have been formed in successive strata, influenced by the Earth's gravitational forces and other external factors, which contribute to the development of various physical properties within these rock layers. It is assumed that the stress in the depth direction (or vertical direction) from any given starting rock layer is Pz1, and the stress in the horizontal direction (or lateral direction) is Ps1. Recent engineering testing and research findings (such as those from the China Coal Research Institute, Beijing

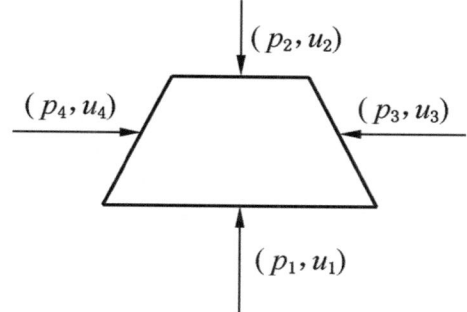

**Fig. 1.11** Schematic diagram of displacement and deformation tendency of surrounding rock layers

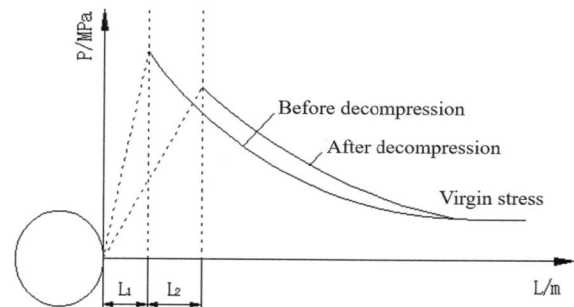

**Fig. 1.12** Natural equilibrium effect of geostress and decompression equilibrium principle of surrounding rock

University of Science and Technology, and other academic institutions) indicate that in the Z-depth direction, the vertical stress Pz1 is less than the horizontal stress Ps1, meaning that the stress field in the original rock layer, which has not been influenced by mining activity, generally follows the objective relationship of Ps1 > Pz1. Each rock layer possesses its own intrinsic cohesion, strength, elastic–plastic properties, and resistance to deformation. However, mining activities cause the surrounding rock in the radial direction of the roadway to undergo a redistribution of the original rock stress field at certain depths or thicknesses. The stress components and state of the original stress field change rapidly, even under slow creep, ultimately leading to a situation dominated by unidirectional (depth Z-direction) compressive stress, i.e., Pz > Ps. This trend is especially noticeable in large-section roadways. This phenomenon can be described as the result of the decompression effect occurring in the original coal-rock mass. This decompression effect, in terms of morphology, often results in horizontal displacement and deformation of the surrounding rock being greater and faster than the vertical (longitudinal) displacement and deformation. The displacement and deformation of the surrounding rock toward the roadway space or radial direction lead to an increase in the thickness of the loosened zone surrounding the roadway.

Based on this, the process of roadway excavation for the support of newly mined roadways, such as the use of bolt-mesh cable support systems, must pay special

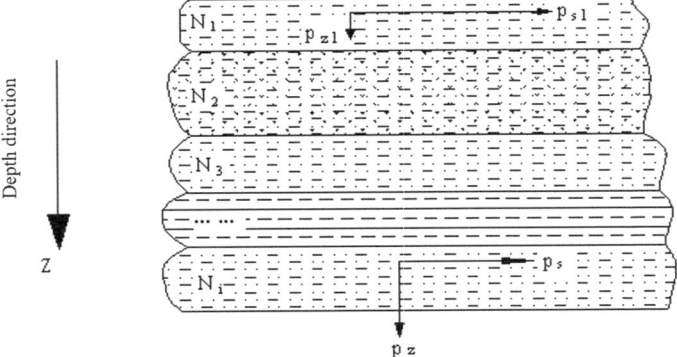

**Fig. 1.13** Diagram depicting the basic composition and stress of sedimentary rock layers

attention to adapting to the physical and mechanical properties of the rock layers and the temporal and spatial scale of the original stress field distribution. This is mainly reflected in the distance between excavation and support, as well as the time required for the redistribution of the original rock stress field, which helps form a general loosening zone in the surrounding rock structure (as shown in Fig. 1.14). Subsequently, corresponding support schemes and support technologies should be developed. As illustrated in Fig. 1.14, the loosened zone has a large thickness of fractured and loosened rock, while the elastoplastic zone has a reduced strength, creating cracks but still maintaining a certain thickness of the elastoplastic zone. The elastic zone represents minimal changes in the original rock layer's properties. For extremely soft surrounding rock, it is nearly impossible to grasp and apply stress field distribution equilibrium theories for support control, as detailed in Fig. 2.6. Its characteristic is that the shape of the surrounding rock undergoes significant changes, with large creep ranges and rapid speed; that is, the transition from creep to instability is extremely fast. Through monitoring or calculation, determining the scale of the surrounding rock loosening zone is very difficult. Therefore, timely and proactive deep preventive support methods must be employed.

## *1.2.2 Production Service Lanes*

The production service-type roadway refers to the roadway that, after completing the support, is designed to serve the production activities of the mining area, such as the working face transport roadway and the return air roadway. These roadways are entirely under the control of artificial support systems. After being supported by bolt-mesh-rope systems, the support structure formed by the bolts, mesh, and ropes effectively integrates the surrounding rock within the loosened zone with the

1.2 Characteristics of Creep Destabilization of the Rock Layers … 19

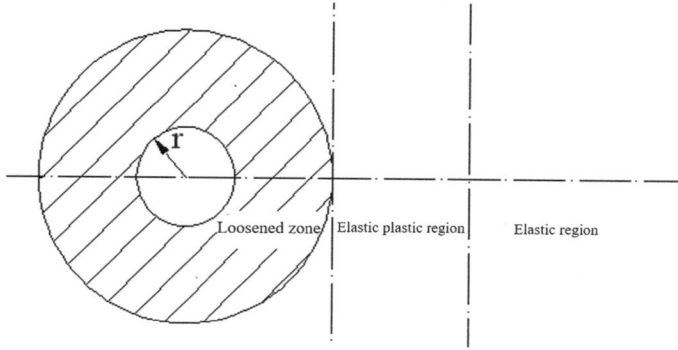

**Fig. 1.14** Schematic decompression and loosening rings for diagram of bare-channel perimeter rock

original rock mass (elastic zone) to form a cohesive structure. This support structure provides excellent overall support performance for the surrounding rock of the roadway, displaying distinctive "shell-arch-beam" characteristics. However, after the roadway has been supported by a bolt-mesh-rope concrete system, it will still experience the formation of new loosened zones due to the dynamic impact from mining activities and the movement of the overlying strata. The formation mechanism of these new loosened zones is illustrated in Figs. 1.12 and 1.14.

In recent years, with the gradual improvement of coal mining equipment and technological processes, large-scale spatial structural mining areas have been progressively adopted. After mining operations commence, the surrounding rock of the roadway in the working face undergoes severe deformation and instability failure. The overall stability of the bolt-mesh-rope support structure of the roadway is thereby threatened. The common disaster phenomena that follow can be categorized into the following types:

(1) Soft rock roadway: due to the roadway gang and the overall support of the deep rock body to a certain extent limited to the roadway gang surrounding rock arbitrary development of deformation, the roadway roof overlying rock layer (including bedrock) to the high rock layer to the surrounding rock compressive stress (compressive stress is the dominant role), this compressive stress induced a certain depth (thickness) range of the rock layer by the strong squeeze, and ultimately triggered the bottom plate to become a channel to unloading and decompressing the channel, the bottom of the roadway to produce Bottom drum phenomenon;

(2) Large-scale spatial structure of the roadway: thick coal seam long-wall once (big) mining full-height working face, adjacent to the mining hollow area of the overlying rock layer subsidence movement induced by the working face roof abnormal to pressure and roof, triggering the roadway surrounding rock affected by mining to expand the scope of the impact, especially thick and hard

roof formed by compressive stress strong impact and vibration, dominantly induced by the roadway deep surrounding rock to occur in the damage;

(3) The effect of the subsidence movement of the overlying rock layer in the mining airspace on the perimeter rock support of the adjacent working face roadway is very easy to be ignored. The movement of the overlying rock layer causes morphological changes in the neighboring production roadway. The top plate pressure is characterized by rapid development from creep to drastic change, with a wide range of impacts, resulting in a wide range of decline in the anchoring force of anchor rods and anchoring cables, and serious deformation of the roadway.

The roadway supported by anchor-net-slope support, including the roadway with anchor-net-slope shotcrete layers, exhibits the tough characteristics of an arch shell structure. The advantage of this support method lies in the overall high disaster resistance and failure prevention capability of the supporting structure. Although the anchorage strength of 20% to 30% of the anchor bolts within a certain section of the roadway may decrease, as long as there is no significant reduction in anchorage strength over a wide range, the roadway's self-stabilizing ability remains intact.

## 1.3 Creep Stability Characteristics of Large Roof Slabs

In recent years, with the continuous innovation and development of mining equipment and technological processes, such as the application of large mining height support systems, ultra-strong anchor mesh cable support technologies, and the goaf-driven (cutting or leaving) roadway technology, as well as the development of trackless mining transport vehicles in mines (especially in working faces), the mining space structure has been significantly expanded, and the advancement of working faces has been accelerated. However, without fully understanding the geological conditions of the roof, an excessive pursuit of the design production capacity and high-speed advancement of large-scale mining faces can lead to roof-related issues in the working face that are associated with technical management. These may include peak values of pressure that differ from the normal pressure, instability in the periodic pressure pattern, or a lack of bearing capacity in the supports.

This book refers to the phenomenon of unstable roof pressure patterns, frequent high-pressure peaks, and continuous pressure increase to peak values as "abnormal roof pressure phenomena." The movement of overlying rock layers caused by the goaf above large mining height working faces creates difficulties for both support and roof management in the working face. The settlement of the overlying rock layers in the goaf and the adjacent goaf areas affects the roof pressure pattern in the working face. Frequent abnormal roof pressure phenomena result in uneven load distribution on supports, leading to poor stability of the supports. These roof pressure issues, which are related to roof control, create a complex challenge for roof management. Particularly under the influence of creep settlement-induced pressure from thick and hard roof layers, large-scale abnormal roof pressure in the working face is highly likely to occur.

It can be said that the occurrence of large-scale abnormal roof pressure in the working face is extremely rare and constitutes an occasional disaster phenomenon. From the current abnormal pressure and catastrophic situations, it is observed that the primary causes of abnormal roof pressure in the controlled roof area are the presence of normal or reverse faults or artificial fractures. Once abnormal roof pressure forms, the high-position roof loses its lateral cohesion constraint, thereby increasing its degree of freedom. This causes the high-position roof pressure to become imbalanced, leading to the generation of high-intensity pressure on the overlying rock layers of the high-position roof, which subsequently impacts the supports. This can result in widespread damage to supports, such as support breakage, collapse, or even catastrophic failure of the support cylinders. The destructive situation is highly dynamic and has the characteristics of a shock wave. This catastrophe is related to mining technology and roof management techniques, representing a dynamic pressure disaster formed by the interaction between the working face roof and the overlying rock layers in the goaf. The triggering factors of this phenomenon are one of the challenges faced by modern mining engineering, as this dynamic instability, from creep to catastrophic change, is highly sudden and random.

## 1.4 Characteristics of Creep Destabilization of the Overlying Rock Layers in the Extraction Zone

In the process of subsidence and deformation of the overlying rock layer in the mining area, the vibration and impact disaster formed by the impact collapse of a wide range of rock layers is a rare catastrophic damage phenomenon triggered by the destabilization of the roof plate in the mining area.

The catastrophic damage caused by roof destabilization in the mining zone can be divided into two situations according to the time:

(1) For a considerable period of time (sometimes as long as several years) after the end of mining in a quarry or mining area, the extensive movement of the rock layers overlying the mining area triggers a large-scale impact collapse of the roof rock mass;

(2) Impact collapse of large rock masses caused by subsidence movement of the overlying rock layer in the mining area during the process of mining back to the quarry or mining area.

The causes of disasters in the two cases mentioned above should be composed of the following factors:

(1) The overlying rock layers of the mining area are endowed with thick hard rock layers and there are large fault tectonic zones in or near the area of subsidence disturbance;

(2) Coal mining face that adopts the artificial roadway gang method such as small coal pillar or leaving along the empty roadway, its hollow area is easy to form a large area of exposed (roof) space without filling;
(3) Thick coal seam one time mining full height working face and rapid advancement of coal mining process conditions.

The combined effect of three factors can easily trigger unstable sudden subsidence movements in the overlying strata of the mined-out area, inducing the sudden collapse of large blocks of high-level roof, which forms a dynamic impact on the floor. The large block layers collapse due to their own weight, impacting the floor and causing intense vibrations. In severe cases, this can trigger vibrations in the mining area, mine, or even the entire mining district. This phenomenon, where the dynamic instability and collapse of the overlying strata of the mined-out area induce vibrations, is referred to as mining-induced seismicity.

If the quantifiable relationship between the overlying strata of the mined-out area and the exposed space area (S), height (h), and collapse time (t) is defined as the spatiotemporal scale of the mined-out area, mining-induced seismicity occurs due to the overlying strata undergoing an intermittent equilibrium reaction process under the action of self-weight (W), which leads to surface intermittent subsidence movements. With the continuous increase in self-weight (W), the overlying strata of the mined-out area suddenly move and collapse onto the roadway floor. In an instant, the impact energy is released through the roadway floor, inducing vibrations.

The impact can sometimes be significant enough to be felt across the entire mining district (mining-induced seismicity), causing phenomena such as the tilting of supports, opening of pressure relief valves, deep rib spalling, and increases in gas and dust concentrations in neighboring working faces. In processes such as the recovery of island coal pillars, coal pillar extraction, and other mining activities, the suspended roof can also suddenly lose stability and collapse, leading to dynamic impact phenomena. The dynamic influence region is accompanied by the sounds of coal body fragmentation, large-scale rock mass collapse, and gas wave shockwaves, which induce phenomena such as coal block ejection, dust flying, and elastic rebound vibrations. The occurrence of mining-induced seismicity seems to lack intuitive early warning signs, making early prediction and identification difficult. Currently, there is no mature prediction method in the academic community, and related techniques are still in the exploratory phase. Unlike dynamic pressure and gas outbursts, mining-induced seismicity is a release of elastic energy that does not involve the expulsion of large amounts of coal and rock. The fall of large structural overlying rock masses upon reaching the floor triggers intense seismic vibrations and damage. This type of damage typically has an early-stage development phase characterized by a large range of overlying rock mass potential energy, or gravitational potential energy, such as during the period of strata subsidence movements. It most often occurs during the final phase of mining (including island recovery, coal pillar extraction, etc.) or after production has ceased.

## 1.5 Characteristics of Creep Destabilization of Impacted Ground-Pressured Coal Rock Bodies

In coal mining, dynamic loads, or coal bursts, are commonly observed in thick coal seams and are rarely seen in thin seams. Coal seams that are prone to dynamic loads or have a tendency to experience such phenomena are typically characterized by coal that tends to break into blocks, has a high hardness, and exhibits elastic–plastic properties, showing both plastic-brittle characteristics and highly elastic strain properties (referred to as elastic-plasticity). These coal bodies have a strong tendency to generate and store elastic potential energy, which is an inherent property of the coal's structural organization.

Furthermore, within coal rock masses (layers), the existence of high-stress geological zones (regions), such as high-pressure stress caused by thick, hard overlying strata (or floor strata), is a necessary condition for the occurrence of dynamic loads. If one of these conditions is met, dynamic loads or a tendency for such loads to occur can be triggered, and they are related to geological factors. Geological stress zones, particularly compressional structural zones, are unique geological formations. These zones store (or harbor) extremely high elastic potential energy, which is a natural attribute of the original coal rock mass (layer). When high-stress geological structures are influenced by mining activities, the latent elastic potential energy within them can be excited by the space effects of mining, propagating, transferring, and releasing energy, thus causing sustained impact failure within the rock mass. This energy effect occurring within the coal rock mass is referred to as the impact effect.

This paper refers to the dynamic loads occurring under such geological conditions as inherent spontaneous coal seam dynamic loads or inherent spontaneous coal seam dynamic load tendencies. Another type of dynamic load occurs when extremely high pressure stress from thick, hard overlying strata acts upon a coal rock mass (layer) with a high capacity to store elastic strain potential energy. Under these conditions, the coal rock mass generates (or stores) extremely high elastic strain potential energy. When this energy is excited by mining activity, it propagates, transfers, and releases energy. The interaction between elastic energy and the particles (or crystals) of the coal body within the coal rock mass induces impact effects, resulting in the destruction of the coal rock mass's structural properties and further propagation of energy. This type of dynamic load is referred to as an inherent induced coal seam dynamic load or inherent induced coal seam dynamic load tendency.

## 1.6 Characteristics of Creep Destabilization in Gas-Abrupt Coal Seams

Coal and gas outbursts (referred to as outbursts) exhibit distinct physical characteristics inherent to the coal seam itself. Gas outbursts are commonly observed in thick coal seams, but rarely in thin ones. Coal seams capable of outbursts generally have low coal hardness, with a Proctor hardness of $f = 0.1 \sim 1.0$. The coal's texture exhibits a scaly granular structure (or scaly crystalline form), with well-developed fissures and pores in the coal rock (verified by gas coal sample water absorption tests). These coal seams display brittle and friable properties along with high rheological characteristics.

Due to the large amount of gas molecules adsorbed for extended periods within the fissures and pores of the coal rock (seam), these coal seams demonstrate pronounced brittleness. A considerable amount of gas is stored within the pores, fractures, and structural formations of the coal seam. The long-term frictional interaction between these gas molecules results in a high concentration of both gas and dust within the coal seam's fissures and pores. Overlying strata above high-gas coal seams are generally composed of rigid or thick layers of fine mudstone. These rock layers form a sealing and compressive effect on the underlying gas coal seams, preventing the natural diffusion and release of gas molecules from the coal seams. These inherent characteristics of gas coal seams belong to the fundamental coal quality features.

However, it is precisely these inherent characteristics of the coal seams that, due to the sealing and compressive effects of the overlying strata, lead to extremely high gas pressure in the coal seam. This results in large-scale gas coal seams exhibiting highly elastic or rheological deformation characteristics. Once the elastic and rheological properties of the gas coal seam are disturbed by mining activities, the gas molecules stored in the pores and fractures are activated, generating kinetic energy. This leads to a vibration impact effect between the coal particles or crystals and the gas molecules, forming a rheological kinetic energy dominated by gas airflow. This subsequently induces the coal particles to generate extremely high elastic potential energy. The combination of kinetic and potential energy promotes the migration of gas within the coal seam, expanding the permeability of the coal seam's pores and fractures, leading to dynamic failure. At the same time, the transfer and accumulation of energy is a latent process. Due to the continuous shock effects of the gas's kinetic energy, the existing fissures and pores in the coal seam are expanded, forming more developed pores and fractures, and new pores and cracks continue to emerge, creating a larger volume for gas flow channels. Gas continually enters these channels, causing the energy of the gas to increase. This cycle leads to the continuous expansion of the gas flow channel's volume, enhancing the kinetic energy within the gas transfer pathways.

The ongoing latent accumulation process accelerates the accumulation of gas and the development of gas flow energy. This process is characterized by creep evolving into a more intense and dynamic change. It demonstrates the rheological motion characteristics of coal and gas, particularly with gas as the dominant factor, reflecting the rheological behavior of energy flow. The rheological kinetic energy of the gas continues to move towards the excavation space, accumulating in a region near the surrounding rock, where it forms regional elastic potential energy. This energy can trigger a sudden release if the energy level is high, resulting in the rupture of the surrounding rock and the ejection of coal and gas, causing an outburst disaster.

Overall, (1) The material properties and bearing strength of the coal rock in the mining face are diverse, variable, and unstable, making it an indeterminate material.

## 1.6 Characteristics of Creep Destabilization in Gas-Abrupt Coal Seams

(2) The deformation and failure trajectory of the coal rock under compressive stress is highly complex. Due to the influence of multiple elements, it is nearly impossible to express this using a single mathematical model. Therefore, various theoretical models based on assumptions have emerged in academic circles, such as fuzzy, nonlinear, neural network, and hierarchical methods, along with simulation techniques using different computational software. A significant amount of research results on fuzzy mechanics have been published.

The deformation and failure process of the coal rock is primarily driven by the formation and development of internal pores, fractures, and cracks. As the strength peak is approached, cracks and fractures begin to develop locally in the coal rock. Once the strength peak is exceeded, the developed fractures start to cause damage, forming a generalized strain region where the properties of the coal rock material undergo significant changes. The strength resisting deformation nearly disappears, and the coal rock becomes a "strain-softened" material. This process also represents the reduction in pressure and strength weakening of the coal rock. According to the material stability criterion proposed by Drucker in plasticity theory, the medium in the strain-softened zone is considered an unstable material. Strain-softened media lack the capacity to resist further deformation, so additional deformation will continue to occur near the strain-softened zone.

Rock bursts and outbursts are caused by the release of energy from high elastic strain potential accumulated in the internal coal rock. This energy transfer results in the weakening of the coal rock strength due to strain-softening (strength reduction) and continuous energy release, leading to dynamic failure. The energy release generated by the internal accumulation and development of energy results in a series of strain-softened destruction trajectories as it moves towards the excavation space. This is a process where a weakened coal rock body loses stability and strength under dynamic forces, and the failure mechanism is consistent with the general principle of creep instability in coal rocks. However, the frequency of changes (stability and fluctuations) and the intensity of the dynamic forces involved in the process of rock bursts and outbursts are more intense.

In other words, when the coal rock experiences strength or strain softening, it typically refers to a situation where the original rock body, under the influence of energy or disturbance forces, expands or deforms towards the excavation space, reducing its strength (or strain) in the decompression direction. As a result, the area affected by strength softening increases, causing the loose region to grow thicker in relative terms and expand in volume. Both from a microscopic perspective and in terms of academic consensus, coal rock is generally considered a medium with well-developed pores and fractures. Therefore, after strength softening occurs in the coal rock, if displacement deformation happens, the dominant direction of this deformation can only be in the spatial direction. Within the coal rock, the inherent pores and fractures become compressed and fractured, producing dust. However, due to the continuous action of elastic energy, the overall volume is compressed and becomes denser, which does not lead to an increase or expansion in volume within the rock.

**Open Access** This chapter is licensed under the terms of the Creative Commons Attribution-NonCommercial-NoDerivatives 4.0 International License (http://creativecommons.org/licenses/by-nc-nd/4.0/), which permits any noncommercial use, sharing, distribution and reproduction in any medium or format, as long as you give appropriate credit to the original author(s) and the source, provide a link to the Creative Commons license and indicate if you modified the licensed material. You do not have permission under this license to share adapted material derived from this chapter or parts of it.

The images or other third party material in this chapter are included in the chapter's Creative Commons license, unless indicated otherwise in a credit line to the material. If material is not included in the chapter's Creative Commons license and your intended use is not permitted by statutory regulation or exceeds the permitted use, you will need to obtain permission directly from the copyright holder.

# Chapter 2
# Principles and Prediction of Creep in the Perimeter Rock of the Quarry Roadway

## 2.1 Overview

Assuming that the roadway in the mining area is not affected by mining-induced disturbances, the surrounding rock of the roadway under the original in-situ stress, based on the principles of temporal-spatial effects and stress release expansion, exhibits a phenomenon where the loosening and deformation of the surrounding rock increases slowly over time. This phenomenon is referred to as the creep behavior of the surrounding rock in the roadway. Compared to mining-induced stress, the influence of geological stress primarily occurs at greater depths, and its impact is generally slower. Therefore, the displacement and deformation caused by geological stress can mainly be observed. Mining engineering faces a complex mechanical interaction between changes in the original rock stress induced by mining and the dynamic stress of the excavation, leading to damage typically caused by a combination of shear forces and compressive stresses. Ultimately, the shear and normal stresses impact the underground engineering space and its support system.

Under the continuous influence of mining-induced stress, the surrounding rock of the roadway undergoes a process of gradual development from creep to significant sudden change, particularly influenced by prominent factors such as large-section roadways, small coal pillar working faces, artificial roadway ribs along goafs (without coal pillars), longwall faces with large mining heights, and subsidence movements of the overlying strata near the goaf. These factors can amplify the effects of mining-induced stress on the roadway surrounding rock, triggering severe deformation of the surrounding rock. As illustrated in Fig. 2.1, the subsidence movement of the overlying rock strata in the neighboring mining area triggers the asymmetric movement of, forming the dynamic load compressive stress on the side of the mining are $P_{Z'}$ the overlying rock on the upper part of the coal seam $P_Z$ is larger than that of the working face area $P_{Z'}$, i.e., $P_Z>$, which results in the coal columns in the mining area unstable destruction of the and the roadway serious deformation of, and in serious cases, it makes the coal seam form the impact tendency or the protruding tendency. Complex geological conditions of coal seams (such as impact and protrusion inclined coal seams) overlying thick hard top plate, the its disaster types and damage-causing state diversity of increases, and use conventional means significantly it is difficult to it and methods effectively control. Therefore, it is difficult to effectively control them by conventional means and methods. Therefore, scientific and objective analysis in the pre-disaster period the research of mastering the disaster-causing mechanism of the roadway rock and effective prediction and forecasting methods have become the faced by the large-scale quarry and deep mining., main theoretical and practical problems.

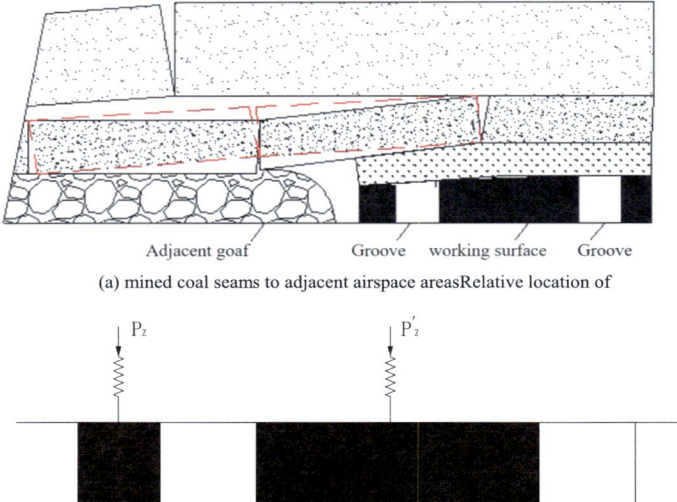

(a) mined coal seams to adjacent airspace areasRelative location of

(b) Principles of the mechanical effects of subsidence of adjacent air-mined areas on the coal pillar at the working face

**Fig. 2.1** Principle of mining coal seam destruction induced by the movement of overlying rock strata in the adjacent airspace area

## 2.2 Creep and Phenomena of the Perimeter Rock of the Roadway

### 2.2.1 Mechanisms and Types of Perimeter Rock Damage in Deep Quarries

When classifying the surrounding rock mass of a roadway into two primary categories—loose and soft rocks—loose refers to the structure of the coal rock, which is characterized by a loose texture, low bulk density, and high porosity. Soft, on the other hand, refers to rocks with low strength that are prone to plastic deformation. In some instances, the same rock layer may exhibit both loose and soft characteristics. Loose and soft rock masses typically display expansivity and fragility. Loose coal rock, especially soft rock, is particularly susceptible to moisture and disturbance caused by mining activities, which can lead to more severe deterioration, ultimately triggering the collapse and fragmentation of the coal rock in the mining face. Drawing upon current research both domestically and internationally regarding the control of fragmented loose surrounding rock, dynamic pressure, and the prevention of coal and gas outbursts, as well as the theoretical and practical exploration of deep mining and surrounding rock disaster control, the dynamic control technology for surrounding rock disasters in deep mining faces has been proposed. This technology is based on the principles of "force, material, and control" in engineering contexts. Specifically, the disaster in the surrounding rock is induced by a combination of three dynamic factors: the changes in the in-situ stress environment and stress field of the surrounding rock, the variations in the physical and mechanical properties of the surrounding rock, and the formulation, implementation, and adjustment of surrounding rock control schemes. Under the dynamic interaction of these factors, the creep instability and instantaneous failure of the surrounding rock and support system occur, leading to catastrophic events.

#### 2.2.1.1 Stress Environment in the Deep Surrounding Rock

Before mining, there already exist various structural stresses within the original rock mass. The mining engineering primarily focuses on the stress changes and influencing factors resulting from the vertical and horizontal stresses caused by mining. These two stresses are the root causes of the mining-induced rock pressure in the mining area. The characteristics of the in-situ stress field prior to mining, including the magnitude and direction of the principal stresses within the original rock and the ratio of vertical to horizontal stress, determine the pattern and distribution of the surrounding rock stress after mining. The formation of stress at each point in the original rock mass is mainly the result of the combined effects of overburden gravity, tectonic forces, and the expansive extrusion of the rock mass. Geological structures and rock stress field examples are illustrated in Fig. 2.2.

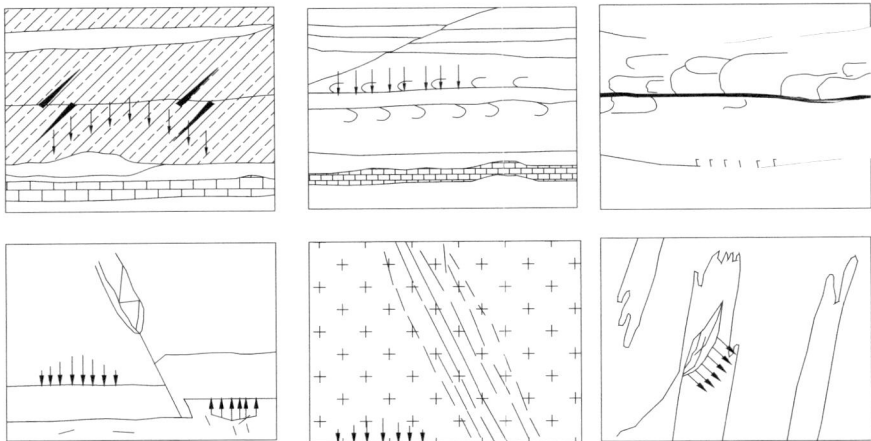

**Fig. 2.2** Geologic formations and rock stress fields

The stress environment surrounding deep mining workings is complex. It transitions from being primarily dominated by structural stress at shallow depths to being dominated by vertical gravity stress at greater depths. The stress environment of the surrounding rock includes both the composition of stress and the characterization of compressive stress. The stress composition refers to the proportion and magnitude of stresses caused by the self-weight of the surrounding rock, structural stress, and concentrated compressive stresses induced by mining activities. The characterization of stress refers to the state of the surrounding rock under the influence of uniaxial or multiaxial stress, as well as the trend of stress environment changes. Various support structures can alter the bearing state and mechanical characteristics of the surrounding rock, which will affect and control the distribution of deep stress in the surrounding rock and the stress composition environment (such as compressive stress and shear force).

Existing coal seam testing data from mining areas indicate that vertical stress increases linearly with depth. This inevitably leads to higher stress in the surrounding rock of deep mining workings, as well as a change in the relationship between the surrounding rock stress and strength, as shown in Fig. 2.3.

Research conducted by the Donbas Coal Scientific Research Institute indicates that starting from a depth of 600 m, for every additional 100 m of mining depth, the roof and floor convergence increases by 10% to 11%. Measured results from the Sun Village Mine's inclined shaft and the Si Coal uphill roadway of the Xinyi Mining Bureau show that, starting from a vertical depth of 570 m, for every additional 100 m of mining depth, the roof and floor convergence increases by approximately 5 cm. Additionally, statistics from the Kailuan Mining Bureau indicate that as burial depth increases, the proportion of roadway length affected by floor heave also increases.

## 2.2 Creep and Phenomena of the Perimeter Rock of the Roadway

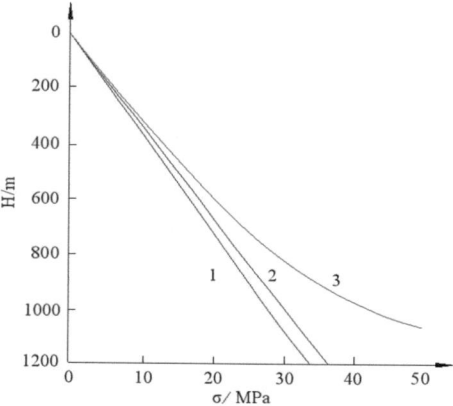

**Fig. 2.3** Raw rock stress as a function of depth

When the mining depth increases from 400 to 1000 m, the floor heave ratio in the mined roadways increases from 32 to 70%, the preparation roadways increase from 24 to 57%, and the development roadways increase from 20 to 42%. Deep shaft roadways' floor heave has become a significant factor influencing roadway stability and secondary repair operations.

### 2.2.1.2 Types of Deep Quarry Perimeter Rock Catastrophes

The deformation, movement, and failure phenomena that affect the normal use of roadways during the process of surrounding rock and its support are referred to as surrounding rock disasters. From the perspective of the characteristics of surrounding rock disasters in roadways, these can be divided into two types: abrupt failure and creep-induced large deformation. Abrupt-type disasters are characterized by dynamic properties, primarily referring to phenomena such as rock burst, coal and gas outbursts, the rheology of extremely soft rock, deformation of thick and hard roof strata, and violent movement and subsidence of overlying rock layers in goafs. The dynamic source of these disasters originates from the properties of the deep-seated original rock mass of the surrounding rock, with sudden and unexpected manifestations. Creep-induced large deformation disasters, on the other hand, are characterized by quasi-static behavior, mainly involving shear or compression failure in the surrounding rock, which leads to localized deformation of the surrounding rock, such as sidewall spalling or local delamination and fragmentation of the roof. The characteristic of this type of disaster is that the local rock mass failure is triggered by its self-weight, and the destruction process develops from the surface inward. According to the manifestation characteristics of these disasters, deep mining surrounding rock disasters can be further classified into three types: roadway surrounding rock cross-section creep convergence, surrounding rock morphology abrupt changes, and dynamic surrounding rock sudden failures.

1. Roadway Surrounding Rock Cross-Section Creep Convergence.

The deformation and failure (fragmentation) of surrounding rock in roadways continue to expand, with the roof, sidewalls, and floor creeping towards the roadway space. Conventional support methods are often ineffective in controlling this deformation, leading to significant surrounding rock deformation and failure. This phenomenon exhibits clear signs of deformation failure, requiring periodic refurbishment to maintain the normal use of the roadway. A typical example of roadway surrounding rock instability is shown in Fig. 2.4. This type of rock mass performance failure is a typical example of brittle (plastic) failure, characterized by loose fragmentation, minimal displacement changes, and rapid expansion fragmentation.

2. 2. Surrounding Rock Disaster and Impact Effects

The occurrence of disasters in the surrounding rock and roof structure of the mining roadway leading to deep-seated impact effects within the rock mass:

(1) Weak Impact and Creep Effects of Original Rock's Elastic Potential Energy in Deep Surrounding Rock: This phenomenon is manifested as decompression-induced expansion, which triggers displacement and local rock mass extrusion

**Fig. 2.4** Typical tunnel perimeter rock destabilization

## 2.2 Creep and Phenomena of the Perimeter Rock of the Roadway

failure. Such damage is often referred to as damage or strength degradation. The creep phenomena occurring within the rock mass are sometimes accompanied by distinct dynamic impact noises and microseismic events, indicating the occurrence of noticeable impact effects within the rock mass (intensity magnitude 1–2).

(2) Moderate-Intensity Impact Due to Elastic Potential Energy in Deep Surrounding Rock: After localized creep occurs within the rock mass, the stored elastic potential energy results in damage to the surrounding rock, causing failure from the interior to the surface. This process is accompanied by energy channels and breakthrough points, leading to widespread damage to the roadway roof (intensity magnitude 2–3.5).

(3) Severe Impact and Failure Due to Deep Elastic Potential Energy: The extremely high elastic energy can cause catastrophic roof collapse in the roadway. This type of damage is typically associated with energy channels and breakthrough points, with large volumes of coal (rock) mass and dust being ejected through these breakthrough points, forming gas waves and causing extensive roadway damage (intensity magnitude 3.5–5).

(4) Catastrophic Energy Impact from Elastic Potential Energy: This results in the collapse of the roadway roof in a single production level or even across an entire mining area, and in some cases, it may affect the entire mine, leading to the shutdown of the mine. The intensity magnitude exceeds 5. Due to improvements in prevention technologies and management levels, such events are now extremely rare.

3. Roadway Surrounding Rock and Gas Outburst (Dynamic Impact Effects)

The destructive behavior of coal bodies with a tendency for gas outburst is similar to that of coal bodies with a tendency for dynamic impact. The key difference lies in the fact that the dynamic impact caused by a gas outburst is driven by gas pressure, which induces deformation in the deep coal body, resulting in a series of failures that degrade the strength of the coal and rock mass. The energy for gas outburst comes from the accumulation of gas in the coal seam.

When the gas content of a coal seam is extremely high, i.e. the static pressure of the coal seam gas is extremely high. The impact energy comes from the potential energy of the gas accumulated and saved in the coal body. The gas in the holes, fissures and large and small structures of the coal seam is affected by the equilibrium effect of the mining space, which will trigger a large amount of gas to accumulate along the holes and fissures in the direction of the mining space, and then form the elastic potential energy together with the coal rock compressive stress. In coal rock body the development process from creep to violent rheological movement of, the coal rock body internal formation from inside to outside (space) of the energy propagation and release channels, continuous accumulation of energy and internal impact, touch friction, and then constantly produce the impact effect, the formation and expansion of the gas energy transfer channel, reach a very high energy is formed break through toto promote a considerable amount of coal and dust break through the barrier of the surrounding rock, spraying to free space. This protruding damage has obvious gas power impact transportation characteristics, spewing out a large amount of dust, the protruding channel of the outer orifice is often pear-shaped, the mouth is small belly and accompanied by a large amount of gas gushing out.

The study on the mechanism of peripheral rock creep should take into full consideration of the mechanical elements and induced deep rock displacement and peripheral rock deformation characteristics and laws, the development trend of the overlying rock layer in the neighboring mining area, as well as creep the stage of creep development of impact pressure and coal and gas protrusion the change of information in, and other phenomena. On the basis of the categorized research on the above phenomena (such as the quality of support, roof slab, small coal pillars, and the range of the movement of roof subsidence in the mining hollow area), the research on the mechanism of peripheral rock creep and catastrophic change is carried out.

### 2.2.2 Enclosed Rock Deformation and Damage Patterns and Basic Mechanisms

Overall, the creep failure mechanism of newly mined roadways is fundamentally similar to that of production service roadways, with the primary difference being the change in the triggering factors. In production roadways, the dominant failure factor is the disturbance effect caused by mining-induced forces. The direction of the compressive stress or pressure applied to the coal-rock mass, or the direction of elastic energy release, defines the direction in which the surrounding rock of the roadway deforms and develops. Displacement and deformation within the coal-rock mass always propagate in the direction of the free mining space, resulting in the creeping propagation of cracks. The crack development gradient typically follows a pattern of increasing fracture density from the interior to the surface, which is also a consequence of the time–space effect.

(1) Roof deformation failure. Deep roadway roof deformation failure can manifest in two forms: One occurs when the roadway is in a high-stress environment. After the excavation of the roadway, the surrounding rock stress and the state of the surrounding rock (microscopically, its crystalline structure or particles) redistribute to achieve equilibrium. This is particularly noticeable with the horizontal (lateral) stress (elastic energy), which, when balancing, induces the lateral expansion and relaxation of the rock layers. This expansion moves toward the free space, creating decompression-induced fractures. As the thickness of these fractures in the surrounding rock increases, stress concentration phenomena occur on the rock face, causing the local roof to enter a plastic state, forming a plastic failure zone. The formation of this plastic failure zone leads to a reduction in compressive stress (i.e., a decrease in stress concentration), resulting in a temporary steady-state equilibrium. In some cases, even slight mining disturbances can alter the thickness of the plastic zone, causing cracks to propagate

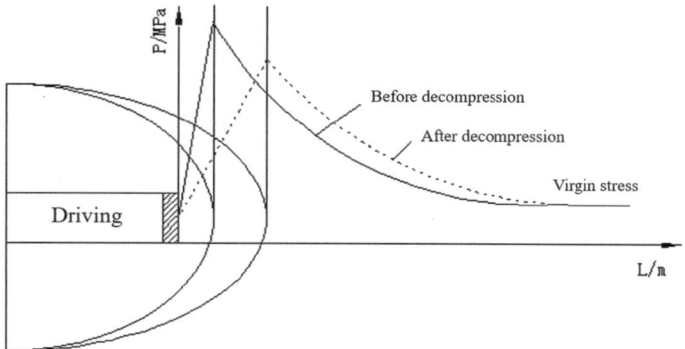

**Fig. 2.5** Schematic diagram of the principle of extractive-induced decompression equilibrium effect

deeper. This type of failure induced by mining activities, which causes decompression equilibrium effects, is referred to as mining-induced decompression equilibrium effects, as shown in Fig. 2.5. As stress (or elastic energy) equilibrium continually changes, new plastic zones are formed, leading to the creation of a loosened failure zone. Generally, the effect of horizontal stress equilibrium is much more significant than that of vertical stress, and in severe cases, this can lead to roof strata buckling failure, manifesting as breakage phenomena.

(2) Deformation and Failure of Roadway Sidewalls. After the excavation of the roadway, the coal-rock mass (strata) transitions from a three-dimensional stress state to a two-dimensional longitudinal and transverse stress state. Due to the existence of uneven factors, the longitudinal and transverse stresses acting on the roadway sidewalls are highly uneven. Under the combined action of the two-dimensional stresses, the original rock stress zone (elastic zone of the original rock) experiences a decompression and expansion effect in the direction of the roadway's free space, or a relaxation effect of elastic energy transfer and release, which is the impact effect. This forms the mechanical elements of the original rock stress field from inside to outside, resulting in the formation of a loosened zone (loosened circle), elastic–plastic zone, and elastic zone (or original rock stress zone) from inside to outside. The extension of sidewall failure further induces the failure of the roof and floor. Ma Nianjie and other scholars conducted numerical simulations on the deformation and failure characteristics of coal roadways at a depth of 1000 m in Xingdong Coal Mine. The results indicated that after the excavation of the roadway, the thickness of the surrounding rock failure gradually expanded from the sidewalls and floor towards the roof. The appearance of plastic zones and spalling in the coal mass surrounding the roadway causes an increase in the roadway span. The maximum tensile stress in the roof rock layer is proportional to the square of the roadway span. When the roadway span becomes too large, the roof may collapse entirely along both sides of the roadway, leading to a caving-in accident.

(3) The serious phenomenon of bottom plate damage is mostly seen in soft rock and very soft rock roadway. Roadway bottom plate damage mainly refers to the bottom dropsy damage, which has become research in the field of mining engineering difficulty. For the type of dropsy, according to the Heman Zhao, Jianyong and other scholars' point of view can be divided into four basic types: extrusion flow type, flexure fold type, shear misalignment type and water swelling type. The current dropsy management method of research results is more, but the attention on the formation mechanism of dropsy is still relatively weak. The formation of dropsy is related to the lithology of the surrounding rocks of the roadway (such as water absorption, siltstone), very soft rock roadway are located in the rock layer are soft rock or mudstone, also has a moisture-absorbing hydrophilicity, the rock body does not exist in the organizational structure, the strength of a small, very easy to soften and muddy, soft and easy to displacement and deformation, usually using the rheological function equations to describe its characteristics. Therefore, the disturbance effect and spatial deformation effect of the excavation process or mining power are very significant, involved in inducing deformation of the roadway stress field depth (or thickness) range is relatively large, i.e., a very large range (such as high rock layer) within the formation of compressive stress in the rock layer $P_2$, $Q_2$ involved in triggering the displacement and deformation of the rock body (rock layer) movement. As Figure shown in, the compressive stress field in the direction of the roadway roof 2.6influence range of is larger than the minimum deformation range of the base plate, and when the deformation range of the longitudinal and transverse compressive stresses of the surrounding rock reaches a certain $P_2$ and $Q_2$ degree, the base plate of the roadway will be deformed or damaged. In many cases, the damage of the bottom plate is preferred to and heavier than other areas, which is due to other parts of the support reinforcement, resulting in a change in the direction of the action (transfer) of stress (energy). When the horizontal stress of the surrounding rock is greater than the vertical stress, or the vertical stress is greater than the horizontal stress, the roadway bottom dropsy will occur (see Fig. 2.6).

The failure of the roadway's "roof—two sides—floor" exhibits significant interactive interdependence: the failure of the two sides will exacerbate the failure of both the floor and the roof; the compressive stress induced by the overlying rock layers on the roof leads to shear failure in the roadway's two sides; meanwhile, the failure of the floor further influences the rate of failure of the two sides. This coupled interaction implies that, in deep floor control, it is crucial to consider how to effectively regulate and control the creep and sudden failure of compressive stresses on the roof and two sides. Therefore, a comprehensive approach should be adopted in support, monitoring, and prediction, taking into account the interactive interdependence between the "roof—two sides—floor."

**Fig. 2.6** Principle of formation of bottom dropsy and the range of large-scale displacement and deformation

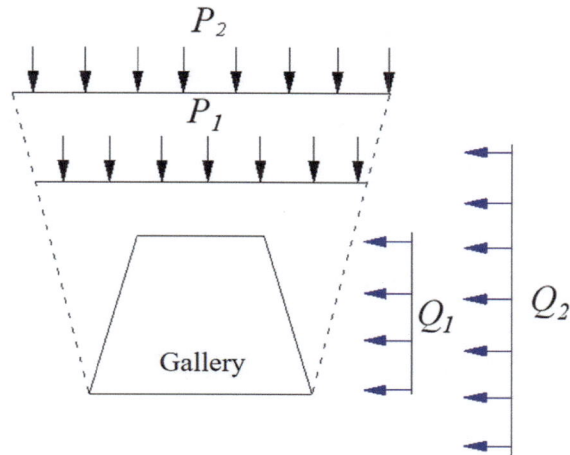

## 2.2.3 Factors Affecting Perimeter Rock Stability

### 2.2.3.1 Stability Analysis of the Roadway Perimeter Rock

As the mining depth continues to increase, the likelihood of engineering disasters in coal and rock masses will gradually rise. Issues such as high rock pressure, significant deformation of the roadway surrounding rock, rapid deformation, and severe damage, as well as the increasing difficulty in roadway support, are becoming more prominent. Changes in mining methods, such as the use of small or no coal pillars and extra-long face transport roadways (panel roads), have exacerbated the impact of subsidence caused by mined-out areas. Due to the effects of high temperature, high surrounding pressure, and high pore pressure, rheology becomes the primary mechanism for roadway deformation. Intense mine pressure and support structure instability are likely to lead to destructive rockburst-induced pressure. According to relevant statistical data, as mining depth increases, the creep instability and catastrophic failure of coal and rock bodies in roadways significantly increase, and the likelihood of safety accidents caused by coal seam gas content also rises sharply. Under deep mining conditions, high ground temperature and deteriorating working environments contribute to additional challenges. Due to the effects of the "three highs" (high temperature, high surrounding pressure, and high pore pressure), the surrounding rock deformation in deep roadways exhibits characteristics not found in shallow roadways. Therefore, research on the stability of deep roadways has become particularly important. Many scientists both domestically and internationally have conducted extensive research on this issue and have achieved rich results. However, there are still many deep mining challenges that require integrated solutions, such as clarifying the spatiotemporal effects and interactions of various factors during the

**Table 2.1** Statistics of deformation of surrounding rock in 800 m deep roadway

| Rock | Approach of two gangs (mm) | Base plate deformation (mm/d) |
|---|---|---|
| Limestone | 20–33 | 0.203 |
| Siltstone | 40–65 | 0.406 |
| Shale | 113–265 | 0.722 |

creep-to-catastrophic failure phase, as well as achieving an objective understanding and consensus on the mechanisms of surrounding rock failure.

### 2.2.3.2 Influence of Rock Properties on Tunnel Stability

Table 2.1 presents the measured amounts of rib convergence and floor deformation of the surrounding rock under different rock mass conditions from a stone tunnel at a depth of 800 m in a mine.

The data in the table clearly show that the surrounding rock in three different rock mass conditions is in stable, unstable, and highly unstable states, respectively. The data also indicate that, at the same mining depth, the stability of the surrounding rock is directly related to the rock mass properties.

### 2.2.3.3 Effect of Depth of Burial on the Stability of the Roadway

According to research from the former Soviet Union, within the pressure influence zone ahead of the working face, there is a relationship between the roof convergence of the roadway and the surrounding rock strength as well as the mining depth, as shown in Table 2.2. Over the years, most technical managers and researchers addressing this issue have focused on analyzing the relationship between displacement (deformation) and time, without giving sufficient attention to the synchronicity of pressure and deformation.

It is evident that the impact of mining depth varies with the performance of the coal rock mass. In the case of soft rocks, the effect of mining depth on surrounding rock deformation is more pronounced than on hard rock layers. The primary cause lies in the comparison between the surrounding rock stress and the surrounding rock strength, specifically the interaction and synchronicity of the compressive stress that induces displacement (deformation).

### 2.2.3.4 Influence of the Strength of the Surrounding Rock on the Stability of the Tunnel Enclosure

The real-time bearing strength of the surrounding rock in the roadway (i.e., the surrounding rock after support installation) is higher, which results in better elastic

## 2.2 Creep and Phenomena of the Perimeter Rock of the Roadway

**Table 2.2** Relationship between the depth of the roadway, the amount of roof movement and the strength of the surrounding rock

| Strength of surrounding rock mass (MPa) | Roadway burial depth (m) | Top plate approach (mm) | Incremental increase in mining depth for every 100 m of removal (mm) |
|---|---|---|---|
| 30 | 400 | 750 | 212 |
|    | 600 | 1200 |   |
|    | 800 | 1600 |   |
| 50 | 400 | 250 | 115 |
|    | 600 | 470 |   |
|    | 800 | 710 |   |
| 90 | 400 | 80 | 35 |
|    | 600 | 140 |   |
|    | 800 | 220 |   |

strain performance. When the real-time compressive stress on the surrounding rock in the roadway is lower than its inherent long-term bearing stress strength (the minimum stress value at the point of rock mass creep failure), the surrounding rock remains in an elastic strain state, preventing the formation of broken and loosened zones. The surrounding rock support structure exhibits good overall integrity with characteristics of a "shell and arch" structure. The deformation of the surrounding rock decreases over time and with increasing spatial extent, as indicated by the characteristic curve shown in Fig. 2.7.

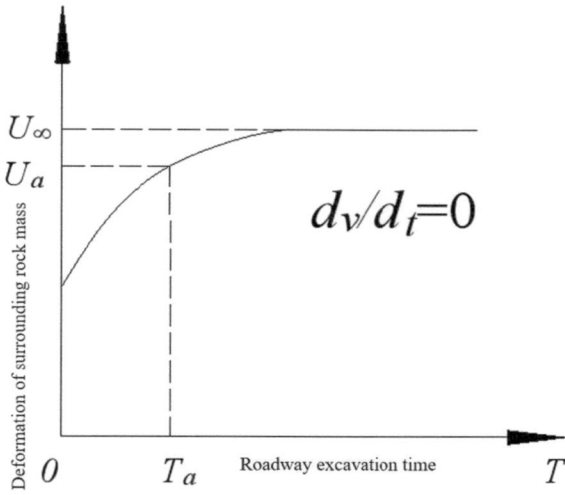

**Fig. 2.7** Enclosure deformation characteristics when the enclosing rock stress is lower than the long-time strength of the enclosing rock

When the real-time compressive stress on the surrounding rock in the roadway exceeds the inherent long-term bearing compressive strength of the surrounding rock, but is less than the instantaneous compressive strength, plastic displacement occurs on the inner side of the surrounding rock, and surface fracturing develops. The compressive stress decreases, or energy is released with a corresponding increase in volume. However, at this point, the surrounding rock still maintains a certain degree of strength and self-stabilizing capacity, with the compressive stress gradually decaying to equilibrium. Changes in the compressive stress state and the loosening and fracturing stop within a certain time range. At the same time, the deformation rate of the roadway stabilizes and gradually becomes consistent as time progresses.

When the real-time compressive stress on the roadway exceeds the inherent instantaneous compressive strength of the surrounding rock, significant damage to the surrounding rock will occur. After the creep to catastrophic failure transition, both from the interior to the surface, a balance effect is reached, and the surrounding rock sequentially experiences loosening, plastic, and elastic zones. Existing academic viewpoints suggest that surrounding rock deformation manifests significant rheological behavior, as shown in Fig. 2.8. This perspective is applicable to soft and extremely soft rock masses but does not adequately address the creep to catastrophic failure in elastic–plastic and brittle-plastic rocks. Therefore, the geological origin and engineering significance of coal rock mass properties, as well as the variation trends of creep and catastrophic failure, are characterized by a diverse representation. In practical engineering, this often manifests as a gradual increase in compressive load on the roadway surrounding rock, with long-term instability, severe deformation of the surrounding rock, and significant heaving of the floor. Observations from the Huainan Jiulonggang Coal Mine's horizontal main roadway show that the roadway displacement increases sharply or gradually over a period of 60–90 days, with some cases extending to 180 days without stabilizing.

### 2.2.3.5 Influence of the Rock Structure on the Stability of the Tunnel Enclosure

Fissures and joints in the coal rock body are collectively called the structural surface, and sometimes the fissures are used to evaluate the two-rock body performance or susceptibility to damage. The structural surface of the rock body can be divided into three categories according to the geological causes: ① The structural surface of the coal rock body formed in the process of rock formation, the primary structural surface; ② The structural surface formed by the earth's crustal movement of the coal rock body (faults, joints, and rock strata misalignment), called the geological tectonic structural surface; structural surface; ③ The surface subsidence movement of the formation of the gravitational effect of the mechanics of the environment, due to the external forces (the pressure becomes smaller, man-made action, etc.) under the action of the rock body to produce a strength weakening structural surface, usually called the secondary strength of the weakened structural surface, usually called the secondary structural surface. There are a large number of structural surfaces in the

## 2.2 Creep and Phenomena of the Perimeter Rock of the Roadway

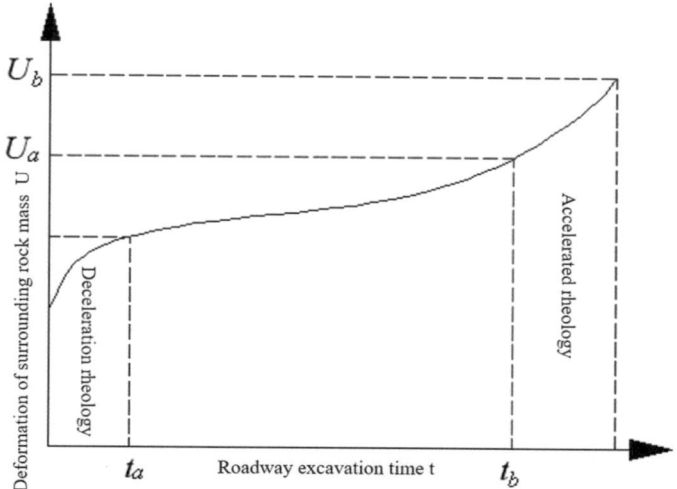

**Fig. 2.8** Deformation characteristics of the surrounding rock when the surrounding rock stress is higher than the instantaneous strength of the surrounding rock

peripheral rock body of the quarry tunnel, but despite this, the peripheral rock body is still able to withstand and decompose higher stresses the action. The high stress is decomposed by the rock mass cut by the structural plane and the structural plane itself. In the rock mass decompose the development of structural surfaces is like a network, which makes the strength of the rock mass weakened to a certain extent, the damage and the bearing strength are uneven, and at the same time makes the rock mass show strong anisotropy. A large number of research results show that the damage to jointed rock bodies is usually due to the slip and expansion of jointed fissures within the rock body and the formation of damaged surfaces, which ultimately leads to the destabilization of the rock body. Therefore, stability analysis of the surrounding rock the deformation characteristics of the rock mass and the evolution effect of joints under mining conditions should be highly emphasized.

1. Damage dynamics and morphology of structural surfaces

The damage patterns of the structural face are mainly slip deformation and dilatational deformation, and we only discuss slip deformation. As shown in Fig. 2.9, assuming that the rock formation there is Ina structural surface with an inclination angle of $\alpha$, the positive and shear stresses acting on the structural surface are respectively:

$$\sigma_n = \sigma_1 \cos\alpha + \sigma_2 \sin\alpha \qquad (2.1)$$

$$\tau = \sigma_1 \sin\alpha - \sigma_2 \cos\alpha \qquad (2.2)$$

**Fig. 2.9** Schematic diagram of forces on structural surfaces

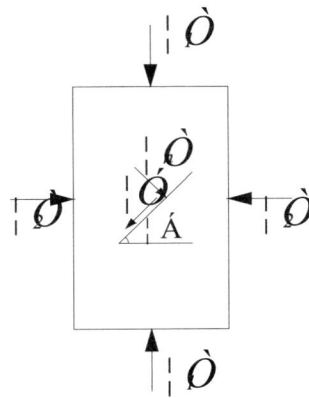

If the structural surface has cohesion C and angle of internal friction φ, by Mohr's law the shear strength of the structural surface is:

$$\tau = \begin{cases} C + \sigma_n tg\alpha & \sigma_n \geq 0 \\ 0 & \sigma_n < 0 \end{cases} \quad (2.3)$$

Thus, the structural face the residual shear stress at after the shear force is:

$$\tau' = \begin{cases} \tau \geq \tau_n = \sigma_1(\sin\alpha - tg\alpha) + \sigma_2(\cos\alpha + tg\alpha\sin\alpha) - C & \tau \geq \tau_n \\ 0 & \tau < \tau_n \end{cases} \quad (2.4)$$

That is, the slip deformation of the closed structural surface is produced by the action of the residual shear stress $\tau'$.

The slip deformation characteristics of the structural surfaces within the rock formation are analyzed below.

When the structural plane is in the influence area of the bearing pressure, the deformation of the structural plane is the largest, which is the object of our discussion. When the structural face is located in the zone of influence of support pressure, the stresses are respectively:

$$\begin{cases} \sigma_1 = K\gamma H \\ \sigma_2 = \lambda K\gamma H \end{cases} \quad (2.5)$$

where: K is the stress concentration factor.

γ is the gravity density of rock strata under the condition of established coal seam.

H is the buried depth of rock strata under the condition of established coal seam.

$\lambda = \frac{\mu}{1-\mu}$ is the lateral pressure coefficient.

Integration of the above equations yields the positive and shear stresses acting on the face of the structure, respectively:

## 2.2 Creep and Phenomena of the Perimeter Rock of the Roadway

$$\sigma_n = K\gamma H(\cos\alpha + \lambda\sin\alpha)$$

$$\tau = K\gamma H(\sin\alpha - \lambda\cos\alpha) \tag{2.6}$$

Further collation can be obtained:

$$\sigma_n \geq \arcsin\frac{C}{K\gamma H\sqrt{(1+\lambda^2)(1+tg^2\phi)}} + \arcsin\frac{\lambda + tg\phi}{\sqrt{(1+\lambda^2)(1+tg^2\phi)}} \tag{2.7}$$

That is, the critical angle of sliding deformation of the structural surface of the rock layer under the supporting pressure condition when the inclination angle of the structural surface is less than $\alpha$ only produces compression deformation, not slip deformation.

When the structural surface lies outside the zone of influence of the support pressure, the stresses are respectively:

$$\begin{cases} \sigma_1 = 0 \\ \sigma_2 = \lambda\gamma H \end{cases} \tag{2.8}$$

Then there are the positive and shear stresses on the structural surfaces, respectively:

$$\sigma_n = \lambda\gamma H \sin\alpha$$

$$\tau = -\lambda\gamma H \cos\alpha \tag{2.9}$$

Then the residual shear stress on the face of the structure is:

$$\tau' = -\lambda\gamma H(\cos\alpha + tg\phi \sin\alpha) - C \tag{2.10}$$

It can be seen that the smaller the inclination angle of the structural surface $\alpha$ is, the greater the possibility of generating slip deformation, which explains that the rock layer at the bottom of the roadway is prone to slip deformation under the condition of a small-angle structural surface, leading to delamination and the formation of bottom drums.

### 2.2.4 Physical Characteristics of Surrounding Rock and Control Program—Integration of Support Structure

The physical characteristics of the perimeter rock of the quarry tunnel mainly include the lithological composition, structure and spatial scale of the rock layer in which the perimeter rock is located, mechanical performance parameters, and the strength of the carrier of the perimeter rock support structure body. Deep mining surrounding rock undergoes elastic–plastic deformation, decompression and expansion displacement to produce fissures and their expansion, expansion and extrusion, energy impacts in the deep protolithic area and gas energy release and other processes. Obviously, the structure, mechanical parameters and strength of the surrounding rock in different deformation and destruction stages have changed accordingly. With different stress states, the surrounding rock presents different strength characteristics; support such as anchor net cable and grouting will improve the mechanical structure parameters, and the surrounding rock then improves the overall support bearing capacity of the surrounding rock. The tunnel support and its post-wall filling will keep the surrounding rock in a good stress state, which significantly improves the bearing strength of the surrounding rock body. This effective support fully embodies the holistic support structure of the tunnel peripheral rock ideology, which greatly extends the service life of the tunnel. Therefore, the control scheme proposed here aims to reflect the holistic concept of the support structure and systematic.

(1) Control program. According to the performance characterization and disaster type of different roadways surrounding rock bodies, different surrounding rock disaster control program is designed. At the same time, they should be in real-time according to the changes in the stress environment in which the roadway is located, as well as the creep and drastic changes in physical characteristics and mechanical properties. Control program adjusted Improving the overall bearing strength (elastic energy) of the surrounding rock is the purpose of controlling the surrounding rock disaster in deep mining. The support and reinforcement structure strength of the surrounding rock, that is, the surrounding rock support structure composed of support structure body and surrounding rock dynamic overall strength, as well as the reinforcing effect of filling and sealing after racking are the to prevent excessive deformation of the roadway and disaster control of preferred solutions. The idea of decompression and energy reduction is an effective control method to prevent the local extremely high elastic energy (coal rock elastic energy and gas dynamic elastic energy) to stimulate the impact effect and induce the surrounding rock disaster.

(2) Construction time. Paying attention to the construction time is the key to fully considering the time effect, construction quality and process effect. For example, the full sealing of the surface holes and cracks of the roadway and the reinforcement of the surrounding rock should be completed within the self-stability time of the surrounding rock exposure; for the extremely broken surrounding

rock that is exposed to collapse, it is necessary to take advanced curtain pre-reinforcement measures; the protective layer mining needs to be ahead of the scale and time effect of the protected layer mining; the technical measures to prevent impact and outburst should first consider the timeliness, effectiveness and creep of the predicted target.
(3) Construction quality. Under the premise of the symptomatic and good time effect of the scheme, the construction quality is the fundamental guarantee to realize the control function of the technical scheme, and also the guarantee of the control effect of the surrounding rock catastrophe, including the strict implementation of the technical scheme and the timely adjustment with the abnormal changes of the construction site conditions. It requires the site construction to have a high technical management level, and fully grasp the diversity of elements that may induce catastrophe and creep and catastrophe.

Obviously, if the stress environment, physical state and control program interact with each other and fail to maintain the overall stability of the perimeter rock and support structure during the service period of the roadway, the perimeter rock and support system will be subject to sustained creep damage or instantaneous rapid deformation damage, and then various types of perimeter rock catastrophe accidents occur, which will significantly reduce the service period of the roadway.

## 2.3 Mechanics of Anchor Mesh Cable Supported Structural Body

Practice shows that the anchor network cable or anchor network cable concrete support structure is the mining roadway support most scientific and effective method. With other support methods compared, this kind of support method is favorable to the reform of coal mining technology, which for coal mining method gives a larger development space for the development of equipment. So far, the main engineering problems are mainly reflected in the bearing strength of the support structure and the improvement of the service period. Summarizing and analyzing the time effect of the change of service life of the anchoring force of a single engineering anchor rod (or anchor cable), and the deformation and destruction law and time effect of the anchor net cable support structure body, it is not difficult to find that the support strength is the key technology to resist the deformation of the surrounding rock, as shown in Fig. 2.10. The variation law of Figures (a) and (b) or the support characteristic curve reveals that the deformation of the surrounding rock of the roadway and the weakening of the anchoring force of the bolt and anchor cable are the inherent creep characteristics and deformation development law of the roadway. Due to the combination of time–space effect and creep effect, both the deformation of the roadway and the anchoring force of the bolt will undergo the failure process of creep to drastic change. Therefore, it is of great significance to take active measures to improve the

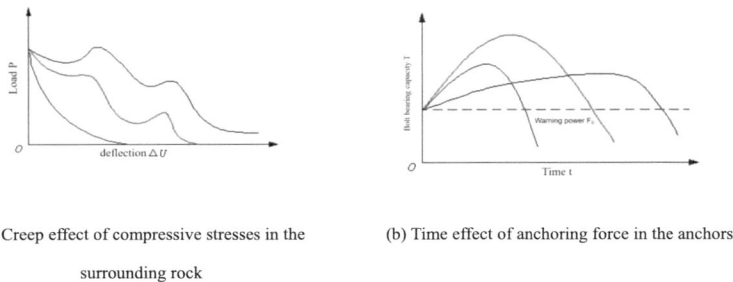

(a) Creep effect of compressive stresses in the surrounding rock

(b) Time effect of anchoring force in the anchors

**Fig. 2.10** Effect of service life of the roadway versus (anchoring force, deformation)

strength of the bolt-mesh-cable support to extend the service cycle of the roadway and support engineering.

## 2.3.1 Anchorage Strength of Anchor Cable

The maximum anchorage strength of the anchor rods and cables is reflected in the axial direction of the anchor rods and cables of the surrounding rock control ability, the engineering physical meaning of the anchorage strength mentioned in the engineering practice is that artificially through the combination of the anchor rods to give the anchor rods in the axial direction of the anchor force generated by the control of a certain range of the surrounding rock, and to make the control of the surrounding rock to achieve the predetermined strength. To make the loose surrounding rock in the control range produce the required strength, with elastic deformation capacity or elastic deformation modulus. The design of reasonable density of anchor rods and anchor cable row spacing and anchor installation angle direction can improve the whole roadway rock strength and deformation resistance (elastic deformation modulus), that is, the role of the anchor network cable support structure is to actively achieve the realization of the anchor network cable support structure and loose surrounding rock combination of a whole structure, to achieve the overall strength and capacity of the perimeter of the rock bearing requirements. Anchor axial force (anchoring force) generated by the different anchoring methods: mechanical anchoring, the axial force is mainly generated by the friction between the anchor and the borehole wall; bonded anchoring, axial anchoring force (load bearing capacity) by the anchoring agent formed by the anchor and the surrounding rock (the borehole wall) between the bond and friction and the joint action.

The numerical size, distribution characteristics, and control range of the anchoring force of the anchor rods and anchoring cables depend on the size of the displacement and deformation between the licensed anchor rods and anchoring cables and the surrounding rock layers, as well as the coupling performance between the anchor rods and the surrounding rock (strength, elastic deformation modulus). Anchor strength of

## 2.3 Mechanics of Anchor Mesh Cable Supported Structural Body

the anchor and surrounding rock is the premise of anchor anchoring force generation, because the coordinated coupling performance between the two under the action of the same force field is the key; coordinated coupling between the anchor and surrounding rock is the guarantee of anchoring strength, the anchoring method and the performance of the anchor combination components and installation angle, and its configuration determines the degree of coordinated coupling, the difference in the anchoring method will result in the overall structural coupling stiffness between the anchor and the surrounding rock, and strength Parameters of the different; installation of the anchor (or anchor cable) after the roadway perimeter rock structure to withstand the continuous mining power (and the overlying rock subsidence movement impact, elastic potential energy of wave energy impact effect) of the perturbation of the influence of the peripheral rock structure will be caused by creep triggered by the anchorage strength of the weak changes, resulting in a series of displacement and deformation, anchor cable anchorage force decline, rock fragmentation and other damage phenomena. If the time period of this failure phenomenon is allowed, it can meet the production requirements. The strength and deformation resistance of the whole support structure system formed by the combination of the anchoring method, anchoring components and anchoring structure (anchor net cable support structure) and the surrounding rock layer is the to ensure that the length of the damaged time cycle meets the production safety requirements technical key.

The anchoring force refers to the pre-tightening extrusion combined force applied to the surface of the surrounding rock by the auxiliary components of the bolt (tray, rigid beam, anchor net, etc.), eliminating the crack cracks of the existing loose circle and improving the overall strength of the surrounding rock anchoring. The magnitude of the anchoring force is equal to the axial force of the bolt at the tail of the anchor, and the anchoring force is one of the key elements to ensure the anchoring strength of the surrounding rock support system of the roadway. The distribution characteristics of the axial force of the anchor rod are related to the properties of the anchor rod and the surrounding rock, the anchoring method, the change of the anchoring strength (or stress field) of the surrounding rock (the influence of mining factors), etc., and also related to the magnitude of the prestress and the configuration and performance of the auxiliary components of the anchor rod. Generally, under the condition that the maximum axial force of the bolt is equal, the anchoring force of the end anchoring bolt is greater than that of the full-length anchoring.

The axial compressive stress of the roadway surrounding rock, that is, the lateral bearing capacity formed by the anchor network cable is the anchor network cable support structure body to restrain and prevent the surrounding rock layer from slipping and breaking, generating new shear damage to the nodal surface as well as the overall axial twisting and deformation of the roadway surrounding rock or asymmetric deformation damage.

Usually, the anchor mesh cable and the surrounding rock combination structure body long continued to withstand mining, roof subsidence motion power, elastic potential energy wave energy of the impact of the formation of the integrated compressive stress field under the action of the surrounding rock layer will produce the following forms of displacement and deformation development trend: ① Between the parts of the weak surface along the relative misalignment; ② Along the specific direction of the generation of a new shear damage surface and along the destruction of the surface of the misalignment; ③ Fragmented rock mass movement, the anchor The structural body of the anchor cable will restrain and prevent the development of this kind of displacement damage, and the longitudinal and transverse force of the anchor cable will be generated. The longitudinal and transverse force generated by the anchor cable and the surrounding rock layer can be summarized as three conditions: ① The shear deformation and relative displacement of the surrounding rock within the anchorage range. Anchor net cable support is a kind of active support mode, through the auxiliary components can make the anchor produce a certain (artificially predetermined) prestressing force (initial anchoring force), so as to make the loose within the scope of anchoring control (loose circle) perimeter rock to form a compressive stress field to achieve the role of support. The support of the whole piece of anchor network cable forms the overall strength of the surrounding rock structure to improve. ② Anchor rods in close contact with the surrounding rock will produce good strength and mechanical properties. ③ Anchor has shear strength and elastic stiffness, so that the anchor (anchor cable) effectively controls the longitudinal and transverse deformation of the surrounding rock damage and forms a strong integral control. According to the conditions of the longitudinal and transverse force of the anchor rod, it can be seen that different anchoring methods will lead to different longitudinal and transverse forces, and different anchoring methods make the overall strength and mechanical properties between the anchor rod and the surrounding rock different. In full-length anchorage, the anchor rod will produce strong longitudinal and transverse force in the whole effective length range, while in end anchorage, the anchor rod only produces a strong control effect on the surrounding rock between the anchored segments. Improving the overall anchorage strength of surrounding rock is not equal to increasing the preload force of anchor rods and cables (initial anchorage force), the initial anchorage force is too large to produce too high a local stress between anchor rods, anchor cables, anchors and borehole walls, which is easy to trigger the anchorage force to fall very early and affect the service life of the roadway. Attention should be strengthened to the details of the support structure in terms of integrity and coordinated coupling (angular direction, density, control range, standardization of pre-tensioning force by area, timeliness of process links, etc.).

### 2.3.2 Mechanical Effect of the Anchor Mesh Cable Structure Body on the Surrounding Rock

Under the condition of different surrounding rock properties, anchor net cable shows different supporting strengths and different mechanisms of action, such as the combination of layered rock mass, the squeezing effect of loose rock strata, the wedge effect of jointed rock mass and the suspension effect of weak rock mass. Although the

## 2.3 Mechanics of Anchor Mesh Cable Supported Structural Body

mechanism of action of an anchor cable is different due to the different performance conditions of the surrounding rock, its essence can be attributed to improving the overall stress state of the surrounding rock, improving its strength index, and forming a whole surrounding rock anchorage with high strength index and strong deformation adaptability. Therefore, the effect of anchor net cable can be measured by the improvement of the overall mechanical property index of the combined structure of anchor net cable and surrounding rock. Enhance and ensure the overall compressive strength of the combined structure of the anchor network cable and perimeter rock, need to interact and coordinate the following four aspects of the interconnection: ① The size of the pre-tightening force (also known as the initial anchor force) of the installation of the anchor cable, the size of the pre-tightening force is generated by the combination of the anchor cable, the lock, the tray, the anchoring agent and the surrounding rock; ② The formation of anchoring force by the installation of the anchor bar preload, anchor, drilling, peripheral rock body (layer) of the interconnection of the interactive formation of the anchorage; ③ The mechanical properties of the surrounding rock mass (layer) reflect the interactive coupling relationship between the compressive strength and the anchoring strength;④ The matching coupling coordination between the anchoring strength reflected by the anchoring force and the inherent strength of the original rock. As shown in Fig. 2.11.

So far, domestic and foreign views on the structural action of anchor network cable can improve the strength of the surrounding rock integrity have converged, but there are still different discussions on the understanding of the way, mode and mechanism of its action. This book carries out the following five aspects of analysis and discussion.

(1) The change rule of anchoring force

The tangential anchoring force generated by the anchor bolt and cable, that is, the resistance and binding force of the anchor bolt and cable to the shear deformation of the surrounding rock and the displacement deformation of the surrounding rock, is essentially to increase the overall shear strength of the anchored surrounding rock mass, that is, to improve the cohesion of the anchored surrounding rock mass.

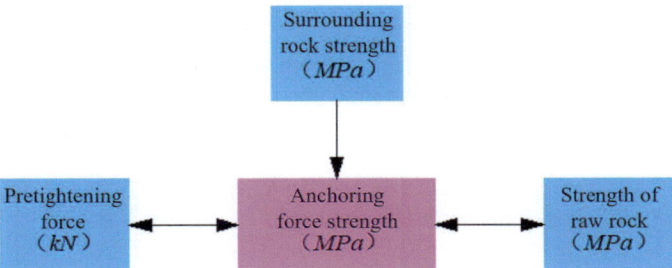

**Fig. 2.11** Principle of interactive and coordinated coupling of the combined structural body of the anchor mesh cable and the surrounding rock

According to the mechanism of tangential anchoring force, the maximum value is the shear strength of bolt material, namely cohesion. If the cohesion of the anchor rods and cables (mainly refers to the maximum tensile strength) is recorded as $C_b$, the cohesion of the anchored perimeter rock body is recorded as $C_r$, and the cohesion of the anchoring solid is recorded as $C$, and it is approximated that the mechanical parameter of the anchoring agent is the same as the mechanical parameter of the anchored perimeter rock body, then there are

$$C = C_r + nS(C_b - C_r) + \sigma \tan\phi \tag{2.11}$$

where: n is the density of anchor arrangement;

$S$ is the cross-sectional area of the anchor;

$\sigma$ is the compressive stress on the surrounding rock caused by the axial action of the anchor;

$\phi$ is the angle of internal friction of the enclosing rock mass.

The increase in cohesion of the anchor solid compared to the cohesion of the surrounding rock mass in the absence of anchors was

$$\Delta C = C - C_r = nS(C_b - C_r) + \sigma \tan\phi \tag{2.12}$$

It can be seen that the magnitude of the improvement of the cohesion of the anchored surrounding rock body depends on the size of the difference between the cohesion of the anchor rods, anchor cables and the surrounding rock body and the proportion of the cross-sectional area of the anchor rods in the total area of the anchored solids (the cross-sectional area of a single anchor rod and the density of the anchor arrangement), as well as the strength of the axial force of the anchor rods, anchor cables. Here for illustration, in the actual engineering of the mechanical properties of the surrounding rock and the mechanical properties of the formation of anchors are different, this difference causes the mismatch and uncoordinated coupling of the performance of the anchor solid and the surrounding rock body, which can greatly reduce the weakening of the support structure system of the anchoring force or anchoring strength.

(2) Resistance effect of anchor cable anchoring force on surrounding rock

Similar to the principle of variation of cohesion, the magnitude of the friction angle within the anchored solid is determined by the internal friction angle of the anchor rods and anchor cables, the magnitude of the friction angle within the unsupported enclosing rock body, and the state of stress on the friction surface. If the stress state generated by the anchor rod and the anchored enclosing rock body are the same, the internal friction factor of the anchored solid is

$$f = f_r(1 - nS) + f_b nS \tag{2.13}$$

where: $f$ is the internal friction factor of the anchor solid;

## 2.3 Mechanics of Anchor Mesh Cable Supported Structural Body

$f_r$ is the internal friction factor of the enclosing rock mass;
$f_b$ is the internal friction factor of the anchor.

The difference between the internal friction factor of the anchor solid and the surrounding rock $\Delta f$ is

$$\Delta f = f - f_r = (f_b - f_r)nS \tag{2.14}$$

Usually, $f_b < f_r$, so the internal friction factor of the anchor solid is reduced compared with that before anchoring. However, since the cross-sectional area of the anchor rods in the anchored solid accounts for a small proportion, this effect can be ignored in practical calculations, that is, the internal friction angle of the anchored solid is considered to be equal to the internal friction angle of the peripheral rock layer before anchoring. It is worthwhile to pay attention to a problem, anchor rods anchor cable anchoring force control under the action of the formation of the anchored surrounding rock body and the uncontrolled surrounding rock performance is different (the formation of friction factors is different), it is very easy to cause the two stresses are different and the formation of destructive fissure cracks, resulting in the phenomenon of detachment from the anchoring control.

(3) Strength of the structural body supported by the anchor network cable

The above analysis shows that the essence of the axial anchoring force on the anchor solid is to change its stress state, that is, to enhance the range of axial compressive stress, so that the anchor solid is changed from the approximate unidirectional compression state before anchoring to a bidirectional compression state, so that the transverse perimeter compressive strength produced by the anchor solid can be enhanced and improved. According to Mohr–Coulomb strength theory, the transverse (perpendicular to the direction of the anchor) compressive strength of the anchor solid can be expressed as follows

$$\sigma_{\max} = \frac{2C\cos\phi}{1 - \sin\phi} \tag{2.15}$$

$\sigma_{\max}$ is the lateral (or circumferential) compressive strength of the anchor solid, which increases compared to the circumferential strength in the unsupported condition $\Delta\sigma_{\max}$

$$\begin{aligned}\Delta\sigma_{\max} &= \sigma_{\max} - R_c = \frac{2C\cos\phi}{1 - \sin\phi} - \frac{2C_r\cos\phi}{1 - \sin\phi} \\ &= 2(C - C_r)\frac{\cos\phi}{1 - \sin\phi} + 2\sigma\frac{\sin\phi}{1 - \sin\phi}\end{aligned} \tag{2.16}$$

where: $R_c$ is the uniaxial compressive strength of the surrounding rock mass (rock layer) before anchoring.

According to Griffth's strength theory analysis, the annular compressive strength of the anchor solid can be expressed as

$$\sigma_{max} = 4R_t\left(1 + \sqrt{1 + \frac{\sigma}{R_t}}\right) + \sigma \tag{2.7}$$

Increase compared to the unanchored perimeter rock $\Delta\sigma_{max}$

$$\sigma_{max} = 4R_t\left(\sqrt{1 + \frac{\sigma}{R_t}} - 1\right) + \sigma \tag{2.18}$$

where: $R_t$ is the uniaxial tensile strength of the surrounding rock mass.

The results of both strength theory analyses show that the axial force of the anchor rods and anchor cables actively changes the stress state of the surrounding rock body (layer), that is, the strengthening effect of strength, so that the lateral (the axial surrounding rock of the roadway) compressive strength of its overall support structure is improved.

(4) Compressive strength of the anchor web cable

Combined beam theory that, when the rock layer is a thin layer, through the anchor rods and anchors will be several layers of rock layers combined as a whole, can make the rock layer's lateral bearing load action produced by the bending stress reduced to the combination of the former $\frac{1}{n}$, that is, equivalent to the strength or load-bearing capacity increased by n times. However, this effect is produced on the premise that the integrity of the combined beam does not break down, that is, the shear capacity of the interlayer should be sufficient to withstand the destruction of the shear stresses generated during the bending deformation. However, the improvement in interlayer bonds caused by anchors is sometimes limited, so the change in flexural strength of anchored solids should be measured not only by taking into account the variability in the deformation characteristics of the combined beams versus the stacked beams; but also by taking into account the changes in the interlayer shear stresses as well as the shear capacity.

From the theory of elasticity, when the combined beam undergoes transverse bending, the distribution of shear stresses in it are

$$\tau_{max} = \frac{Q_{max}S_{max}}{I} = \begin{cases} \frac{3qb}{2n_rh} & (n_r \text{ when even}) \\ \frac{3qb}{2n_rh}\left(1 - \frac{1}{n_r^2}\right) & (n_r \text{ when odd}) \end{cases} \tag{2.19}$$

The shear strength possessed between the layers after anchoring is:

$$[\tau] = nSC_b + \sigma\tan\phi \tag{2.20}$$

where: $Q_{max}$ is the maximum shear force in the cross-section;
$S_{max}$ is the maximum static moment;
$q$ is the transverse uniform load;
$b$ is the span of the combined beam;

## 2.3 Mechanics of Anchor Mesh Cable Supported Structural Body

$n_r$ is the number of layers in the rock formation;
$h$ is the thickness of layering in the rock formation;
$n$ is the density of the anchor arrangement;
$S$ is the cross-sectional area of the anchor;
$C_b$ is the cohesion of the anchor (cohesion embodies the combined deformation strength of shear strength and the ultimate tensile force of the anchor);
$\sigma$ is the normal compressive stress in the layer;
$\phi$ is the friction factor of the interlayer.

Only when $[\tau] \geq \tau_{\max}$, the deformation of the rock formation will exhibit the performance characteristics of the combined beam. Otherwise, the anchorage will fail and the bending condition of the rock layer will return to the state of the stacked beam before combination. It can be seen that the realization of the overall bending effect of the combined beam of anchor rods and anchor cables is conditional. The above analysis shows that the cohesion and confining pressure of the rock mass in the anchorage body are improved due to the action of the anchor rod, so that the transverse compressive strength, bending strength (or deformation strength) and shear strength along the weak surface are improved to a certain extent. The magnitude of the increase depends on the difference between the mechanical parameters of the anchor rod and the mechanical parameters of the surrounding rock mass (rock stratum), as well as the coordinated coupling relationship between the increase of the overall compressive strength of the surrounding rock supporting structure and the anchor force of the anchor cable. It can be seen that the looser and weaker the surrounding rock layer or the rock layer with good laminar morphology, the more obvious the role of the anchor rods and the characteristics of the combined beam.

(5) Compressive strength of the structural body of the anchored mesh rope enclosure

One of the main functions of anchor rods and cables is to improve the stress state of the surrounding rock by pre-tightening and squeezing the existing surrounding rock loosening circle, to improve its compressive strength and to enhance its deformation resistance or elastic deformation modulus. If the anchor rods and cables and the surrounding rock body under their action are regarded as a whole, the reduction of the deformation of the surrounding rock body within the control range of the anchorage is equivalent to the improvement of the elastic modulus of the anchored solid. If the modulus of elasticity and Poisson's ratio of the anchor solid are $E$, $\mu$, the modulus of elasticity and Poisson's ratio of the unanchored perimeter rock layer are $E_r$, $\mu_r$, and the modulus of elasticity and Poisson's ratio of the anchor rods are $E_b$, $\mu_b$, then it can be deduced that

$$E = \frac{E_r}{1 - \lambda \mu_r} \qquad (2.21)$$

$$\mu = \frac{\mu_r - \lambda}{1 - \lambda \mu_r} \qquad (2.22)$$

Among others, $\lambda = \left(\frac{\mu_r}{E_r} - \frac{\mu_b}{E_b}\right) / \left(\frac{1}{E_r} + \frac{1}{nSE_b}\right)$.

From the above equations, it can be seen that if $\lambda = 0$, that is, there is no difference between the deformation properties of the anchor and the surrounding rock body, the modulus of elasticity of the anchored solid is the same as that of the anchored solid without anchor; if $\lambda > 0$, the modulus of elasticity of the anchored solid, E, will be improved and the Poisson's ratio, $\mu$, will be reduced. This analysis shows that the increase of elastic modulus can also have positive strength effects on improving the strength of the surrounding rock, for example, choosing the optimal anchoring agent and size parameters such as the anchoring agent drug ring and the diameter of the drilled hole, as well as increasing the diameter of the anchor rods and the scale of the anchors are ways to increase the modulus of elasticity of the anchored solids. It is also positive to understand the engineering meaning of elastic deformation modulus or compressive strength as elastic strength.

## 2.4 Monitoring Rationale and Methodology

The deformation and destruction process of creep to drastic change occurs after the coal rock body is subjected to the action force or impacted by the elastic potential energy. The driving force for the creep of the roadway surrounding rock and the functional failure of the supporting structure is the result of the joint action of the high compressive stress of the roof plate and the decompression and release of elastic energy from the elastic zone of the original rock in the deep part of the roadway gang. However, there are multiple elements behind this result change, and the quarry roadway supported by the anchor network cable structure is subjected to impact damage from both impact forces or elastic strain energy during the service period.

(1) Working face mining and mining hollow area (including adjacent to the hollow area) overlying rock movement induced deformation and destruction of the roadway surrounding rock or roadway local form of asymmetric deformation and other phenomena, resulting in the roadway loose circle of the surrounding rock extrusion and swelling, when serious, induced the release of the deep original rock area coal rock energy decompression resulting in the rock loosening and expansion, expanding the scope of the circle of loosening. This dynamic process produces impact damage to the anchor network cable support structure, causing the rock body within the anchoring range to loosen and break, reducing the anchoring force or completely losing the anchoring effect, which in turn reduces the overall impact resistance of the anchor network cable structure. By installing the anchor force gauge, it can be found that the reading of the measuring instrument decreases or the display value remains unchanged or the display value is zero (however the monitoring value of the multi-point displacement meter or off-layer meter, becomes larger). It indicates that the load force of the anchor cable has abnormal changes, the anchoring force has different degrees of decline or no anchoring force phenomenon, the real-time

## 2.4 Monitoring Rationale and Methodology

anchoring force is lower than the initial anchoring preload, and it is not possible to realize the original intention of the implementation of a unified anchoring preload standard for the installation of the anchor cable.

(2) Coal seam protruding tendency with impact ground pressure and, the roadway, in addition to bearing the impact of mining and the neighboring hollow area roof subsidence movement, but also to bear the coal rock body high-stress area to the support system to transfer the release of elastic strain energy impact. In the process of elastic stress wave release and transmission, the impact effect of wave energy expands the original rock holes and fissures, reduces the performance of the supporting structure system, leads to the uneven anchoring force of the bolt (cable), and then weakens the anti-disaster ability to a certain extent.

As shown in Fig. 2.12, the compressive stress transmitted by the roof plate (or overlying rock layer) $P_Z$ and the elastic energy released from the original rock zone in the deep part of the gang $\phi$ induce the original rock body to decompress and expand to displace and create cracks. The roadway side is subjected to the shear action of the horizontal compressive stress $P_S$ and $P_Z$ transmitted by the deep rock mass, which reduces the bonding force between the bolt anchoring agent and the rock mass (that is, the anchoring force is reduced), resulting in the failure of the anchoring effect of the bolt (cable). The peripheral rock on the top plate of the roadway began to creep(ropes) and developed to the state shown in Fig. 2.12a: $P_Z > P_S$, such an extremely unbalanced mechanical relationship, exacerbated the roadway damage. Especially adjacent to the side of the hollow area of the roadway surrounding rock to bear the hollow area overlying rock layer subsidence movement is more serious, by the impact of subsidence dynamic load force, the bearing strength of the supporting structure is the more common phenomenon of reduction. As shown in Fig. 2.12b, due to the anchor installation preload is too large, resulting in anchor solid (anchoring agent and drilling hole wall around) around the local stress concentration and producing cracks, and in serious cases, producing the phenomenon of de-anchoring.

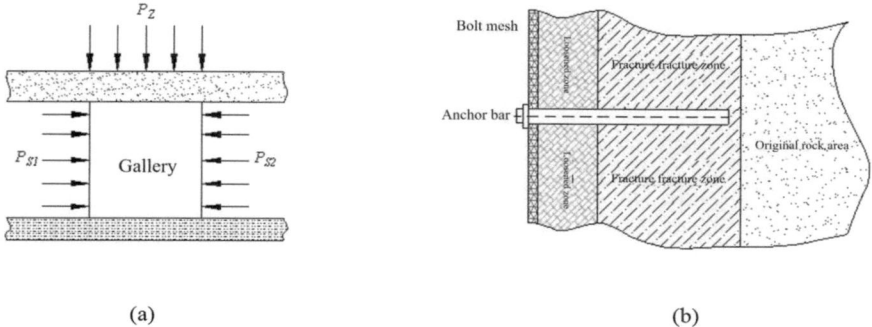

**Fig. 2.12** Sketch of the damage failure principle of anchor network cable engineering

## 2.4.1 Alternative Information

In the process of creep to the drastic change of the supporting structural body of the roadway perimeter rock, corresponding information signals, which will be released can be called physical information signals or physical quantity signals. Objectively speaking, the strength of the information signal generated by each link is different from the bare excavation of the roadway to the production service roadway. For some reason, the surrounding rock structure sometimes creeps and develops, and sometimes drastically changes or destabilizes. In terms of monitoring methods and the difficulty of recognizing information signals alone, four factors should be emphasized in the monitoring methods of quarry roadway perimeter rock:

(1) The amount of absolute or relative change in the signal is large and intense;
(2) Continuity of information change (from nothing to the best) and ease of recognition and utilization;
(3) The environmental and mechanical-structural relationships are adapted to the monitoring site whichever method of monitoring the signal;
(4) The information signal is associated with compressive stress, and the signal appears when there is compressive stress or elastic energy acting.

The commonly used information that is currently found to be recognized is shown in Table 2.3.

The monitoring of the information of the creep development process of the roadway surrounding rock should focus on distinguishing different environments and different parts of the information signal generating triggers and geomechanically triggering factors. When analyzing the monitoring information, we should

**Table 2.3** Information table for monitoring deformation and damage in the lane

| Typology | Type of information | Location of the quarry | Signal strength | Monitoring utilization degree of difficulty | Relationship between quarry stress and signal generation |
|---|---|---|---|---|---|
| Acted upon by force or elastic energy | Peripheral rock deformation | Roadway | Creep rise | General analysis | Significant, direct |
| | Deep displacement (physics) | Roadway | Slate | Continuous monitoring and analysis | Direct |
| | Compressive stress | Anchor | Creep | General analysis | Significant, direct |
| | Acoustic emission | A coal seam | Sore | Continuous monitoring and analysis | Significant, indirect |
| | Compressive stress | A coal seam | Sore | Continuous monitoring and analysis | Associated, direct |

consider: the influence of mining on the stability of the surrounding rock, the environment of the surrounding rock of the roadway in the first mining face, the change of the support method and parameters, the movement of the rock layer on the roof of the adjacent mining area, the influence of the overlying rock layer of the mining area on the surrounding rock of the roadway, the mechanical effect of the thick hard top plate on the surrounding rock of the roadway, the mining height and the size of the pillar, the cross-section of the roadway, and the size of the roadway, etc. There are a lot of affecting the mechanical structure factors. Changes in each element may induce changes in monitoring information or disaster signals.

## 2.4.2 Evaluation of the Application of Monitoring Methods

Roadway perimeter rock support structure body deformation damage directly induced power for the deep surrounding rock stress (two-way stress) or strain elastic potential energy. Deep surrounding rock mutation type power disaster breeding evolution process, rock decompression effect induced rock expansion displacement, fissure expansion will be energy and release in the form of acoustic wave, electromagnetic and other information signals, resulting in a series of physical information changes. The information collection and monitoring method should also fully consider whether the information released by the coal and rock body is suitable for continuous real-time monitoring or random monitoring. The preliminary stage of geological disasters such as impact ground pressure, coal and gas protrusion, and water burst will produce corresponding physical information signals, for example, acoustic emission or microseismical phenomena, displacement and deformation, and pressure changes. For the monitoring method and instrumentation development, there is no essential difference, the only difference is the identification and utilization of monitoring signals, that is, the difference of identification methods in different fields of expertise, such as the existence of karst water in front of the roadway excavation, the water pressure can also induce the displacement and deformation of the rock body, triggering the acoustic emission phenomenon and the change of the temperature of the coal and rock body. Based on this, the use of acoustic emission or microseismical monitoring technology, ultrasonic detection technology, electromagnetic radiation monitoring technology and other geophysical methods to evaluate the dynamical change rules of the mechanics of the deep coal rock body has become a research hotspot in recent years.

#### 2.4.2.1 Monitoring of Pressure, Displacement and Deformation of the Perimeter Rock of the Roadway

In engineering practice, the parameters such as pressure, displacement and deformation of the roadway perimeter rock are usually simply understood as the roadway stability parameters. The evaluation of compressive stress is used to illustrate the direct destructive force that triggers the creep to drastic instability of the anchor mesh cable perimeter rock support structure body, and the evaluation of deformation and displacement is used to analyze the surface deformation and deep rock displacement induced by the compressive stress on the anchor mesh cable perimeter rock support structure body.

(1) Top plate pressure P1 monitoring. The deformation force or compressive stress of the roof plate within a certain range is analyzed by monitoring the bearing force (or load force) P1 of the anchor rods and anchor cables. Accurately speaking, P1 should be the load force and compressive stress of the surrounding rock mass (layer) in the anchorage range of the bolt (cable), not including the deformation compressive stress (pressure) of the roof given by the upper strata above the anchorage range (the transfer force of the subsidence movement of the overlying strata adjacent to the goaf).

(2) Roadway gang compressive stress monitoring. By monitoring the bearing force P2 (or compressive stress) given to the anchor by the surrounding rock body within the anchoring range (anchoring section) of the inner and outer gang anchors in the roadway, the value of this stress reflects the deformation and destructive force directly induced to the gang. In fact, P2 is a combined force, it is the top plate on the rock layer transferred to the gang of the compressive stress and the gang of the horizontal compressive stress of the combined force, the direction of its action is the radial direction of the roadway, as to which side of the transfer of compressive stress for the dominant, should be analyzed according to the specific circumstances.

(3) Tunnel deformation monitoring. Surface displacement of the roof plate is monitored by using the peripheral rock top plate detachment meter, which is used as an index for evaluating and analyzing the stability of the peripheral rock top plate and the depth stress damage. This kind of displacement and deformation that occurs in the deep part of the perimeter rock of the roadway from the inside to the surface also indirectly reflects the distribution of destructive force around the perimeter rock of the roadway, which provides judgment and evaluation for determining the key parts of the support (such as the length, installation site and angle of the anchor rods) and the reliability of the support.

(4) Deep displacement monitoring of the roadway surrounding rock. By installing displacement meters of different depths in the drill holes, the development and distribution of fissures and cracks inside the coal and rock bodies are analyzed. Judge the stability of the rock body through the deep displacement information, evaluate the rationality of the support design and the adaptability of the rock body stability, and analyze the destructive effect of the stress and the depth of the original stress area through the location of the fissure cracks.

### 2.4.2.2 Strength Monitoring of Coal Pillars in Quarries

Based on the mine pressure perspective, the role of the quarry coal pillar in the back-mining face is to carry the subsidence pressure of the rock layer on the upper part of

## 2.4 Monitoring Rationale and Methodology

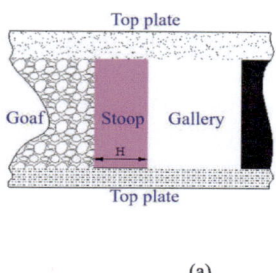
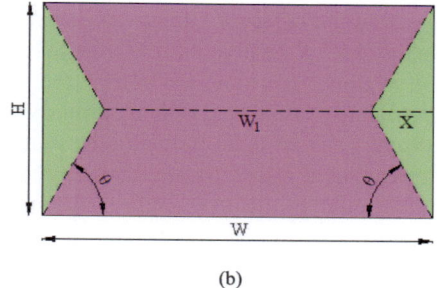

(a)                                          (b)

**Fig. 2.13** Coal pillar height versus damage domains

the roof plate and play the role of supporting the roof plate to come under pressure. The triangle or symmetrical triangle of the rupture area of the coal pillar, the rupture width X of the coal pillar and the height H of the coal pillar are positively related, as shown in Fig. 2.13.

$$\frac{W}{H} = 0.5\left(\frac{W_1}{X} + 2\right)\cot\theta \qquad (2.23)$$

where, $W$ width of the ore column;
    $H$ Pillar height;
    $X$ Pillar rupture width;
    $W$ Width of the core zone of the pillar;
    $\theta$ Corner of rupture;

Consider the generalized formula for calculating the strength of a mining column, as shown in the following equation

$$\sigma_c = \sigma_{cl}\left[A + B\left(\frac{W}{H}\right)^\alpha\right] \qquad (2.24)$$

where: $\sigma_c$ is the compressive strength of the ore column;
    $\sigma_{cl}$ is the compressive strength of the cubic dwelling specimen;
    A, B, $\alpha$ are constant factors.

Joining the above two equations yields the following equation:

$$\frac{\sigma_c}{\sigma_{cl}} = A + B\left[0.5\left(\frac{W_1}{X} + 2\right)\cot\theta\right]^\alpha \qquad (2.25)$$

The above equation shows that the compressive strength of the column increases with the increase in the ratio of the width of the column core zone to the rupture zone.

The stability and damage degree of the coal pillar in the quarry can reflect the roof plate of the quarry and the pressure strength of the roof plate of the mining hollow area and the adjacent mining hollow area. Especially the stability of the coal pillar on the side of the neighboring mining hollow area has a non-negligible influence on the working face roof, and in turn, the pressure strength of the roof plate will affect the stability of the coal pillar. Currently recognized methods: the monitoring of coal columns can be done by using the roadway gangs off the layer meter, multi-point displacement meter to continuously observe the damage depth; and the application of hydraulic pillow to monitor the deformation pressure of the drill holes.

### 2.4.3 Coal Pillar Support Pressure Monitoring

The main functions of coal columns include waterproofing, isolating the hollow area and supporting the roof, etc. Among them, the coal columns, which play the role of supporting the roof, are subject to the combination of dynamic and static pressures during the period of mining. After the working face stops production, the coal pillar is in a relatively constant static pressure field, and its destruction is mainly affected by the compressive stress field. If the compressive stress field in the coal pillar is continuously monitored, the pressure changes in the coal pillar can be understood in time, so that it is easy to react dynamically to the changes in the stress field in the coal pillar.

Nowadays, most of the monitoring of coal pillar compressive stress adopts the method of drilling stress gauges into the coal pillar measurement point, and then the pressure field in the coal pillar is continuously monitored through online monitoring, and the changes of the compressive stress field are reflected in the mining engineering diagram in a timely manner.

## 2.5 Application and Outlook for New Monitoring Technologies

### 2.5.1 Acoustic Emission Monitoring Technology

Application of guided wave technology to monitor the acoustic emission (AE) signals of the tunnel enclosure and the development of a new type of acoustic emission transducer (often referred to as acoustic transducer) applying the guided wave principle. Acoustic emission or microseismical monitoring are both dynamic monitoring techniques. When a rupture occurs within the tunnel enclosure, seismic waves will be generated to propagate around. Setting up microseismical sensors in different directions in space can record the arrival time, propagation direction and other information of these microseismical waves, and then utilize various calculation methods to determine the rupture point of the coal rock body, that is, the spatial location of the seismic source. The location of the seismic source, the moment of vibration, and the intensity of the seismic source in microseismical monitoring are all unknown, and determining these factors is precisely the primary task of microseismical monitoring.

Microseismic monitoring techniques at home and abroad can be categorized into 3 main groups:

Type 1 is a system that mainly monitors the vibration of rock layers in a wide range of mining areas, with a vibration frequency of 100 Hz or less, focusing on monitoring the mining earthquake, and with a positioning accuracy of 100–500 m in general;

Type 2 is a system that mainly monitors the vibration of the rock formation around the working face, with a vibration frequency of 20–300 Hz, focusing on monitoring the rupture of the rock formation, and with a positioning accuracy of 5–10 m in general;

Type 3 systems that are based on monitoring a small area (such as roadway surrounding) of rock rupture with a vibration frequency of 300 Hz or more are often referred to as ground-sound systems, which have a smaller monitoring area.

In recent years, China has achieved fruitful work in using microseismical technology to monitor the impact of ground pressure in coal mines, rock explosions in metal mines, and sudden water discharge in mines. It is foreseeable that the use of microseismical technology for real-time monitoring of the evolutionary process of the roadway surrounding rock power disaster breeding is an important means of disaster prevention and mitigation.

## 2.5.2 Acoustic and Electromagnetic Wave Monitoring Technology

The application of ultrasound and high-frequency electromagnetic wave monitoring technology to develop acoustic wave transducers (probes) suitable for monitoring the depth or thickness of the perimeter rock loosening circle, that is, the transmitting acoustic wave transducer and the receiving acoustic wave transducer; the application of this technology in other fields has achieved great success and significant results. The basic principle of acoustic monitoring technology is the use of ultrasonic propagation in the medium, if the acoustic wave encounters the anisotropic medium layer when the reflection wave occurs, this reflection wave can be used to receive the acoustic wave receiver transducer to receive. From the launch of sound waves to receive sound wave time (sound time) and known wave speed can be known wave away from the distance, that is, to get the depth of the perimeter rock loosening circle, acoustic wave monitoring principle as shown in Fig. 2.14. Electromagnetic wave monitoring technology is the use of electromagnetic waves emitted to the surface of the rock layer, the use of high-frequency waves to penetrate the loose rock layer, when the wave arrives at the dense coal rock body is the same as the reflected wave, monitoring the time of the reflected wave will be easy to calculate the thickness of the circle of loosening, the principle is basically the same as that shown in Fig. 2.14.

**Fig. 2.14** Principle of acoustic and electromagnetic wave monitoring of the thickness of the loose ring of the surrounding rock

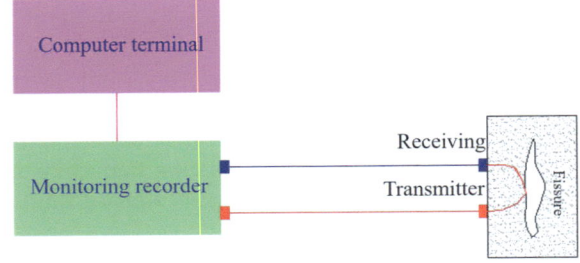

### 2.5.3 Acoustic Monitoring Applications

#### 2.5.3.1 Loose Circle Testing of Enclosed Rock

As a result of excavation and vibration, the integrity and strength of the perimeter rock of a certain thickness in the radial direction of the roadway will be reduced, resulting in the formation of stress reduction zones, or loosening zones. The thickness of the loose zone is an important basis for evaluating the stability of the rock mass and for designing the parameters of the support structure.

According to the sound velocity and the integrity of the rock body relationship between stress conditions, etc. It can be determined, in the same section of the surrounding rock under the same conditions of the original nature of the change in the wave velocity can be determined according to the thickness of the loosening circle. The test of the loosening circle is generally arranged in the cross-section of the roadway under test, and the arrangement of the inclination angle of each part of the test hole is in Fig. 2.15 shown. The angle of the holes is usually 90° for the top of the arch, 45° for the foot of the arch, and −5° for the side wall, and can be tested by double-hole or single-hole method. In order to understand the anisotropy of the rock mass, a number of test holes can be added along the direction of the roadway axis. Figure 2.15b shows the principle of probe (transducer) monitoring, T is the acoustic wave transmitting probe and R is the acoustic wave receiving probe.

**Fig. 2.15** Principle of sonic determination of loose rings

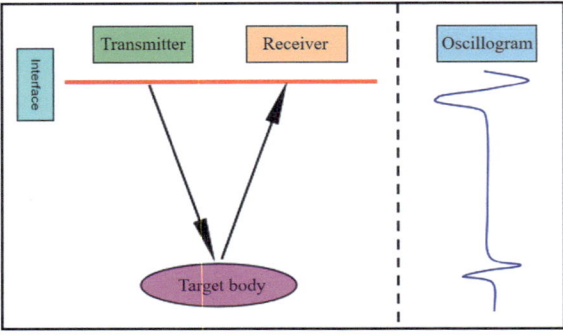

## 2.5 Application and Outlook for New Monitoring Technologies

It is generally believed that the range of wave velocity in the surrounding rock of the roadway that is less than the wave velocity of the original rock is called the loosening circle or stress reduction zone. The measured longitudinal wave velocity versus borehole depth can be summarized as follows:

(1) A zigzag curve. The curve of wave velocity versus hole depth basically stays at the original rock wave velocity value, indicating that the rock body is intact, and there is no obvious change in the integrity of the surrounding rock and stress after the roadway excavation. Therefore, it can be assumed that the surrounding rock has not been loosened and damaged.
(2) Factory-shaped curve. The wave velocity is lower in the front part of the curve and higher in the back part, and it is close to the value of the wave velocity of the original rock, which indicates that there is a loosening of the surface of the surrounding rock, and a stress reduction zone has been produced.
(3) Decaying curve. The front part of the curve is higher than the original rock wave velocity, and the back part is gradually close to the original rock wave velocity value, indicating that the rock body is complete and hard, the surrounding rock has no loosening circle, and there is a temporary stress increase area (the increase area will be balanced with the original rock body stress area eventually).
(4) Peak type curve. The wave velocity in the front part of the curve is low, while the middle part is higher than the original rock wave velocity value, and then close to the original rock wave velocity value, which indicates that there is a loose circle on the surface of the surrounding rock and the stress decreases, while the middle part is a pressure-dense zone, that is, an area of increased stress, and then an area of pressure in the original rock.

When the joints and fissures are more developed, the wave velocity and depth curves will have more complicated patterns, and the general trend should be noted. When there are several rock layers in the probing depth, attention should be paid to the influence of lithology and anisotropy on the wave velocity, and the division of enclosing rock loosening circle, compression and density belt and original rock belt should be correctly determined, which can provide the basis for judging the stability of enclosing rock and designing and constructing accordingly.

### 2.5.3.2 Acoustic Pore Detection Method

The application of acoustic sounding (or logging) can identify the structure and fracture zones of the underlying stratigraphic range, as well as the degree and depth of weathering of bedrock, and the physical and mechanical parameters of each stratum. The detection method adopts the "single-hole height difference synchronization method", in which the transmitting contact transducer and receiving contact transducer are placed in the same hole filled with water or mud, and a certain distance is maintained between them, as shown in Fig. 2.16. In the figure: T- transmitting transducer; R- receiving transducer; C- acoustic wave meter, along the hole wall at the same time moving up and down, measured acoustic wave propagation velocity,

**Fig. 2.16** Schematic diagram of the principle of acoustic hole measurement

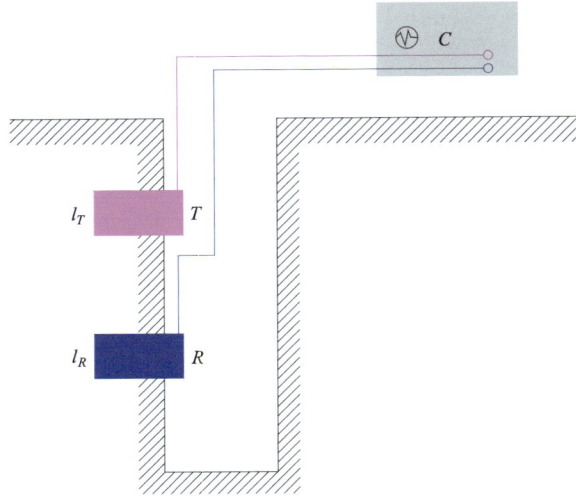

according to which the geological profile is plotted to analyze the engineering geology of the rock layer. The net distance between the transmitting and receiving contactors should be satisfied:

$$\frac{l}{l_{p(water)}} = \frac{l + l_T + l_R}{l_{p(rock)}} \tag{2.26}$$

It has been proved that sonic logging has a higher stratification accuracy of the drilling profile, especially when the core is very incomplete and difficult to drill, which shows the advantages of sonic logging.

### 2.5.3.3 Rock Strength Assessment

The relationship between wave velocity and compressive strength of rock specimens, it is not difficult to find a large number of experimental research results in a number of fields, representative of some of the formulas are as follows:

$$\sigma_c = 0.098 \left[ \left( \frac{V_r - 1.385}{124} \right)^2 + 15 \right] \tag{2.27}$$

$$\sigma_c = 10 V_r^3 \tag{2.28}$$

$$\sigma_c = 7.252 V_r^{1.61} \tag{2.29}$$

where: $\sigma_c$ is the compressive strength of the rock specimen, MPa;

## 2.5 Application and Outlook for New Monitoring Technologies

$V_r$ is the rock specimen wave velocity. m/Ms.

Although many different formulas have been obtained by researchers, the positive correlation between the wave velocity of a rock mass and its compressive strength is consistent.

A more influential method for estimating rock strength using wave velocity is the quasi-rock strength formula proposed by Japanese scholars:

$$R_c = \sigma_c \left(\frac{V_m}{V_r}\right)^2 \qquad (2.30)$$

The formula for calculating the strength of the rock body, which was calculated for the disturbed rock body in the medium stress state in the NGI classification system: was obtained from the statistical relationship between the RMR and Q quality classification index values of the rock body and the wave velocity of the rock body as well as the Hoke-Brown strength criterion

$$R_c = \sigma_c \sqrt{\exp(2.4945 V_m - 17.2973)} \qquad (2.31)$$

Since the law of variation of rock body strength with structural surface yield is different from the law of variation of wave velocity with structural surface yield, it is difficult to use wave velocity value to predict the rock body strength in engineering practice for the rock body with obvious anisotropy. For example, for a rock mass containing a group of through joints, as shown in Fig. 2.17, acoustic testing practice shows that the wave velocity along the direction of the maximum principal stress is the largest at, and $\alpha = 0°$ the wave velocity is the smallest at. The strength of the rock mass shows two high and one low changes with the change of the structural surface yield, and the lowest point of strength appears at $\alpha = 90° \alpha = 60°$. At this time, no matter which formula is used for the positive correlation between the strength of the rock mass and the wave velocity, the predicted strength of the rock mass along the direction of the maximum principal stress will not be consistent with reality.

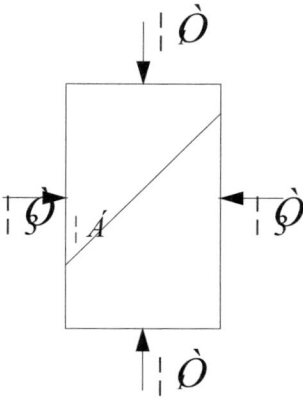

**Fig. 2.17** Model of a rock mass containing a set of penetrating structural surfaces

**Open Access** This chapter is licensed under the terms of the Creative Commons Attribution-NonCommercial-NoDerivatives 4.0 International License (http://creativecommons.org/licenses/by-nc-nd/4.0/), which permits any noncommercial use, sharing, distribution and reproduction in any medium or format, as long as you give appropriate credit to the original author(s) and the source, provide a link to the Creative Commons license and indicate if you modified the licensed material. You do not have permission under this license to share adapted material derived from this chapter or parts of it.

The images or other third party material in this chapter are included in the chapter's Creative Commons license, unless indicated otherwise in a credit line to the material. If material is not included in the chapter's Creative Commons license and your intended use is not permitted by statutory regulation or exceeds the permitted use, you will need to obtain permission directly from the copyright holder.

# Chapter 3
# Principle and Prediction of Large Area Roof Plate Coming to Pressure in Working Face

## 3.1 Overview

With the continuous innovation and development of coal mining technology, such as the application of super-strong anchor net cable support, small coal pillar (no coal pillar) in gob-side entry driving (retaining) roadway and other technologies. The spatial change brought about by the development of trackless transport vehicles in the mining area has made the application of large-scale and strong mining face a reality. It has not given sufficient time for the dynamic balance of subsidence movement in the overlying strata, resulting in rare and unique.

Mine quake induced by the roof plate of the hollow zone and large area of the roof plate coming under pressure is a recent year to the mining technology and technique new type of dynamic disaster phenomenon, and its triggering factor is one of the important problems faced by the modern mining engineering; this dynamic destabilizing phenomenon is very sudden and random. This dynamic instability phenomenon is very sudden and random. It is difficult to explore the environment of the mining area, and the analysis of the triggering factors is very ambiguous, which brings great difficulties to the application of prediction methods and control techniques. The goaf environment is difficult to spy and the analysis of inducing factors is very vague, which brings great difficulties to the application of prediction method and control technology.

## 3.2 Principles and Phenomena of Large-Area Roof Coming Under Pressure

### 3.2.1 Large Roof Coming Under Pressure and Dynamic Characterization

The large area of roof plate pressure, impact pressure and gas protrusion are all coal rock power instability disaster phenomena, with strong power instability characteristics, and when serious, it can trigger mine quakes in the quarry and even in the mining area. In recent years, with the continuous innovation and development of mining technology, the mining intensity and working face advancement speed have rapidly increased, and major unique sudden mining power instability phenomenon occurs from time to time. Especially the phenomenon of large area roof weighting disaster in working face, its occurrence process is difficult to predict. The coal rock power phenomenon (or power disaster) in the process of mining is one of the important problems faced by mining engineering. The coal rock dynamic instability phenomenon is very sudden and variable, due to the limitations of the environment, it is difficult to analyze the triggering factors of dynamic disasters, and the corresponding prediction and control technology faces great challenges.

In recent years, large-area roof pressures have occurred in many mines, including Shenhua Dalriata Mine, Yinzhou Coal Company's Jining No. 3 Coal Mine, Xinglong Zhuang Coal Mine, and Huaibei Qidong Coal Mine. Based on the analysis of on-site events and the study of literature, the authors believe that: the large-area roof pressure belongs to a series of dynamic disasters triggered by large-scale roof laminar breakage and destabilization under specific conditions, and the roof rock layer is completely damaged in the form of gravity impact on the stent, that is, the gravity impact damage is the first factor.

A certain area of the roof rock layer suddenly exceeds the normal pressure rule and the definition of the scope of articulated rock beam action and loses the regular braced controllable roof pressure cycle. The whole of large roof formations (direct roof and old roof) loses control of transverse and longitudinal transfer rock beams. The development of gravity-type action of the rock layer reaches the limit value, and in a moment, the degrees of freedom and gravity associated with the rock layers overlying the roof plate (longitudinal and transverse) are instantly transformed into maximization, and the large body of the rock layer is completely transformed from gravity-type potential energy into impact kinetic energy to act on the stent. This process is a large area of the top plate rock layer from the potential energy instantaneously converted into impact kinetic energy, the impact energy of the strong impact force on the support body, and with the support body in the form of elastic energy release energy. The impact force formed by the strong impact kinetic energy is up to thousands of tons, which is more than 2–3 times of the normal value. It has an impact, distortion and disturbance instability damage to the support, resulting in a large area of the support column being crushed or the rigid body bursting, and the

## 3.2 Principles and Phenomena of Large-Area Roof Coming Under Pressure

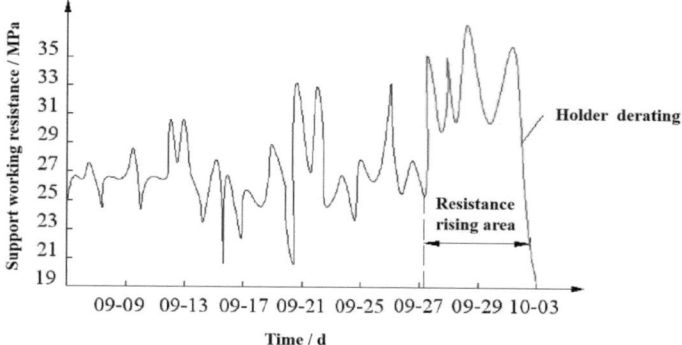

**Fig. 3.1** Working resistance curve of a certain number of brackets

support collapsing in pieces. In this series of development processes, the impact damage formed by kinetic energy is the second factor.

From the observation data, the roof rock layer continuously acts on the stent before the large-area roof comes under pressure, which makes the stent's working resistance show a rising trend. Figure 3.1 shows that the seepage of water from the upper part of the working face of a mine causes the intrinsic strength of the local old top of the next face to be lowered and damaged, and the direct and old tops are destabilized in a holistic way (especially the lateral destabilization of the roof plate of the working face). Before the large area of the roof plate came to pressure, the pressure recorder monitored the abnormal fluctuation of the bracket working resistance and showed a trend of continuous increase.

### 3.2.2 Large-Area Top Plate to Pressure Principle

#### 3.2.2.1 Primary Geotectonic Factors

From the point of view of the mines where large-scale roof pressurization disasters have occurred, the occurrence of large-scale roof pressurization events has remarkable objectivity. Based on a large number of research analyses and discussions, it is believed that: specific geological and tectonic factors are the internal cause of the event, which is also the necessary condition; under the premise of having the necessary conditions, the formation of specific spatial–temporal positional relationship between the roof rock layer and the bracket, that is., spatial–temporal positional constitutive condition is the sufficient condition. Specific reference:

① In rock formations, primary geotectonic zones (including fissures, faults, faults, etc.), or large-scale structural surfaces (including primary structural surfaces,

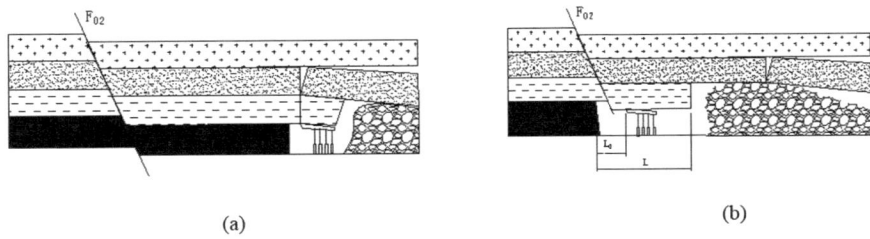

**Fig. 3.2** Schematic diagram of the temporal and spatial positional composition of the top rock support

tectonic structural surfaces, and secondary structural surfaces) related to geological genesis are present in the rock layers above the top plate, as shown in Fig. 3.2 (a);

② Direct top or old top rock formations are capable of forming and constituting massive rock masses, that is, discrete bodies with a certain thickness and a certain extent;

③ The top plate has hard or rigid mechanical properties. These three basic elements are necessary conditions for the influence of geological formations.

In addition to the necessary conditions, the corresponding temporal and spatial structural relations must be satisfied. When the support advances to close to the geological structure zone, and close to the time of cycle pressure, that is, at the end of the previous normal roof cycle pressure, after the workforce advances for a certain distance close to the next roof cycle pressure, the geological structure zone of the roof rock layer happens to be in (or close to) within the range of the support roof control distance. Due to the disturbance of the shifting process (including the participation of dynamic pressure) and the self-weight of the top plate rock layer, it presents a sufficiently free end face at the fault and tends to form a sufficiently large blocky layer. The large-scale block-shaped layer body loses the role of rock layer (beam) longitudinal and transverse binding force and is almost in the state of unconstrained free structural system, and its gravitational potential energy acts completely on the brace, instantly forming a large area of the top plate to press the mechanical structure of the relationship, that is, the large area of the top plate on the brace to produce a sufficiently large and strong twisting, disturbing and impact action. The brace is in the critical state of a short-lived critical load (pressure). At this moment, the top plate rock layer forms a sufficient gravity shaking impact situation, which develops into a spatial–temporal structural situation of large-area top plate pressure under the action of the potential energy of the top plate shaking impact movement, and the synthesized mining stent and the top rock layer are in the transient and instantaneous destabilized state, as Fig. 3.2b shown in.

## 3.2 Principles and Phenomena of Large-Area Roof Coming Under Pressure

### 3.2.2.2 Roof Secondary Structural Factors Caused by Coal Mining Technology

Coal mining technology triggers secondary tectonic factors of the roof mainly refers to the application of small coal pillars, artificial road gangs (without coal pillars) along the open road, and large mining height and long-wall mining method, which has changed the traditional mode and concept of roof management, and of course, includes the application of artificially forced roof releasing process similar to the thick and hard top plate. First of all, the coal pillar supports the strength and support life of small coal pillars, artificial roadway gangs (without coal pillars) along the empty stay roadway, and the coal pillar support strength and support life of the large mining height longwall mining method have changed qualitatively, and the mechanical structure model formed by this process is Fig. 3.3 shown in. Figure (a) is the working face structure schematic diagram, and Figure (b) is the working face force model. Obviously, the support strength of the small coal pillar or artificial coal pillar to the roof is smaller than that of the coal body, or the compression deformation of the roof allowed by the small coal pillar and the artificial roadway is larger than that allowed by the coal body. Under such conditions, the top plate of the working face is very likely to produce an unbalanced and asymmetric pressure tendency that is not favorable to the stability of the comprehensive mining support, and the top plate or the rock layer overlying the mining area produces an unbalanced and asymmetric subsidence movement, especially the side adjacent to the mining area bears a larger supporting pressure, so that the load borne by the support body is close to the supercritical load state. The compressive stress of the roof has a linkage damage effect on the coal pillar, artificial roadway side, roof and support, including the rate and degree of subsidence movement of the overlying strata adjacent to the goaf, which will have different degrees of damage effect. Even relying on the filling method can not fully solve such problems.

For example, in one mine, the artificially forced roof release of thick and hard roof slabs and the formation of abnormal roof pressure by the coal seam through the roadway when the work face was faced caused rows and rows of damage to the stent. Figure 3.4 shows the principal model of the large area coming pressure formed by the working face coal seam crossing in this mine. The working face should pay attention to avoid destroying the integrity of the roof plate, especially the use of poor performance of the synthesized mining support, otherwise, when the working face advances close to the coal seam through the lane, it is very easy to form an abnormal roof pressure, the support bearing pressure suddenly rises to reach the critical state, resulting in large-area roof pressure and causing the shutdown of the production accident. The damage to the roof plate caused by the artificial forced roof release method is also prone to produce similar abnormal pressure situations.

Therefore, the use of such coal mining technology should pay special attention to the monitoring of the working resistance of the fully mechanized mining support, the reasonable layout of the support monitoring area, and the monitoring density of the support should be strengthened, and the change law and development trend of the bearing capacity of the support should be fully grasped, as well as the stability change

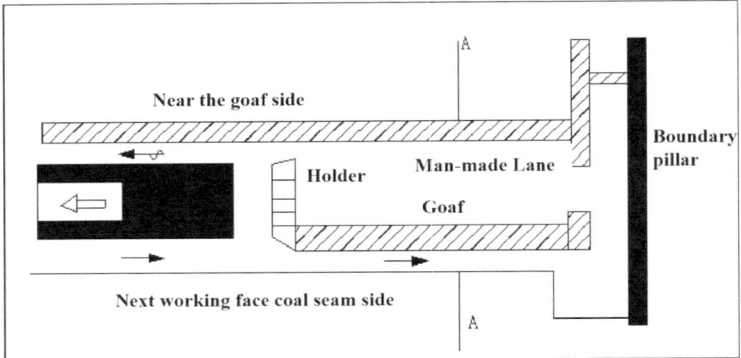

(a) Schematic diagram of the working face plan structure

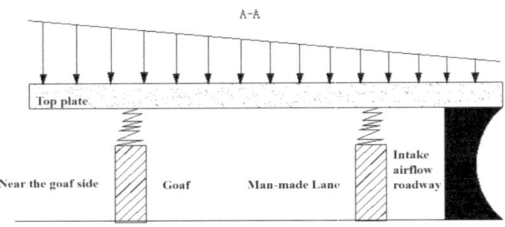

(b) Schematic diagram of the mechanical modeling

**Fig. 3.3** Mechanical structure model of small coal pillar and artificial roadway

**Fig. 3.4** Schematic diagram of the principle of abnormal roof pressure caused by coal seam through the roadway

law in the operation of the support. In addition, it is also necessary to strengthen the following two aspects of key technology management and analysis and research.

(1) On the basis of scientific proof of adopting small coal pillar or no coal pillar or large mining height mining technology, carry out monitoring of mining pressure, fully and carefully grasp the characteristics of mining pressure manifestation in the mining field and the study of the change rule of mining pressure, especially the understanding of the cyclic pressure law of the roof plate and the characteristics of the roof plate collapse, and the characteristics of the interaction and coupling relationship between the stent and the roof plate and the coal pillar. Reasonably estimate the development height of the "three zones", and grasp the

impact effect and influence of the mechanics of the subsidence movement of the overlying rock layer on the roof pressure in the mining area. Reasonably determine the top control distance between the support and the working face, and prevent changes in the tilt of the support triggered by changes in the top plate pressure or the bottom plate. Monitor and study the influence of the movement of the overlying rock layer in the neighboring mining airspace on the mining pressure law of the quarry.

(2) Combined with the development of large mining height mining technology in large mining face, the performance parameters of fully mechanized mining support are designed, and the new support design is carried out in combination with the research results of mine pressure. The design concept is innovated, and the previous design concept of focusing on increasing the design of cylinder body to obtain high support strength and strong support force is changed. Ignoring the value of small support surface of fully mechanized mining support, the corresponding relationship between support force and support area will be reconsidered, and reasonable support strength (floor specific pressure) will be selected through various experimental research methods to design and produce new fully mechanized mining support with high stability.

## 3.3 Monitoring Principles and Methods

### 3.3.1 Monitoring Modeling

Combined with the content of the previous section, when the working face advances to the near geological tectonic zone, the relative position of the support changes constantly, and it can be considered that the potential energy (gravitational potential energy) of the roof rock layer is in the process of continuous gathering and rising, and within a certain stage, the roof rock layer will produce a stronger than normal roof pressure subsidence vibration phenomenon. That is, under the action of potential energy, the overlying rock layer will produce a strong acoustic emission phenomenon, and the acoustic emission signal of this process is two to three times higher than that of normal roof pressure. In a certain time and space scale, such as the continuous increase of work resistance of the stent, the stent is not stable, the quarry perimeter rock and the bottom plate vibration, the working face dust, gas concentration increases, the liquid injection safety valve overload opening and other phenomena. These phenomena are the basic information for the prediction of large-area roof plates coming under pressure.

Based on the above analysis, it can be seen that the occurrence of the large-scale roof instability to pressure event in the quarry belongs to the conditional instability state. The event formation model is shown in Fig. 3.5, and the logical relationship of its dynamic characterization can be explained by the correlation diagram shown in Fig. 3.6. Based on the logical relationship of large-area roof plates coming under

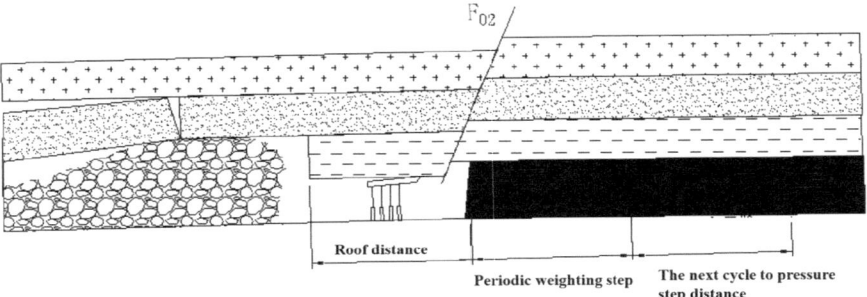

**Fig. 3.5** Schematic diagram of the large-area top plate coming under pressure model

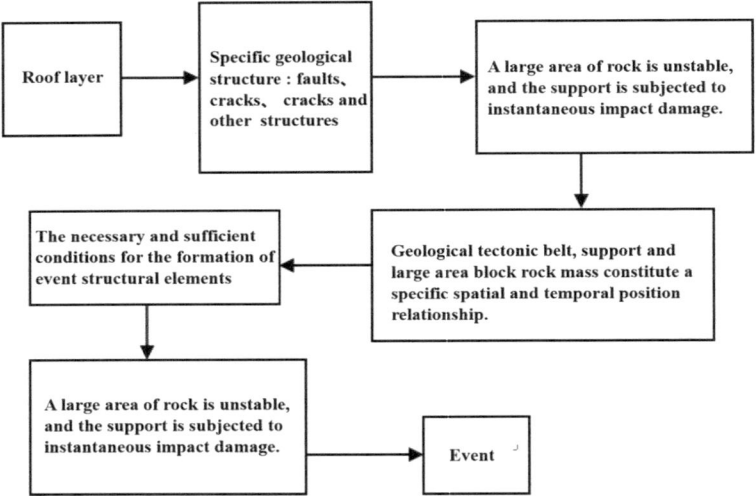

**Fig. 3.6** Schematic diagram of the logical relationship between the large top plate coming under pressure

pressure shown in Fig. 3.6, the technical route of preventing large-area roof plates from coming under pressure in the working face is derived.

## 3.3.2 Discriminative Method for Geological Survey Data

By analyzing the existing geological data, we can grasp the geological and tectonic characteristics of the roof plate and even the overlying rock layer in the mining area, as well as the temporal and spatial location of the quarry where the faults are located. During the advancing process of the working face, we always observe and analyze

## 3.3 Monitoring Principles and Methods

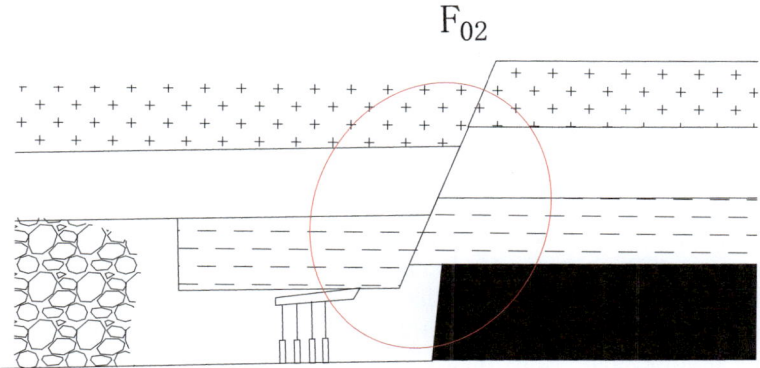

**Fig. 3.7** Prediction of anomalous roof incoming pressure reaction zone

the reaction degree of the roof plate affected by mining, clearly grasp the normal reaction law and phenomenon of the roof plate's cycle of pressure and pressure step, maintain a high degree of sensitivity to abnormal phenomena, and analyze the relevant geological data in a timely manner.

Specifically, the focus should be on controlling the following three key aspects:

① Visual observation must be carried out at regular intervals to determine the controllable roof plate coming under pressure status and the location of the abnormal reaction zone. Comprehensively analyze the monitoring data and visual observation, and draw the results of the analysis into a diagram, as shown in Fig. 3.7.

② In the production process, the relevant technical departments should be established to coordinate the production management of the mining area, to achieve data and information sharing, especially production, technology, geodetic survey, district teams and other departments under the leadership of the chief engineer to promote the production process. Timely discover the unexplored geological defects and adjust the technical management strategy. Relevant departments should actively coordinate to solve and respond to the geological tectonic phenomena in the anomalous regional zones. Establish the production operation mechanism. As shown in Fig. 3.8.

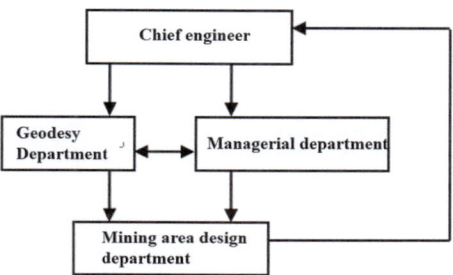

**Fig. 3.8** Logic schematic of top plate abnormal coming pressure analysis

③ Real-time monitoring of the changes in the working resistance of the stent, through the analysis of the working resistance of the stent to determine the development and change rule of the roof cycle to pressure, the initial support force, the working resistance, the end resistance. From the monitoring data, under the premise of normal operation of mining equipment and relatively stable change of geological parameters such as the thickness of coal seam, the change of work resistance of stent mostly shows a stable fluctuation range.

Therefore, once the abnormal fluctuation phenomenon of work resistance of the bracket occurs in the spatial and temporal scales, it should cause the technicians to pay full attention to it, analyze and study the reasons for the occurrence of work resistance abnormality, and point out whether it is the change caused by the operation of the equipment or the abnormality induced by the change of the geological parameters of the quarry.

### 3.3.3 Instrumented Real-Time Continuous Monitoring Methods

(1) Adopting equipment with continuous recording functions such as computer online monitoring and infrared recorder continuous monitoring system, technicians analyze the characteristics of changes in resistance parameters of stent work in real-time, analyze the anomalies produced by large areas of roof plate coming under pressure by integrating the geological exploration data, and if necessary, take underground site monitoring and observation and other means to obtain auxiliary information for analysis.

(2) Application of acoustic emission and microseismical monitoring methods.

The mechanical behaviors that can induce the acoustic emission phenomenon from coal and rock bodies in mining engineering are different, so different monitoring targets should have different monitoring modes and focuses, such as the coal and rock bodies that have been proven to have the tendency of impact pressure or protruding, the prediction and forecasting system should be set up with the guidance of the corresponding inducing mechanism. Long-term research results show that no matter which kind of acoustic emission source is generated by mechanical behavior, the key technology of its monitoring should reflect how to select the appropriate sensor arrangement and signal strength identification method under the guidance of a professional technology system.

Coal-rock dynamic instability phenomena such as impact ground pressure, gas protrusion, mine tremor, roof cycle pressure and coal pillar instability, although sudden, have precursors, such as increased gas and dust concentration, the steep increase in bracket pressure, sound of coal cannon, destruction of surrounding rocks and tremor. Coal rock body is a non-homogeneous elastic body, there are a variety of original micro-fractures, pores and other morphological defects, so that in the disturbance of external forces (such as mining), the area around these defects will occur to

## 3.3 Monitoring Principles and Methods

varying degrees of stress concentration, which in turn produces acoustic emission or micro-seismic phenomena, to release elastic energy in the form of stress wave mode, resulting in a sudden impact on the coal rock body to form the impact effect. The accumulated energy in the coal rock body is generally released or transferred along the easiest and fastest path to propagate. In the process of deformation to subsidence and collapse of the overlying rock layer in the mining area, the accumulated strain energy propagates to the surrounding area in the form of an elastic wave through the coal pillar and bottom plate.

Acoustic emission and microseismic are two characterization methods of homologous associated events. Accurate and timely acquisition of acoustic emission signals is of great significance for studying the formation mechanism, failure scale and prediction of induced dynamic instability. Therefore, monitoring mode is the key problem of acoustic emission detection technology. Long-term investigation and research results show that the installation location of acoustic emission or microseismic sensors and the geometric parameters of buried boreholes are one of the key technologies for acoustic emission signal location and early prediction. If the installation position is improper, it will lose and delay the propagation of a large number of early acoustic emission signals. The acoustic emission signal released by the roof pressure (abnormal pressure) in the initial goaf is weak, so the sensor should be set in the dangerous area confirmed by the evaluation of professional and technical personnel. The field monitoring engineering practice shows that the sensor is directly installed in the coal pillar and floor of the goaf for waiting monitoring, which can effectively capture the information available for the prediction and analysis of the subsidence of the overlying strata in the goaf.

The monitoring equipment should meet the requirements of continuous recording and data communication functions, and the signal recording should meet the requirements for use in computerized prediction and forecasting analysis. The basic information set based on professional analysis needs can be recorded as: signal timing, signal frequency and amplitude frequency, signal threshold setting, etc., which can be used in computerized prediction and forecasting analysis software systems. The computer comprehensive analysis system needs to carry out timely or real-time spectrum analysis, signal judgment, waveform recognition and parameter calculation of monitoring data. Based on the positioning timing calculation of the forecast sound source signal, it realizes the substantial prediction forecast and pre-evaluation analysis of power disaster prevention.

When the roof is pressed in a large area, the sound of coal and rock mass and the perception of microseismic are often heard, which are used as early warning signals for dynamic disasters. Therefore, objective and accurate monitoring of mining-induced coal rock microseismical or acoustic emission phenomenon is also an effective technology for the prevention and control of large-area roof power disasters.

The frequency of sound waves perceived by human hearing is between 20 Hz and 20 kHz. The frequency of microseismical waves induced by mining activities is 1 Hz above 10 kHz, and the wave frequency characterized by the gravitational potential energy of the roof rock layer is mostly concentrated in a certain frequency band in this frequency range. The height of the frequency band is directly related to the

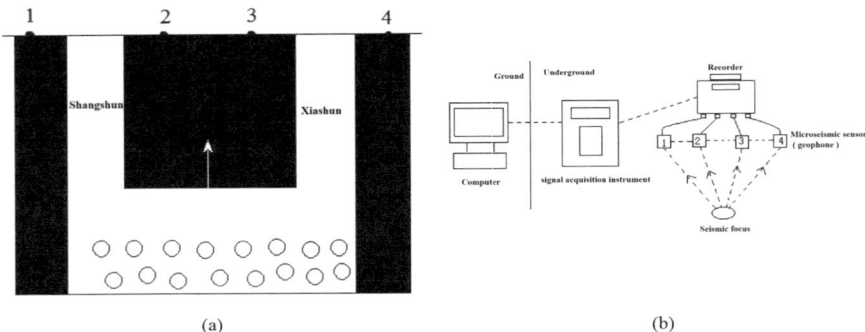

**Fig. 3.9** Schematic diagram of the monitoring principle of the acoustic emission (microseismical) triangular mesh method

stored energy and physical properties of the coal and rock body, and the frequency increases with the increase of energy release, and the energy has the characteristic of decreasing by overcoming the work done by the medium resistance in the process of transferring and transferring.

In the process of large-area roof pressurization, the gravitational potential energy continuously increases and releases extraordinary values of vibrational energy (mechanical energy), and in particular, the roof formation exhibits an acoustic emission propagation rate that increases with the increase in bracing pressure. The increase in emission rate precedes the destabilization of the roof formation. Because there is a time difference in the propagation of the shock wave to each sensor, microseismical sensors (or ground sound sensors) can be applied to monitor the shock wave. Combined with the propagation speed of the wave in the coal rock body, applying the triangular network method (a set of triangles formed by two measuring points and the seismic source), the orientation of the seismic source can be obtained more accurately. The principle of the four-measurement-point deployment triangular network method is used, as shown in Fig. 3.9.

Engineers and technicians with engineering experience and other similar examples to investigate and analyze, determine the location of the measurement point, the depth of the borehole and the distance between the measurement point. Figure 3.9a is a way to locate the measurement point at the outer end of the mining area; as shown in Fig. 3.9b, the monitoring frequency band of the seismic wave sensor is 10–800 Hz, and the wave speed is more appropriate.

The sensor will convert the acoustic frequency signal obtained by monitoring into an electrical signal, and amplify and adjust the electrical signal according to the need. The monitoring host will be timed (the time interval is $\Delta t_1$, $\Delta t_2$.....) By continuously recording the frequency signal of the threshold value in the preset frequency band, the monitoring host monitors the time when the seismic wave reaches each measuring point and uses the geometric relationship and spacing formed by the location of the source and the sensor, as well as the monitoring parameters composed of the propagation velocity of the stress wave in the coal and rock medium. Through

## 3.3 Monitoring Principles and Methods

these basic data, the source orientation can be determined. At the same time, the relative energy level of the fluctuation event is determined by the recorded amplitude and signal duration. By using these recorded events and combining them with the accumulated engineering practice experience, engineers and technicians can judge the source orientation and the corresponding danger degree of the region, so as to take corresponding control measures.

Acoustic emission sensors are an important part of acoustic emission monitoring technology, and acoustic emission signals can be received through the sensors. Therefore, the performance of the acoustic emission sensor directly affects the acoustic emission monitoring results. Combined with the characteristics of coal rock acoustic emission signals, generally speaking, the frequency response range of acoustic emission sensors should be between 100 and 3000 Hz, the amplification should be between 100 and 5000 times, and the sensitivity should be not less than 250 mV/(cm-s$^{-1}$).

The acoustic emission monitoring system (Patent No.: ZL200920226992.4, ZL200910019022.1) independently developed by Shandong Kedar Mining Monitoring Equipment Co., Ltd was used to monitor the large-area roof activity law of the 19,160 fully mechanized mining face of Wu$_{9-10}$ coal seam in Pingdingshan No. 4 Coal Mine. One of the key technologies of this monitoring is the installation and arrangement of acoustic emission sensors. The key technologies include: delineating the monitoring area and the effective receiving range of the sensor, the receiving direction of the sensor, the installation position of the sensor, the installation structure of the sensor, the installation horizon and installation depth of the sensor, etc.

For the monitoring of large-scale roof collapse in the mining area, the installation of acoustic emission sensors should be chosen to facilitate the transmission of acoustic waves of roof deformation to the coal pillar or bottom plate of the roadway near the stopping line, and the number of installations should be 2–3 groups per roadway. In order to ensure the continuity and reliability of the acoustic emission monitoring, the acoustic emission sensors should be arranged 10–20 m away from the stopping line (inside and outside) and should be installed before the working face is put into production. As a long-term monitoring program, part of the sensors should be arranged in the confining wall after stopping mining, which is used to continuously monitor the activity of the roof plate in the hollow zone, as shown in Fig. 3.10a. Acoustic emission sensors should be installed in the drill hole at a depth of 1.5–2.0 m and outside the loose ring of surrounding rock, as shown in Fig. 3.10b.

During the monitoring process, it was particularly important to successfully receive a number of more active roof damage activities in the mining zone and make timely warnings. The first vibration event lasted about $T_1 = 2.5$ s, and there was another one after 7.5 h, which lasted about $T_2 = 1.5$ s. Typical waveforms of these two vibration events after processing and analysis are shown in Fig. 3.11. After comparing with other observation data, it was determined that these two events were vibrations caused by roof activities in the mining area.

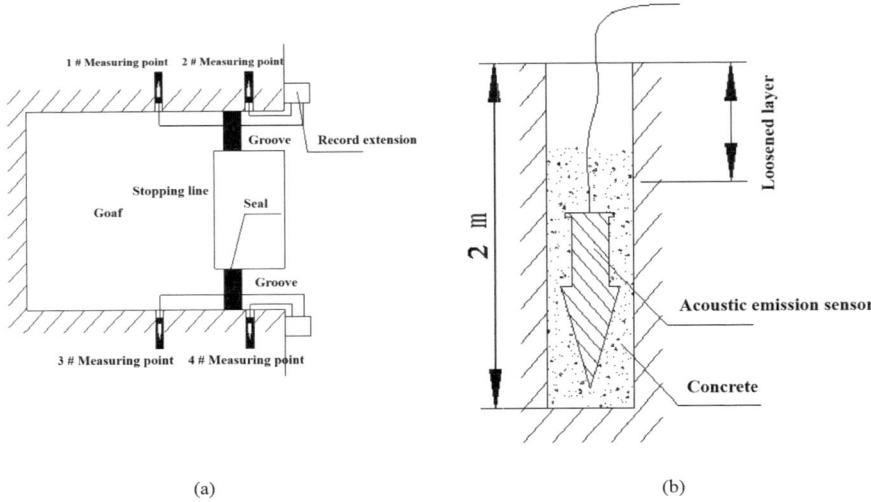

**Fig. 3.10** Schematic diagram of measurement area layout and sensor installation

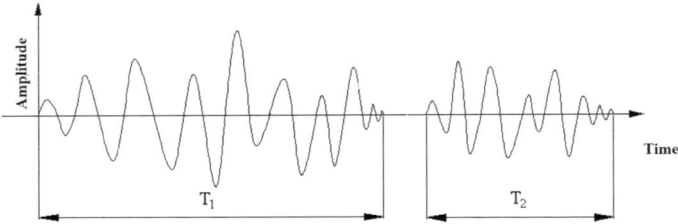

**Fig. 3.11** Waveform of perimeter rock vibration caused by two adjacent roof movements

## 3.4 Techniques Faced by Large Roof Slabs Coming Under Pressure in the Working Face Problems

Under the engineering background of pursuing high productivity and high efficiency, the majority of engineers and technicians and scientists have continued to carry out research work on coal mining technology that is conducive to high productivity and high efficiency. The overall development trend of coal mining technologies, such as long-wall coal mining with large mining height, small coal pillars, and artificial roadside gangs along the open road (without coal pillars), is large scale and large space. The application of these technologies greatly increases the movement speed and influence range of the roof plate and overlying rock layer in the mining field, which creates a serious challenge to the stability of the bracket operation and puts forward higher requirements for the pressure monitoring technology in the mine.

Most of the large-area roof pressurization occurs in the back-mining face where the rock layer of the roof plate is relatively hard, and the roof plate is mostly sandstone or conglomerate, and the roof plate hangs over the top of the face in a large area after mining. When the face is first mined, the initial fall of the roof plate can reach 50–70 m, or even more than 100 m. Such a large area of the roof plate, once the fall, will cause great harm. When mining by the cutter-pillar method, because the area of the hollow zone is too large, the coal pillar is crushed and a large area of roofing occurs. Even if the coal mining face adopts the synthesis mining technology, the phenomenon of large-area top plates coming under pressure will still occur in ten thousand square meters.

It is generally believed that when the hard and hard-to-rock top plate overhangs the roof over a large area, the top plate will bend and sink away from the layer under the action of self-weight. Whether treated as a plate or analyzed as a beam, at this time, the roof rock layer has a considerable elastic deformation, and when the bending stress exceeds its strength limit, a fissure will be generated. Once the fissure penetrates through the hard rock layer, sudden collapse will occur, resulting in a large area of the top plate to pressure disaster. Another situation is that the large area of roof overhang makes the formation of shuttle-type flat microstrip holes in the mining area, and the coal pillar produces continuous shear stress on the overlying roof plate, resulting in the roof plate being cut off, and then the sudden collapse occurs.

In addition, part of the large area of the roof to pressure will occur in the large mining face, large mining face advancing speed is generally faster, sometimes advancing speed is greater than the reaction time of the roof collapse, then often form a large area of hanging roof, resulting in a large area of the phenomenon of pressure, the formation of windstorms to destroy the roadway, overturning the bracket, resulting in casualties and equipment damage.

Preventing large-area roof weighting is mainly to find a balance between high-yield production and roof safety. If all the collapsing method is used to manage the roof in the mining area, enough time should be left for the roof movement on the basis of production safety, so as to let the roof collapse completely and avoid producing large areas of hanging roof. If it is impossible to meet the requirement of roof collapse time, it is necessary to carry out artificially forced roof release (deep hole blasting roof breaking technology).

## 3.5 Principles and Research Directions of Large-Area Roof Coming Under Pressure

Up to now, our understanding of the pressure pattern of the working face and the characteristics of the pressure manifestation is still based on the research results of the mining pressure monitoring of the high-grade general mining working face. However, the control strength (or ability) of the two types of support, that is, synthesized mining support and mono pillar, is different for the roof slab. As a result, the corresponding roof pressure law and manifestation characteristics are also different. The application of integrated mining support brings the evolution of coal mining technology and the promotion of large-scale mining fields, which will inevitably trigger some new mining pressure phenomena, so the following aspects of technical work should be emphasized.

(1) The large-area roof plate coming under pressure is related to the physical properties of the rock stratum, and is extremely sudden and random when it occurs, which is a kind of coal-rock dynamic disaster phenomenon under the modernized high-yield and high-efficiency modernized mining method of the large mining face.

(2) The process of large-area roof pressure is often accompanied by the extraordinary increase of dust and gas concentration, the destruction of surrounding rock and deep rock bodies; microseismical and ground sound, flake gangs, and the abnormal increase of mine pressure, etc. These phenomena are all indicative of the rock dynamics. All these phenomena are the characterization of rock body dynamics, and it is of great significance to extract useful information (anomalous information) from the signals for effective prediction of the overlying rock layer instability, especially for effective analysis of the monitoring information of anomalous deformations and support pressure anomalies.

(3) Prevention and control of large-area roof pressure should attach great importance to the effective use of geological exploration data. For the abnormal area zone, the relevant departments should provide detailed geological data, actively adjust the production technology management strategy, coordinate and solve the abnormal situation, and establish a coordinated operation mechanism for safe production management.

(4) The microseismical or acoustic emission prediction method is more objective has a high degree of credibility in predicting and forecasting information, and should be popularized and applied.

(5) The mechanism of large-area roof pressure is different from that of impact ground pressure and gas protrusion, but they are triggering factors for each other, and the formation process is characterized by abnormal phenomena such as acoustic emission and microseismicity. How to accurately identify the abnormal phenomena from the "unusual information" is the key.

(6) With regard to long-wall large mining face, large mining height, small coal pillars or no coal pillars (artificial roadway gangs) process technology, attention should be paid to analyzing the influence of the sinking movement of the roof plate of the neighboring mining hollow area on the pressure coming from the roof plate of the working face.

## 3.5 Principles and Research Directions of Large-Area Roof Coming Under ...

**Open Access** This chapter is licensed under the terms of the Creative Commons Attribution-NonCommercial-NoDerivatives 4.0 International License (http://creativecommons.org/licenses/by-nc-nd/4.0/), which permits any noncommercial use, sharing, distribution and reproduction in any medium or format, as long as you give appropriate credit to the original author(s) and the source, provide a link to the Creative Commons license and indicate if you modified the licensed material. You do not have permission under this license to share adapted material derived from this chapter or parts of it.

The images or other third party material in this chapter are included in the chapter's Creative Commons license, unless indicated otherwise in a credit line to the material. If material is not included in the chapter's Creative Commons license and your intended use is not permitted by statutory regulation or exceeds the permitted use, you will need to obtain permission directly from the copyright holder.

# Chapter 4
# Principles of Creep in the Overlying Rock Layer of the Mining Zone and the Prediction of Ore Tremor

## 4.1 Overview

The vibration phenomenon induced by the power instability and collapse of the overlying rock layer in the mining area is called mining earthquake, and compared with the phenomenon of large-area roof to pressure, the occurrence of mining earthquake is very occasional. Mine quake is induced by the rock layer overlying the mining airspace, must have a large enough area of the mining airspace, i.e., a large area of overhanging roof as a basic condition. In recent years, due to the development of mining technology and coal mining equipment, such as long-wall high mining, small coal pillar or no coal pillar (along the empty roadway) mining and other advanced technologies, in the absence of filling of the mining void area, a large overhanging roof for the overlying rock strata subsidence movement to provide space conditions, so that the overlying rock strata of the mining void area subsidence movement is very dynamic characteristics, and compared with the general movement laws that have been mastered, it exhibits extremely unstable impact. Under the action of self-weight, after a certain time and space scale balance process, the movement state reaches the critical state will trigger the ground surface rapid subsidence movement of, resulting in the sudden movement of the overlying rock layer of the overlying rock layer of the air-mining area collapsed in the air-mining area on the bottom plate, forming a strong impact in a moment with the help of the air-mining area on the bottom plate to release a large amount of high-frequency energy and induced vibration. Sometimes, the impact energy is large enough to cause the whole mining area to feel the tremor, appearing stent toppling, unloading valves open, the depth of the sheet gangs, gas and dust concentration rise and other phenomena. Mining quake is a dynamic phenomenon that occurs during the production process of overlying rock subsidence and sudden collapse of goaf, as well as the instability and collapse of hanging roof rock layers in corner coal mining faces, isolated coal pillar mining faces, and coal pillar recovery. power phenomenon that sudden destabilization and collapse of the rock layer the dynamically affected area overhanging is coal body breaking accompanied by the sound of, the collapse of large areas of rock to form an air wave shock wave, inducing of coal block ejection, dust flying the phenomena, and elastic rebound, etc. The mine quake has no intuitive early prediction. Mine quake without intuitive early omen phenomenon, early in all direction's prediction is more difficult, there is no mature prediction method in the academic world, is still in the exploration and development stage. Different from rockburst and gas outburst, mine vibration in goaf is an elastic energy release phenomenon that does not involve a large amount of coal and rock ejection. The large structural overburden rock falls to the bottom and triggers a strong impact vibration, and most of such damages have an early stage of the development of a wide range of overburden potential energy, such as the period of rock subsidence movement, which occurs in the mining (including island mining, coal pillar recovery, etc.) near the end of the mine or a period of time after the cessation of production.

There is a direct relationship between the geological structure of the overlying rock strata and the occurrence of mining earthquakes in the mining area. The formation and occurrence of mining earthquakes in the mining area can be divided into two periods according to time:

(1) The first period: in the middle and late stages of mining in the mining area, a large-scale hollow zone has been formed, i.e., there is a large-scale phenomenon of unconfined touchdown between the roof plate and the hollow zone. From the occurrence, the mining shock mainly occurs in a period of time at or near the end of the working face.

(2) The second period: within a period of time (a few months, half a year, a year or even a few years) after the end of mining, a wide range of rock body collapses in the mining hollow area, including filling materials undergoing complex morphological changes, the original loose accumulation of collapsed crushed rock, filling materials, etc. can be sufficiently deposited and densely compacted after the change, a wide range of space and exposed rock layer, for the mining hollow area overlying rock subsidence movement and the This provides conditions for the subsidence movement of the overlying rock layer in the mining area and the accumulation of gravitational potential energy.

The relationship between this type of mine quake and geological factors is mainly manifested:

(1) Within the coverage of the subsidence movement of the overlying rock layer, there exists a relatively large fault structure or a secondary structure or fissure zone triggered by the influence of the subsidence movement.

(2) Sudden and extensive deformation movement of thick and hard rock layers under the environment of large-scale mining and hollowing area endowment.

(3) The presence of thick or hard rock layers in the overlying rock formation, including the old top, under large mining height conditions.

## 4.2 Principles of Creep in the Overlying Rock Layers of a Large Mining Area

### 4.2.1 Overlying Rock Layers and Their Movement Patterns in the Mining Area

The new thinking concept drives the coal mining method we advocate and preferred: long-wall large height mining method. With the advancement of coal mining face, the roof plate (direct roof or direct roof + old roof) of the mining area will produce subsidence and deformation movement and collapse under the pressure of the mine. Usually, according to the damage degree of the rock layer, the deformation and damage of the overlying rock layer can be divided into three zones, i.e., I: the bubble fall zone, II: the fissure zone, and III: the bending and sinking zone, as shown in Fig. 4.1,

When all the roof collapsing method is used to manage the roof, after the coal mining face advances for a certain distance, under the influence of the hollow roof effect, the direct roof firstly undergoes subsidence creep and eventually forms a subsidence movement range that is completely free from the constraints of lateral cohesive force of the rock layer, and this destructive range is commonly known as the bubble fall zone (Fig. 4.1-I). The rock layer in this area firstly collapses due to self-weight, and it is mainly supported temporarily by the synthesized mining support.

A rift zone is a rock formation that is intermediate between a rift zone and a bending zone. The rock strata in the rift zone are characterized by the supporting effects of the integrated mining support and the boundary coal pillar at the front and back of the rock strata, respectively, which can also be called the shear force effect. The center area of the rift zone is subject to the joint extrusion of the vertical

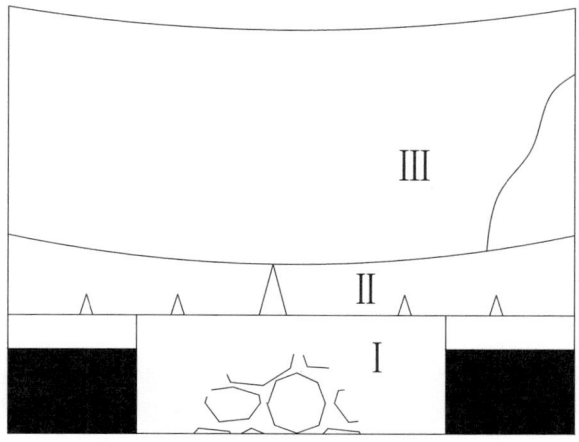

**Fig. 4.1** Schematic diagram of the zoning of the overlying rock layers in the extraction zone

subsidence movement force W and the decompression and expansion stress of the rock layer at the boundary of the subsidence movement. Since the decompression and expansion extrusion precedes the subsidence movement, the cracks developed due to the rock movement are cut off, but the collapsed rock blocks can still be aligned by extrusion (Fig. 4.1-II).

A flexural subsidence zone is generally defined as a rock formation that lies above a rift zone and whose subsidence influence extends upward to the surface. Rock formations within a flexural subsidence zone will remain monolithically laminated (Fig. 4.1-III).

Production practice shows that the overlying rock layer has the conditions for free fall only when it develops into a load through complex movement. During the mining period, after the top plate of the mining zone is collapsed through the cycle, the overlying rock layer can still be suspended above the mining zone by relying on its own transverse bonding (or cohesion) and transverse extrusion, and ultimately form the morphology and structure of the mining zone as shown in Fig. 4.1. This part of the rock layer is equivalent to the total thickness of the fallout zone and fissure zone, generally 6–8 times the mining height.

After the end of mining, the overlying exposed rock layer is in a relatively stable interval of subsidence movement, the length of which is related to the height of accumulation of collapsed rock layers in the mining hollow area and the existence of thick or hard rock layers in the overlying rock layer. Once the thick or hard rock layer is fractured, it is very easy to trigger a wide range of violent movement of the overlying rock layer. The overlying rock layer has two forms of movement, i.e. bending and tensile damage and shearing damage. Even in the temporary stable state of the overlying rock layer of the mining area, the movement of the elements of the long-term creep and subsidence movement, sometimes induced by the subsidence of the boundary area of the microscopic fissure development of large secondary faults, which in turn formed the sudden change of the overlying rock layer collapse conditions.

## 4.3 Mine Quakes Triggered by the Movement of Overlying Rock Layers in the Extraction Zone

A more unified academic view is to shock ground pressure, gas protrusion, large area roof collapse and other rock instability process caused by the mining area-wide vibration damage phenomenon collectively referred to as mining earthquake. In order to facilitate the discussion, the vibration phenomenon induced by the power instability and collapse of the overlying rock layer in the mining area is called mining earthquake. If the quantitative relationship between the overlying rock layer and the exposed space area S, height h, and collapse time t is defined as a spatiotemporal scale, then the mining earthquake in the goaf is caused by the self weight of the overlying rock layer. After a certain spatiotemporal balance process, it triggers surface subsidence movement and reaches a critical state, resulting in the sudden movement and collapse of large rock masses in the overlying rock layer on the goaf floor. In an instant, the strong impact can be released by the goaf floor to induce vibration.

### 4.3.1 Mineral Earthquakes and Their Characterization in Old Mining Areas

There are several main aspects of mine seismicity in old mining areas:

(1) The sudden collapse of a large area of overhanging roof in the mining hollow area caused an impact that triggered a mining earthquake. Due to blindly catching up with the production progress, it is easy to cause a large overhanging roof in the hollow area. If not dealt with in time, the subsequent mining face or mining area of the impact of the disturbance and ground subsidence movement or other forms of vibration impact, may break the hollow area suspended roof equilibrium state, very likely to cause a large area of the roof suddenly collapsed, resulting in a great impact, which triggered the whole mine shaft vibration (mine quake). The main method to prevent this disaster from occurring is to mine the necessary means to deal with the hollow area in the production process, to prevent the roof rock layer from generating excessive power impact and subsidence movement speed. At the same time, the roof of the mining hollow area should be made to fully collapse and fill, and the method of artificial roof release can be used. After the stoppage of the last mining face, the production of the next adjacent successive mining face should be reasonably organized, so that the top plate activity in the old empty area can be stabilized as far as possible before carrying out the subsequent production.

(2) Sudden destabilization of the coal pillars left in the old empty area, as well as subsidence movement of the overlying rock strata induces rapid subsidence of the existing tectonic faults (fissures) in the affected area, resulting in a wide range of impact collapses of the roof plate and triggering a mine quake. As shown in Fig. 4.2, various types of coal pillars need to be left in the coal mining face. After the advancement of the mining face is finished, most of the above coal pillars are left in the mining airspace. As an elastic–plastic material, the coal body will have creep effect over time under constant stress. When the elastic energy of the coal body accumulates to a certain degree and enters the critical state, it may suddenly produce destabilization damage, triggering a series of impact movement or sudden collapse of the upper roof plate, thus triggering a wide range of high-level roof collapse impacting the bottom plate.

**Fig. 4.2** Schematic diagram of the coal pillar remaining in the old air

## 4.3.2 Working Face Mechanism of Occurrence of Mining Earthquake in the Mining Area and Its Characterization

From the documented cases, the power caused by the sudden destabilization and collapse of the rock layer on the roof of the mining area impact phenomenon can be attributed to two mechanical mechanisms, which is in line with homologous concurrent events the of the basic characteristics, and can be referred to as the double power effect.

(1) Destabilization and collapse of the rock layer on the roof of the mining area

In recent years, large mines such as Yinzhou Coal Group's Boudican Coal Mine, Dongtan Coal Mine, Datong Mining Area and Shenhua Dalriata Mining Area have experienced the collapse of overlying rock layers over large structures in open areas. From the statistical data, the power disaster caused by the collapse of overlying rock in the mining area is relatively rare. Case analysis found that there are three elements play a key role in catalyzing: ① mining area overburden rock layer there is a certain thickness of hard rock, which is the basis for the formation of large structural overburden; ② mining area overburden rock layer in a certain direction on the direction of the existence of (or by the subsidence movement triggered generated by) a large fissure structure, such as fault tectonic zones, dislocation of the rock layer, etc.; ③ mining airspace overburden rock layer on the top of the existence of a large range of exposed area. These are the triggering factors for the formation of subsidence collapse of large structural overlying rock formations.

Obviously, as far as the occurrence mechanism is concerned, the influencing effect of the geological structure is the first factor of the event. When the workface was close to the stopping line, the exposed area of the roof plate reached a great value, which formed the exogenous conditions to push the strong subsidence movement of the overlying rock layer in the mining area. The influence accelerated the surface subsidence movement of the geological tectonic zone, which in turn formed a strong movement potential. Under the strong action of subsidence movement, the top plate potential energy increases sharply, and when the deformation and potential energy reaches a great value, the top plate is in a fleeting non-stable state. When the deformation and potential energy reaches a great value, the roof plate is in an instantaneous unstable state. The nearly parallel whole roof plate overhanging a large area in the mining area suddenly breaks and collapses to touch the bottom, which triggers a large spatial range of vibration. Collapse process produces impact air waves, causing dust flying, block gangue to the surrounding projectile, the power mechanism and the phenomenon of concurrency, so the concurrency phenomenon is called double power effect. Large structure roof collapse mechanism, as shown in Fig. 4.3.

(2) Mechanisms of Impact Air Waves

The biggest hazard of the collapse of a large area of the roof in the mining area is the top rock fall accelerated compression of the gas below the impact of gas

**Fig. 4.3** Schematic model of the collapse mechanism of overhanging overburden in the extraction zone

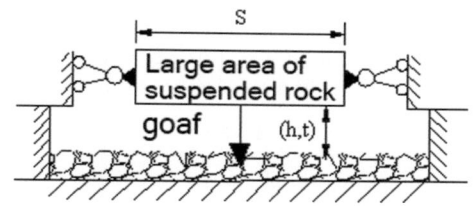

waves (storms, hurricanes), the drop process of the instantaneous formation of high-pressure kinetic energy of the gas, will be piled up in the bottom of the gangue and other miscellaneous as a hurricane like momentum rolled up and thrown to the periphery of the mining area, the impact of the gas wave can be instantly broken through the mining area outside the stop line of the airtight wall, a huge impact of the gas waves and even the airtight brick wall rushed out of the dozens of meters away.

During the collapse of a large overhanging roof, the potential energy of the overhanging rock mass is eventually converted into kinetic energy, which is instantaneously released in the form of a gas-dynamic manifestation. The kinetic energy of the gas (wave) reaches its maximum value, forming a strong impact wave. According to the theory of fluid dynamics, depending on the value of the gas dynamic pressure as the strength of the power impact, it is proposed to establish an instantly when the roof falls to the bottom mathematical model of the gas dynamic pressure of the impact type air wave as Fig. 4.4

As shown in Fig. 4.4, Figure taking cross sections I-I,XIIXII, and establishing the Bernoulli equation for the gas between the two cross sections, the pressure energy of the gas kinetic model (neglecting the effect of potential energy) is: for the

$$P_1 + \frac{V_1^2 \rho_1}{2} = P_2 + \frac{V_2^2 \rho_2}{2} \tag{4.1}$$

where: $P_1$, $P_2$—Static pressure energy of cross-section, Pa, I-I, respectively XIIXII

$V_1$, $V_{(2)}$-, for I-I, respectively gaseous kinetic velocities, m/s cross sections, XIIXII

$\rho_1$, $\rho_{(2)}$—I-I, respectively gas density for cross-section, kg/m XIIXII$^{(3)}$;

**Fig. 4.4** Power impact air wave pressure modeling

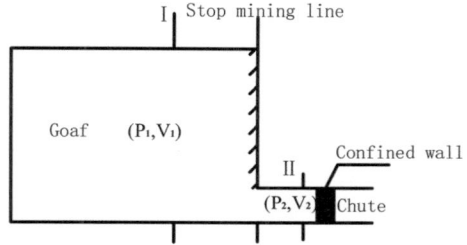

Assuming $\rho_1 = \rho_2 = \rho$, since the gas flow space in the I-I cross section is much larger than the cross section space of, it XIIXII is deduced the principle of energy conversion and conservation in 's equation: there are according to Bernoulli $P_1 \gg P_2$, XIIXII the kinetic pressure energy on the face is much larger than that on the I-I cross section, i.e., $V_2^2\rho/2 \gg V_1^2\rho/2$, Considering the instantaneous of the gas dynamics shock destructive nature, the above equation is simplified as follows:

$$P_1 \approx \frac{V_2^2 \rho}{2} \qquad (4.2)$$

The above equation reveals the mathematical model conservation of energy conversion and the hydrostatic energy generated by the potential energy of the suspended roof plate during fall its. At the same time, it also shows that the hydrostatic pressure of the gas on the I-I face can be almost completely converted into the dynamic pressure energy, which is on the face IIII instantly increased to produce strong agas impact destructive power. In the of the kinetic disaster that has occurred case, its energy is large enough to break down the two brick cement walls. outside the stopping line of the working face Fig. 4.4 and the above two formulas comprehensively reveal the basic principle of the to the far side gas dynamic effect between the I - face, which will break down III face and IIII the sealing wall of the roadway, and the coal gangue and stones will be rolled up and thrown. This is the large area or large structure roof principle of fall to form gas dynamic disaster. Rock Impact ground pressure, coal and gas protrusion and quarry large area roof to pressure collectively belong to the coal and rock power disaster phenomenon, all have strong power instability characteristics. Their occurrence mechanism is not the same, but can be mutual triggering factors.

With the continuous innovation and development of mining technology, such as the application of super-strong anchor cable network support technology, along the air tunneling and along the air to stay the application of technology, so that the application of large-scale and high-intensity mining working face has become a reality. In the face of the future special and large inclination of thick coal seams mining, such as Xinjiang, Inner Mongolia, some mining areas, coal seam features will make the existing coal mining technology changes, will face new problems in the management of the airspace. This will undoubtedly increase the overlying rock layer subsidence collapse induced mine quake opportunity, so we should have a full understanding, grasp and control of the movement and balance of the roof rock layer of the mining area, especially the change rule of the speed of movement, to improve the effect of the effective filling of the mining area, to prevent the occurrence of a major unique sudden coal-rock power disaster. Mine quake induced by the roof plate of the air-mining zone is a kind of coal rock and gas concurrent power shock phenomenon related to the coal mining process that appeared in recent years, and its inducing factor is one of the important problems faced by the modernized mining engineering. This dynamic destabilization phenomenon is very sudden and random. The environment of the mining area is not easy to be explored, and the analysis of the triggering factors is very ambiguous, so the prediction methods and control techniques are facing serious challenges.

## 4.4 Predictive Forecasting Methods and Case Studies

### 4.4.1 Acoustic Emission Monitoring and Fundamentals

A large amount of experimental data shows that acoustic emission monitoring technology can effectively capture the acoustic emission signals (elastic waves) generated during the damage process of coal and rock bodies. The acoustic emission signals generated by the early dynamic behavior induced by mining disturbance are very weak, and the propagation process is very easy to attenuate. However, in engineering practice, the monitoring technology is often required to obtain the prediction signal as early as possible to give engineers and technicians enough time to analyze the signal, which requires the acoustic emission monitoring technology to capture the signal in a timely manner, especially the early weak acoustic emission signal.

An acoustic emission monitor usually consists of three basic parts: an acoustic emission sensor or microseismical sensor, an acoustic emission signal monitoring receiver and terminal computer software. As far as the current level of technological development is concerned, the key to acoustic emission monitoring technology is the sensor's sensing device and the monitoring of the selected point, which seriously restricts the popularization and application of acoustic emission technology, and the research results on acoustic emission mostly stay in the laboratory and thesis. Another problem is the computer real-time forecasting and analyzing software. In short, solving these key technical problems requires the experience of the relevant specialized technicians and the full support of intellectuals.

At present, the commonly used in monitoring instruments are roughly divided into coal mining face two major types. One type is the without record storage function monitoring device, this kind of monitoring instrument to operate is more cumbersome, and the reliability and real-time data collected are poor, which can't satisfy the requirement of real-time forecast for on-site monitoring, and can only satisfy the general macroscopic management. The other type is the real-time online monitoring system, the accuracy of this kind of instrument is relatively high, the with signal monitoring and record storage function micro-controller as the core with continuity is good, it can collect a large amount of data, and the forecasting accuracy can meet the coal mine production and needs of, and scientific research it can be according to the needs of the scientific research and management personnel based on different methods analyzed and utilized, and it is flexible highly.

The breakage and collapse of the overhanging roof plate in the mining area has a developmental process. During the development process, the roof plate forms a fracture zone under the effect of gravity. Acoustic emission occurs during the formation of the fracture zone, and the fracture zone is the source of acoustic emission (or acoustic source), which propagates in the form of elastic waves in the direction where the energy is most easily released. In the process of subsidence of the overlying rock layer in the mining area, the elastic wave mainly propagates to the bottom plate of the quarry through the coal pillar, and the piezoelectric sensors are installed in the bottom plate according to a certain geometric relationship at a fixed point (e.g., drilling hole) to form a sensor array (or array), and the relative time difference of the acoustic emission from the sound source propagating to the sensors is determined. These relative time differences are substituted into a set of equations to satisfy the geometric relationship of the array, the location coordinates of the fracture area can be obtained. To simplify solving the equations for the location of the sound source, the sensors are arranged in a specific regular geometry.

**(1) Installation location selection**

The role of acoustic sensors (or probes) is to capture the hazard signals released when creep occurs in the coal rock mass, i.e. to monitor the target signal. Therefore, the installation location of the sensor should be chosen to be the closest to the expected source of acoustic emission, the easiest location for acoustic wave propagation, and consideration should also be given to setting and anticipating the path of acoustic wave propagation, as well as being less susceptible to interference from non-monitoring signals and damage from the production process.

**(2) Linear positioning method**

Linear localization method is to determine the position coordinates of the acoustic emission source in one-dimensional space, also known as line localization. Line positioning is the most basic method of sound source localization. After a large number of experimental studies and engineering analogies, the line localization method is more suitable for the application of mine seismic prediction technology in the mining area. The line localization method is mostly used to seek the localization of acoustic source signals in the envelope space domain between two points, as Fig. 4.5. Two sensors are placed in one-dimensional space$_1$, $S_2$, and the location of the determined acoustic source must be on the connecting straight line or circular arc line of the two sensors. Take the coordinate origin as the midpoint of the connecting straight line between the two sensors, and take the direction of as the positive direction. If $S_1$–$S_2$ $_1$ receives the acoustic emission signal first, the value counted by the time difference counter takes the negative sign; on the contrary, $S_2$ receives the acoustic emission signal first, and the value of the time difference counter takes the positive sign. Then, the coordinates of the position of the sound source P(x) can be found by the following formula.

$$X = sign(\Delta t)\frac{\Delta t}{2}c \quad \begin{array}{l} sign(\Delta t) = 1 (signal\ finding\ S_2) \\ sign(\Delta T) = -1 (signal\ finding\ S_1) \end{array}, \quad (4.3)$$

where, $\Delta t$—time difference (absolute value) between the arrival of the two sensors, s;

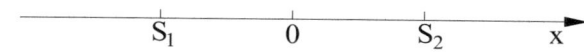

**Fig. 4.5** Schematic diagram of the line positioning method

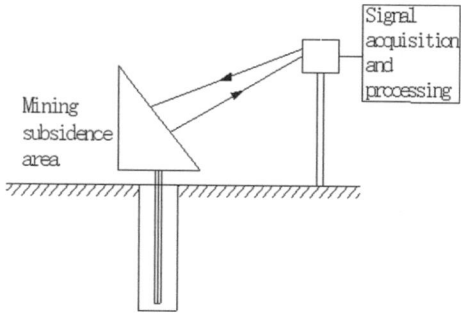

**Fig. 4.6** Schematic diagram of acoustic wave monitoring of surface subsidence deformation

c–Speed of sound, statistically obtained by engineering test method measurements, m/s.

### 4.4.2 Ground Acoustic (or Geophonic) Monitoring Methods

The basic principle of acoustic monitoring is shown in Fig. 4.6. The acoustic wave transmitting probe is utilized to transmit the acoustic wave at a fixed point (set base point) at regular intervals, while the acoustic wave receiving probe is utilized to monitor the reflected acoustic wave. Using the acoustic wave propagation distance and time or propagation speed, the development of subsidence movement can be obtained by calculating the position change of the acoustic wave reflector (surface).

## 4.5 Prospect of Prediction Methods for Subsidence of Overlying Rock Layers in the Mining Area

(1) Mine quake is a dynamic disaster induced by the collapse of the roof plate in the mining area after the mining work is finished. Therefore, it is also necessary to carry out research on the formation mechanism and its prevention and control methods in a categorized and periodized manner.
(2) Reveal the operation mechanism of mine shock induced by the collapse of a large area of overhanging roof slabs in the mining area and its manifestation. Apply the principle of energy conservation and conversion to study the mechanism of instantaneous increase of static pressure energy of gas in the mining area triggered by collapsing roof slabs under different environmental conditions, as

well as the gas dynamics model and the gas dynamics mechanism of gas shock wave formed by converting static pressure energy into dynamic pressure energy.

(3) The theoretical research and application of acoustic emission monitoring technology and monitoring model system have obtained many results. Based on the destabilizing phenomena such as impact ground pressure and gas protruding tendency, large-area roof plate coming under pressure and abnormal pressure of supporting body, there is an urgent need to study the respective occurrence mechanism, monitoring and forecasting method and the development of related computer software.

(4) At present, the research and development and application of acoustic emission monitors are still in the primary stage, and the watchful monitoring of roof activities in the hollow zone has obtained the expected results, which can provide an effective way for monitoring and forecasting of mine earthquakes in the hollow zone, but its performance needs to be further improved.

(5) GPS positioning monitoring: GPS can cover a wider area than traditional surveying methods, with an accuracy of millimeters, and the results obtained from continuous measurements are more continuous and require less hands-on experience on the part of the operator. There is no need for visualization between observation stations and no need for a control network to ensure a good geometric structure, thus making the choice of points more flexible. Short observation time, short baseline (about 20 km), fast positioning, its observation time is only a few minutes. All-weather operation and high positioning accuracy. Many data show that the relative positioning accuracy of short baseline (baseline less than 20 km) can reach $1 \sim 2 \times 10^{-6}$. Through cyclical monitoring of deformation, it can understand and grasp the trend of surface development and change and the mechanism of denaturation and destruction, and realize accurate forecasting and reduction the goals of disaster and prevention.

The observation line of technology can directly pass through the buildings, structures or other obstacles, and the monitoring network can be set up directly in the mining area, and the static operation mode can be used to observe the monitoring network, comprehensively collect the spatial coordinate information of the measurement points PS, and carry out automatic software solving and analyzing of the observation data, so as to obtain the surface movement law and guide the safe production of the mines and the restorative construction of the old mining area, which provides a powerful guarantee for the monitoring of the subsidence of the mines. Guarantee.

(6) Comprehensive monitoring methods refer to the monitoring methods that apply various principles on the ground and underground. For example, ① the acoustic emission monitoring technology is used in the bottom plate on the outside of the stopping line of the working face or the coal body (coal pillar) on the upper hill of the mining area for continuous monitoring of the AE signals generated by the rock movement of the roof plate in the hollow area inducing the coal pillar to be affected by the impact of the subsidence movement of the roof plate, and the AE signals are propagated to the acoustic wave probe on the outside of the hollow area through the bottom plate (or the coal pillar); ② before the stopping of

mining in the working face, the installation of monitoring instrumentation with a continuous signal storage and recording function, such as pressure monitoring (ii) Before stopping mining at the working face, install monitoring instruments with the function of storing and recording continuous signals, such as pressure monitoring, roof dynamic meter, and layer separator; (iii) Ground monitoring instruments, the monitoring mode can be regular or continuous. Regardless of the form of monitoring, special attention should be paid to analyzing the jumping and jittering signals or the continuously increasing signals with a fast rate of change.

**Open Access** This chapter is licensed under the terms of the Creative Commons Attribution-NonCommercial-NoDerivatives 4.0 International License (http://creativecommons.org/licenses/by-nc-nd/4.0/), which permits any noncommercial use, sharing, distribution and reproduction in any medium or format, as long as you give appropriate credit to the original author(s) and the source, provide a link to the Creative Commons license and indicate if you modified the licensed material. You do not have permission under this license to share adapted material derived from this chapter or parts of it.

The images or other third party material in this chapter are included in the chapter's Creative Commons license, unless indicated otherwise in a credit line to the material. If material is not included in the chapter's Creative Commons license and your intended use is not permitted by statutory regulation or exceeds the permitted use, you will need to obtain permission directly from the copyright holder.

# Chapter 5
# Principles and Prediction of Impact Ground Pressure in Coal Rock Bodies

## 5.1 Overview

Seismic Roof Pressure, as one of the dynamic hazards in coal mining, is a dynamic phenomenon characterized by sudden, violent, and intense damage caused by the release of deformation energy from coal and rock mass in mine roadways and working faces. It represents a specific manifestation of mine pressure. The occurrence of seismic roof pressure in metal mines is generally referred to as a pressure bumping. The disaster releases the elastic potential energy of the overloaded coal and rock mass in a sudden, violent, and intense manner. The coal and rock blocks, along with dust, are ejected, resulting in damage to supports, roof falls, roadway blockages, injury to personnel, and the production of loud noise and seismic vibrations in the coal and rock mass. The duration of the vibration ranges from several seconds to tens of seconds, and the ejected coal and rock can weigh from several tons to several hundred tons. The energy released by this ejection ranges from $10^{-5}$ J, associated with the micro-fracturing of coal and rock, to $10^{9}$ J, associated with large-scale destruction of coal and rock, which is equivalent to a seismic magnitude of 6–5 on the Richter scale.

Globally, many countries experience seismic roof pressure phenomena during underground mining. Currently, over 20 countries, including China, Germany, South Africa, Poland, the United States, and Canada, have experienced seismic roof pressure disasters. For example, more than half of Poland's 67 coal mines exhibit seismic tendencies. Between 1940 and 1980, Poland recorded over 2,000 instances of destructive seismic roof pressure. The earliest documented instance of seismic roof pressure in China occurred in 1933 at the Fushun Shengli Coal Mine. As mining depths increased and the scope of excavation expanded in the 1980s, seismic roof pressure disasters occurred at several coal mines, including those of the Beijing Mining Bureau, such as the Fangshan, Daguoyu, Muchengjian, Fushun Laohutai, Tangshan, Chaili, Yingcheng, Gaode, Tiechang, Didao, Tianchi, and Huancheng mines. In the

1990s, mines such as Huafeng, Sanhejian, Quantai, Jining, and Datong also experienced seismic roof pressure disasters. On October 20, 2018, a seismic roof pressure accident occurred at the Longyun Coal Mine of Shandong Energy Group, resulting in 21 fatalities. On June 9, 2019, a seismic roof pressure accident at Longjiapo Coal Mine, Jilin Coal Mining Group, led to 9 fatalities. On August 2, 2019, another seismic roof pressure accident at the Tangshan Mining Branch of Kailuan (Group) Co., Ltd. caused 7 deaths. In 2017, the number of mines affected by seismic roof pressure in China reached 177. The names, distribution, and quantities of these mines are shown in Table 5.1. The distribution of seismic roof pressure mines in China has expanded to more than 20 provinces and autonomous regions. As of June 2019, the number and distribution of mines in China identified as seismic roof pressure mines are detailed in Table 5.2.

The mechanism of seismic ground pressure is highly complex, with numerous influencing factors, gradually becoming a significant topic of research in the fields of mining and safety engineering. Most coal seams currently being mined in China exhibit varying degrees of susceptibility to rock bursts, with the likelihood of occurrence greatly increasing when the coal body experiences stress concentration at certain critical depths. The depth of coal mining in China increases by approximately 10 m annually, and as a result, rock burst disasters are becoming more severe, posing one of the primary constraints on the safety and productivity of the nation's mines. Despite substantial advancements in research on the mechanisms of rock bursts, monitoring techniques, and control technologies, the complexity of rock bursts means that a comprehensive system for their prevention and control has not yet been fundamentally developed.

A review of the research on rock bursts shows that past findings reflect a diversity of mining contexts, technical conditions, and conceptual understandings. Such technical factors have inevitably led to a wide array of research outcomes in China on rock bursts. Due to the complexity of geological structures, the occurrence and disaster characteristics of rock bursts exhibit variability, thus attracting many experts to conduct studies on the relationship between rock bursts and geological factors, leading to numerous valuable theoretical contributions. The most influential and significant theoretical work in this regard is the paper by scholar Zhang Mengtao: "Unified Instability Theory of Rock Burst and Outbursts" (Journal of Coal Science, 1991, Issue 4). This paper delves deeply into the causal relationship between geological structures and rock bursts.

Experts and scholars have categorized rock bursts based on various criteria: by the coal or rock seam in which they occur, classifying them into coal seam rock bursts and rock mass bursts (rock explosions); by stress type and loading method, they are divided into gravity-induced, structural, vibrational, and combined types; based on instability types of coal and rock strata, they can be classified into coal body compression, roof fracture, and fault displacement types; by the source of impact force, they are divided into rock explosion, roof collapse, and structural types; by material and structural instability, they can be categorized as material instability, structural instability, and slip-drift instability; according to the process of occurrence, rock bursts are divided into concentrated static load type and concentrated dynamic

## 5.1 Overview

**Table 5.1** Statistics of impact ground pressure mines in China (as of January 2017)

| Province and city | A mine shaft |
|---|---|
| Liaoning | Fushun Huotai Mine, Fushun Longfeng Mine, Fushun Shengli Mine, Fuxin Wulong Mine, Fuxin Evergrande Company, Fuxing Xinfu Mine, Shenyang Hongyang San Mine, Beipiao Taiji Mine, Fuxin Aiyou Mine, Benxi Niu, Xintai Mine, Shenyang Binggou Mine, Benxi Caitun Mine, Beipiao Guanshan Mine, Fuxin Gao De Mine, Fuxin Dongliang Mine |
| Shandong | Linyi Gucheng Mine, Zibo Tangkou Mine, Feicheng Liangbaosi Mine, Zaozhuang Taozhuang Mine, Zaozhuang Bayi Company, Zao Zhuang Gao Zhuang Mine, Xinwen Suncun Mine, Xinwen Huafeng Mine, Xinwen Xinjulong Mine, Xinwen Pansi Mine, Xinwen Zhangzhuang Mine (Huayuan), Yanzhou Boudian Mine, Yanzhou Dongtan Mine, Yanzhou Nantun Mine, Yanzhou Jiji Mine, Yanzhou Jisi Mine, Yanzhou Jisi Mine, Weishanhu Huancheng No.2 Mine, Weishanhu Chaoyang Mine, Weishanhu Jinyuan Mine, Qufu Xingcun Mine, Xinwen Xiezhuang Mine, Xinwen Liangzhuang Mine, Xinwen Huaheng Mine, Zaozhuang Chaili Mine, Xinwen Ezhuang Mine, Tengdong Mine, Yankuang Xinglongzhuang Mine, Yankuang Zhaolou Mine, Yankuang Beiju Mine, Longkou Beishou Mine, Linmin Wanglou Mine, Weishanhu Huancheng No.1 Mine, Luneng Pengzhuang Mine, Zizi Xufang Mine, Zizi Daizhuang Mine, Zizi Giting Mine, Zaojiang Tianchen Mine, Zaojiang Daxing Mine, Fengyuan Beixuelou Mine, Daizhuang Huxi Mine, Xinwen Longgu Mine, Luneng Guotun Mine, Luneng Pengzhuang Mine |
| Heilongjiang river forming the border between northeast China and Russia | Shuangyashan Jixian Mine, Shuangyashan Dongrong No.2 Mine, Hegang Nanshan Mine, Hegang Fuli Mine, Hegang Xing'an Mine, Hegang Junde Mine, Jixi Chengshan Mine, Qitaihe Xinxing Mine, Qitaihe Taoshan Mine, Shuangyashan Qixing Mine, Shuangyashan Xin'an Mine, Shuangyashan Dongguang Mine, Jixi Xinghua Mine, Jixi Didao Mine |
| He'nan Mengguzu autonomous county in Qinghai | Yima Qianqiu Mine, Yima Yuejin Mine, Yima Changcun Mine, Yima Gengcun Mine, Pingdingshan Ten Mine, Pingdingshan Eleven Mine, Pingdingshan Twelve Mine, Luoyang Xinyi Mine, Pingdingshan Eight Mine, Pingdingshan Four Mine, Pingdingshan Shushan One Mine, Hebi Five Mine, Hebi Three Mine, Pingyu Fangshan Mine, Yima Yangcun Mine, Jiaozuo Jiuilishan Mine |
| Jiangsu | Datun Yaoqiao Mine, Datun Kongzhuang Mine, Xuzhou Sanhejian Mine, Xuzhou Quantai Mine, Xuzhou Qishan Mine, Xuzhou Zhangshuanglou Mine, Xuzhou Zhangxiaolou Mine, Xuzhou Zhangji Mine |
| Beijing, capital of People's Republic of China | Muchengjian Mine, Dayanshan Mine, Chengzi Mine, Mentougou Mine, Datai Mine, Changgouyu Mine, Fangshan Mine |

(continued)

**Table 5.1** (continued)

| Province and city | A mine shaft |
|---|---|
| Shanxi | Datong Coal Yukou Mine, Datong Tongjialiang Mine, Datong Siraogou Mine, Datong Xinzhouyao Mine, Datong Jinhuagong Mine, Datong Baidong Mine, Zhaozhuang Mine, Yanya Coal Company, Datong Yongdingzhuang Mine, Jincheng Zhaozhuang Mine, Datong Tashan Mine |
| Anhui | Jizhong Dashucun Mine, Kailuan Zhaogezhuang Mine, Kailuan Tangshan Mine, Handan Guantai Mine |
| Sichuan | Mianyang Tianchi Mine, Mianyang Wuyi Mine, Mianyang Ronggu Coal Mine |
| Gansu | Huating Yanbei Mine, Huating Mine, Shanzhai Mine, Yaojie Mine (No. 1 shaft, Jinhe Coal Mine), Jingyuan Wangjiashan Mine |
| Also, Jilin prefecture level city, Jilin province | Liaoyuan Xi'an Mine, Liaoyuan Longjiabao Mine, Liaoyuan Taixin Mine, Tonghua Daoqing Mine, Tonghua Ironworks Mine, Jiutai Yingcheng Mine |
| Anhui | Huaibei Luling Mine, Huaibei Hazi Mine, Huainan Panyi Mine, Huainan Zhangji Mine, Huainan Xieyi Mine |
| Jiangxi | Pingle Yangou Mine, Huagushan Mine, Bajing Coal Mine, E4 Wells |
| Chongqing | Nantong Coal Mine No. 1 shaft, Yanshitai Mine, Jianghe Coal Mine |
| Xinjiang | Sulphur Gully Mine, Tiechang Gully Mine (Wudong Coal Mine North), Kuan Gully Mine, Dahong Gully Mine (Wudong Coal Mine South), Alkali Gully Mine (Wudong Coal Mine West) |
| Shaanxi | Xiayukou Mine, Xiashijie Mine, Hujiahe Mine |
| Inner Mongolia | Pingzhuang Gushan Mine |
| Hunan | Shaoyang Shixiajiang Mine, Shaoyang Xindong Mine, Shaoyang Niumasi Mine, Changsha Coal Dam Mine, Loudi Enkou Mine, Loudi Dougasan Mine, Loudi Qiaotouhe Mine, Lianyuan Xiantian Mine, Baisa Nanyang Mine, Chenzhou Kakizhuyuan Mine, Nanyang Inter River Coal Mine, Chenxi Jiangjaping Mine, Hunan Coal Huangnuling Coal Mine, and Longjiashan Coal Mine of Hongwei Company |
| Guizhou | Panjiang Shanfushu Mine, Panjiang Yueyuntian Mine, Liuzhi Shijiaotian Mine, Liuzhi Mine, Liuzhi Liangshuijing Mine, Liuzhi Dayong Mine, Xifeng Nanshan Mine, Shuicheng Dahabian Mine, Kaiyang Mine, Liuzhi Huadizhi Mine |
| Hillsides | Xingning Daxing Coal Mine |

load type; for shallow coal seams, rock bursts are classified into static load, dynamic load, and mixed types; and deep mining rock bursts are classified into strain-type, fault slip-type, and hard roof types.

Reviewing the findings on rock burst research, most engineering and technical experts and scholars agree that rock bursts result from the elastic potential energy of

## 5.1 Overview

**Table 5.2** Number and distribution of impact ground pressure mines in production in China (as of June 2019)

| Province and city | Beijing, capital of People's Republic of China | Anhui | Shanxi | Liaoning | Also, Jilin prefecture level city, Jilin province | Jiangsu | Shandong | He'nan Mengguzu autonomous county in Qinghai | Shaanxi | Xinjiang |
|---|---|---|---|---|---|---|---|---|---|---|
| Quantities | 1 | 2 | 2 | 8 | 2 | 6 | 40 | 4 | 13 | 9 |
| Province and city | Gansu | Guizhou | Inner Mongolia | Heilongjiang river forming the border between northeast China and Russia | | | | | | Add up the total |
| Quantities | 10 | 3 | 10 | 11 | | | | | | 121 |

strain formed by a specific geological structural stress field, which is then developed by the dynamic forces induced by mining operations. Based on the latest research findings, hot topics, and the current development status of rock bursts, the author conducted an analysis of rock burst research results from coal mines such as Shandong Huafeng Coal Mine, Jining Mining Area, Shanxi Tongmei Group, and Henan Yima Mining Area, while also engaging in discussions with on-site technical personnel. Our findings suggest that, for the foreseeable future, conducting research on rock bursts is of great theoretical and practical significance, but it also presents considerable challenges. Based on the research process, we have summarized and analyzed the following aspects of rock bursts: (1) Geological structures and rock bursts, (2) The interaction between roof stress and rock burst occurrence.

Compared with the level of forecasting technology, China has made great progress in the prevention and control of. The authors believe that domestic scholars have focused on the research of prevention and control of impact ground pressure, and the research on its occurrence mechanism is relatively weak. Based on the decompression principle of impact ground pressure, many effective methods of preventing and controlling impact ground pressure have been obtained, such as decompression methods of large-diameter drilling and unloading of coal seams, deep-hole precracking blasting of roof slabs, and hydraulic fracturing of roof slabs. These methods have achieved good prevention and control effects, but there are still many things that need to be supplemented and improved. Therefore, the focus of the research should be to impact ground pressure analyze and study the case impact ground pressure and occurrence mechanism of, pay more attention to the prediction and forecasting technology of impact ground pressure, especially the development of monitoring equipment technology, and pay more attention to the application of the existing monitoring technology of impact ground pressure and the comprehensive application of other monitoring technology. Proposing and establishing on-site monitoring ground pressure and forecasting technology system of impact, and researching and developing monitoring for impact ground pressure equipment are the key research directions in the long term in the future.

## 5.2 Current Status of Research and Development on Impact Ground Pressure

According to relevant documents, rock burst is a dynamic phenomenon characterized by sudden, intense, and violent destruction, caused by the release of deformation energy from the coal-rock mass in the shaft roadway or mining face. Rock burst is a common issue faced by the global mining industry. The phenomenon was first reported in 1738 in the United Kingdom, and later, incidents of rock burst were recorded in the former Soviet Union, Germany, South Africa, the United States, Poland, Canada, France, Japan, India, Hungary, the Czech Republic, Bulgaria, New Zealand, Austria, and other countries. Currently, more than 20 countries and regions,

## 5.2 Current Status of Research and Development on Impact Ground Pressure

including China, have experienced rock burst incidents. This fact suggests that almost all countries and regions involved in mining activities around the world are threatened by rock burst to varying degrees. The most severe rock burst disasters and the most effective prevention and control measures have been observed in the former Soviet Union, Poland, and Germany.

At present, China is the largest coal producer among the major coal-producing countries, and most mines predominantly use underground mining methods. As shallow resources decrease, underground mining will remain the primary method for a long time. Currently, with increasing mining depths and mining intensity, rock burst disasters are becoming increasingly severe. Statistical data shows that multiple rock burst incidents causing casualties and property damage occur annually in China, and the country has gradually become one of the countries facing the most severe rock burst risks globally. Therefore, research on rock bursts should receive full attention from Chinese scholars.

The first recorded rock burst in China occurred at the Shengli Mine, located in the original Fushun Mining Bureau, in 1933. With the increase in mining depth and the expansion of mining areas, rock bursts have occurred in mines such as the Mengtougou Mine, Chengzi Mine, and Fangshan Mine of the former Beijing Mining Bureau, the Longfeng Mine and Laohutai Mine of the former Fushun Mining Bureau, the Taozhuang Mine and Bayi Mine of the former Zaozhuang Mining Bureau, the Tangshan Mine of the former Kaiqi Mining Bureau, and the Gaode Mine and Wulong Mine of the former Fuxin Mining Bureau, as well as the Tianchi Coal Mine in Sichuan. Since the founding of the People's Republic of China, destructive rock bursts have occurred over 4,000 times in key mining areas, with magnitudes ranging from M0.5 to M3.8, resulting in significant roadway damage and considerable personnel casualties.

Due to differences in the geological environment and mining methods, the manifestation of rock bursts in China shows great diversity. Through summary and analysis, the main characteristics of rock bursts in China include:

(1) **Sudden Occurrence**

The occurrence and characteristics of rock bursts are closely related to the mining technology and monitoring technology levels. Often, there are no obvious visible precursors before a rock burst, and most of the known rock bursts are triggered by objective factors such as blasting in the working face, coal exposure in roadways, roof caving, and the instability of coal pillars. However, many rock bursts also occur after mining has ceased in the working face, making it difficult to pinpoint the exact triggering factor. The process of rock burst is generally short-lived and is accompanied by loud noises and vibrations, with the maximum magnitude reaching a seismic level of 5 on the Richter scale, with significant vibrations felt over several kilometers from the site.

(2) **Diverse Occurrence Locations**

Many recorded rock bursts in China are caused by the instability and destruction of coal-rock masses. For example, the Datai Mine, Panxi Mine, Bayi Mine, Taiji Mine, and Chaili Mine have experienced multiple rock bursts triggered by coal-rock mass instability. In addition, several incidents of rock bursts caused by roof and floor movements have been reported, such as the rock burst at the Fangshan Mine, where the roadway floor suddenly bulged and cracked, forming fissures approximately 9 cm wide. Rock bursts caused by coal-rock mass instability usually manifest as fragmented coal-rock suddenly ejected from the coal wall, damaging equipment. In some extreme cases, large coal-rock masses may undergo overall sliding, resulting in equipment damage and production disruption.

(3) **Severe Degree of Damage**

Due to varying mining conditions and methods, the severity of rock burst disasters and associated damages in China also vary. Mines such as Laohutai, Mengtougou, Huafeng, and Taiji have all experienced rock bursts with seismic intensities above level 3. These are considered high-intensity seismic shocks. According to monitoring records, the Laohutai Coal Mine experiences over 300 recorded vibrations monthly; Mengtougou Coal Mine averages over 100 rock burst and vibration incidents per month; Huafeng Coal Mine records over 1,000 instances of underground vibration monthly; and Taiji Mine is a typical case of multiple mining disasters, including rock bursts, coal and gas (rock) outbursts, and high-temperature thermal hazards, in one of China's deep mines. Vibrations caused by rock bursts can lead to failure of support systems, causing severe roadway collapses, roof falls, floor heaving, and broken supports. In severe cases, workers may be injured, and roadways may become completely blocked, forcing production to halt.

(4) **Increasing Frequency**

Before the founding of the People's Republic of China, only two pairs of mines experienced rock bursts. By 1950, this number had increased to seven pairs, and by 1960 to 12 pairs, 22 pairs by 1970, 32 pairs by 1980, and 50 pairs by 1990. In 2008, 121 pairs of mines experienced rock bursts, and by 2017, the number had risen to 177 pairs. In recent years, China has adjusted its coal mining industry by closing several mines with poor safety conditions, outdated technology, and exhausted resources, including some mines prone to rock bursts, resulting in a slight decrease in the number of such mines. However, there are still over 120 pairs of mines experiencing rock bursts. As mining depth and intensity increase, the number of mines at risk of rock bursts and the frequency of rock burst events will continue to rise year by year.

In the early decades, the formation and occurrence of rock bursts were relatively complex, and the variety of destructive forms and phenomena were closely related to the diversity of mining methods at the time, the lack of unified technical standards, and the uneven levels of equipment technology and process design. It can be said that the scientific and technological development and understanding before the 1980s, or even the 1990s, represented a period of technological advancement.

## 5.2.1 Review of and Theoretical Development of the Understanding Impact Ground Pressure

The systematic study of the mechanism of coal and gas outbursts (rock bursts) is of great significance for accurately understanding the occurrence of such outbursts, predicting their occurrence, and reducing the harm caused by these events. The occurrence of coal and gas outbursts is related to geological structures, mining layouts, and changes in mining-induced stresses. Mining activities lead to the redistribution of mining stresses, and the continuously changing mining-induced stresses act on the surrounding rock of the working face, causing stress concentration at certain locations. The superposition of these stresses with structural stresses makes these areas of coal and rock highly susceptible to instability and failure. In severe cases, this may trigger catastrophic accidents. A thorough study of the spatial structure of mining-induced stress and its relationship with structural stress fields and dynamic mining pressures, combined with the use of appropriate monitoring techniques to systematically observe phenomena such as coal and rock body fractures, is an important means to prevent and reduce coal mine outburst accidents.

## 5.2.2 Development of Research Impact Ground Pressure Mechanisms

The phenomenon of pressure bumping in the mining sector has been documented for over two centuries, but it wasn't until the twentieth century that it truly captured the attention of researchers worldwide. During this period, specialized institutions focused on pressure bumping research were established in various countries. Research efforts began with extensive statistical surveys and analyses of pressure bumping, evolving into laboratory theoretical studies and field monitoring research. The most effective countries in studying the mechanisms and prevention of pressure bumping were the former Soviet Union, Poland, and Germany. The research into pressure bumping mechanisms in China began in 1978 with a water injection experiment conducted in the coal seam at the Tianchi Coal Mine. Since then, multiple key scientific projects at the ministerial level have been carried out. Around 1980, foreign monitoring systems were introduced, providing effective technical support for the expansion of theoretical research, and preliminary theoretical and practical experience was accumulated. In recent years, through scientific cooperation and academic exchanges with foreign countries, China has absorbed numerous advanced technologies and ideas.

Simultaneously with foundational work, Beijing Mining Bureau, Zaozhuang Mining Bureau, the Institute of Mining Research at the China Coal Science and Technology Research Institute, and some instrument manufacturers completed two national "Seventh Five-Year"technological research projects in 1990, namely "Pressure bumping Monitoring Devices" and "Pressure bumping Prediction and Prevention." In 1993, the Institute of Mining at the China Coal Science and Technology Research Institute completed a key project of the Ministry of Coal Industry, the "Experimental Research on the Tomographic Imaging Method for Pressure bumping Coal Seams," which was the first attempt to apply seismic tomography methods to the evaluation of pressure bumping hazards in coal mines. In 1997, the Ministry of Energy completed a key project on "Comprehensive Prevention and Control Technology for Pressure bumping at Huafeng Mine," systematically and comprehensively conducting practical research on the causes, prediction, and prevention of pressure bumping.

Entering the twenty-first century, against the backdrop of a significant increase in pressure bumping disasters in coal mines, both the state and enterprises began to pay more attention to the development and investment in pressure bumping monitoring equipment and prevention technologies. In 2005, the Ministry of Science and Technology established the "11th Five-Year" National Science and Technology Support Program project, "Multi-Parameter Identification and Hazard Mitigation Key Technologies and Equipment for Deep Mining Coal and Rock Dynamic Disasters," marking the first time a project related to coal and rock dynamic disasters was included in the "973 Program." In 2010, the relevant national standards for pressure bumping measurement, monitoring, and prevention methods were issued and implemented, signifying a new phase in China's research on pressure bumping. In the area of pressure bumping research, the state increased its investment. In 2010, the "973 Program" project "Fundamental Research on the Mechanism and Prevention of Dynamic Disasters in Deep Coal Mining" was established with Chief Scientist Jiang Yaodong, and in 2011, another "973 Program" project, "Theory of Coal and Gas Co-Mining in Deep Coal Mining," was established with Chief Scientist Xie Heping. In 2016 and 2017, the Ministry of Science and Technology launched two projects related to pressure bumping: "Risk Identification and Monitoring and Early Warning Technology for Typical Dynamic Disasters in Coal Mines" led by Yuan Liang in 2016, and "Prevention and Control Technology for Coal and Rock Dynamic Disasters in Deep Mining" led by Qi Qingxin in 2017.

At the end of 2016, the State Administration of Work Safety and the National Coal Mine Safety Supervision Bureau organized relevant units and experts to restart the drafting of the "Regulations for the Prevention and Control of Pressure bumping in Coal Mines." Based on the 2013 drafting efforts, the work continued through 2017 and the first quarter of 2018, culminating in the release of the "Detailed Rules for the Prevention and Control of Pressure bumping in Coal Mines" by the National Coal Mine Safety Supervision Bureau on May 2, 2018. In 2018, the drafting of Part 3 to Part 14 of the series of standards "Methods for the Measurement, Monitoring, and Prevention of Pressure bumping" reached its final stages. This work, initiated in 2013 and led by the Safety Branch of the China Coal Science and Technology Research Institute, involved over 20 research institutions, universities, and enterprises from the coal industry. As a result, significant progress was made in China's research on pressure bumping in a short period, laying a solid foundation for further in-depth study of the phenomenon.

### 5.2.3 Review and Summary of Impact Ground Pressure Theory and Research Results

Impact ground pressure (rock explosion) mechanism research is: stiffness theory, strength theory, energy theory, "three norms" theory, impact propensity theory, deformation system destabilization theory, fractal theory, mutation theory, expansion theory, time effect theory, dynamic destabilization theory, impact ground pressure and prominent unified destabilization theory, and so on. The above theories play in revealing the formation mechanism of impact ground pressure in depth. A brief overview of the commonly used stiffness theory, strength theory, energy theory, and impact propensity theory will be given below:

(1) Stiffness theory

Stiffness theory was proposed by Cook et al. around 1960 theory states that when the coal rock body by the stiffness of to yielding force |K external $_n$| greater than the stiffness of the top and bottom plate |K bracket or $_s$|, the impact ground pressure phenomenon will occur. On the contrary, when |$K_{(n)}$|<|$K_{(s)}$|, the impact dynamic phenomenon. will not occur here.

The process of uniaxial compressive testing shows that when the stress in the specimen exceeds its strength, the specimen only undergoes destruction and there is no impact phenomenon. Therefore, the strength theory only explains the cause of coal rock body damage, and does not reflect the intrinsic mechanism of the occurrence of impact ground pressure. In order to correct this error, the recent strength theory proposes that the determining factor leading to the impact damage of coal rock body is not only the size of the stress, but the ratio of the stress to the strength. However, the strength theory is not clear about the critical value of this ratio.

(2) Strength theory

Strength theory, proposed by G. Braemer, suggests that a coal rock body is mainly clamped by top and bottom slabs the, and that the clamping characteristics characterize the structural nature of the system formed by the coal and the surrounding rocks. When coal the contact surface between the rock body and the surrounding rock reaches its limit equilibrium condition, the coal rock body will occur be destabilized and damage, which may lead to impact ground pressure. It can also be shown that the stress conditions are required for the occurrence of impact ground pressure: $\sigma \geq \sigma^*$. The clamping theory of coal-rock mass can well explain the limit equilibrium conditions of the mechanical system of coal-rock mass and surrounding rock, but it cannot explain the dynamic characteristics of rock burst caused by coal-rock mass. In some cases, $\sigma \geq \sigma^*$, but there is no occurrence of the phenomenon of impact ground pressure.

### (3) Energy theory

Cook and others in 1960 on the South African region a decade for more than of impact ground pressure collation and generalization found, that the mining support and surrounding rock mechanical system in its mechanical equilibrium state of destruction of energy is greater than the energy consumed, that is, the impact ground pressure occurs. In other words, the energy accumulated in the local coal rock body is greater than the energy consumed by the destruction of the coal rock body, that is, impact ground pressure occurs. From the viewpoint of conservation of energy, part of the energy accumulated in the coal rock body is used to destroy the coal rock body, and the other part will throw out the destroyed coal rock body. Therefore, it is correct to use energy theory to describe the cause of impact ground pressure. However, the early energy theory is only a qualitative description of the mechanism of impact ground pressure occurrence. Afterwards, Petukhov the energy of impact ground pressure studied composition. The energy theory assumes that are the top and bottom plates of coal rock purely elastic, while impact the coal rock body even if it is extremely loose. With low risk may cause impact ground pressure phenomenon under high pressure.

### (4) Impact Propensity Theory

Impact susceptibility is the inherent of a coal rock body to undergo impact damage property. The theory of impact susceptibility based on this interpretation suggests that when the impact susceptibility of a coal rock mass, $K_{(E)}$, is greater than a critical value, $K_{EC}$, impact ground pressure phenomena may be triggered. Existing studies have shown that used to the parameters measure the impact susceptibility of a coal rock body are summarized as follows: the elastic energy index of the coal rock body, the dynamic damage time $W_{ET}$, $D_t$, and the impact energy index, $K_E$.

The determination of the elastic energy index relies mainly on the uniaxial compression test, when the coal rock body specimen reaches about, the is carried out immediately85% of the ultimate pressure unloading, then the elastic index expression: can be derived

$$W_{ET} = \frac{\Phi_{SP}}{\Phi_{ST}}$$

where $\Phi_{SP}$ is the elastic energy, $\Phi_{ST}$ is the loss energy, and the impact energy index can be derived from the following equation:

$$K_E = \frac{F_S}{F_X}$$

where $F_S$ is the coal full course area of the before the peak of the stress–strain curve and $F_X$ is the curve area after the peak of.

To sum up, in the twentieth century, China's impact ground pressure research results are rich and colorful, due to the lack of a relatively unified coal mining methods, various coal mining methods in parallel development, the degree of standardization is in the development stage, thus resulting in the diversification of understanding and thinking, research results are relatively abstract, and did not form a relatively unified academic theory and engineering practice and technology system. It is not difficult for us to find that the research results on impact ground pressure have to be further applied in engineering practice.

## 5.3 Twenty-first Century Impact Ground Pressure Gestation Mechanism and Creep Characterization

In brief, the coal mining methods in the twenty-first century have largely standardized around the fully-mechanized coal mining method, with a high degree of standardization. The focus has shifted toward studying the structural stress and the transfer and release effects of strain elastic potential energy within or at depth in the coal-rock mass. There is now an increasing awareness of the real-time information on the creep to catastrophic change of the coal-rock mass's properties, recognizing the significant engineering importance and value of prediction and forecasting.

The causes of seismic events are generally triggered by factors such as meteorological conditions, tidal forces, geological disasters, and crustal movements. These factors induce intense movements, compressions, and collisions of tectonic plates in the Earth's deep structure, which then cause vibrations and displacement of rock masses, manifesting as destructive phenomena on the Earth's surface. In contrast, pressure bumping pressure (or impact pressure) is caused by underground mining activities. It results from the intense accumulation of elastic strain energy stored in a local region $\Omega$ within the coal-rock mass, which then suddenly transfers and expels dust and coal-rock from the mined space. This process resembles an avalanche, exhibiting intense, impulsive damage with relatively smaller energy levels.

During coal mining, the stress concentration and transfer in deep coal-rock masses can easily lead to the accumulation of high strain elastic potential energy within the coal-rock mass region $\Omega$. This stress, induced by mining activities, is often referred to as secondary stress. Secondary stress can either amplify or counteract the original in-situ stress field, resulting in a composite stress field. The elastic strain energy manifested by the stress field is transferred within the coal-rock mass as elastic stress waves, which, through the action of wave energy, provoke frictional collisions and impact compressions among internal medium particles. This phenomenon is referred to as the impact effect of elastic stress waves. When the elastic strain energy is sufficiently large, it can lead to continuous internal impact effects within the coal-rock mass, eventually resulting in the formation of a damaged medium structure in the coal-rock mass within region $\Omega$. The extent of the damaged region $\Omega$ depends on the magnitude of the elastic strain energy $\Phi$. When the elastic strain energy reaches a critical state, the range of $\Omega$ will continuously expand toward the weak planes of the surrounding rock, with continuous or intermittent energy transfer causing strain energy damage to the medium's properties, ultimately fracturing the coal-rock structure and generating dust and coal chunks. This combination of effects is referred to as the fracture and energy dissipation effect of elastic potential energy. The continuous and intermittent consumption of elastic potential energy is completed through work W0, i.e., under the action of $\Phi$, the coal-rock mass undergoes an impact effect. This is a composite, dynamic change. If the value of $\Phi$ within $\Omega$ is insufficient to allow the elastic energy to propagate continuously along the most easily transmissible path (such as weak planes in surrounding rock or mining spaces), it will continue until $\Phi$ is entirely consumed as work W0, at which point the impact effect ceases within the coal-rock mass.

Impact and outburst effects occur inside the coal-rock mass, at any depth within the surrounding rock. The author believes that, during mining operations, the impact effect of strain energy in the coal-rock mass can occur at any time. The intensity of the impact and outburst effects within the coal-rock mass is related to the magnitude of stress in the coal-rock mass. When these effects are severe, energy transfer and sudden accumulation alternate rapidly, causing mutual impact and compression between the elastic strain energy, the coal-rock mass, and the gas contained in its pores and fractures. As gas fills the fractures, and when the elastic stress waves communicate (or connect) with the fractures, the gas and fracture walls may experience impacts and friction during elastic compression, potentially producing sound and weakening the fracture walls. The theory behind the production of sound and noise is rooted in gas dynamics and acoustic principles, which help explain why, during roadway excavation, coal extraction at coal doors, or the pressure displacement of coal-rock mass, one might hear or detect sounds emitted from the rock mass. The author believes that the phenomenon of impact effects due to the elastic energy generated in deep coal-rock masses during mining is a common occurrence, reflecting the dynamic nature of coal-rock mass properties. The process of elastic energy impact effects is a process of energy attenuation and work production, with phenomena such as displacement, deformation, and pressure changes occurring within the coal-rock mass. This phenomenon is universal, as it can happen or cease at any time. Therefore, one cannot solely rely on the occurrence of impact effects in the coal-rock mass to predict whether pressure bumping pressure will occur or whether it will compromise the stability of surrounding rock support systems or lead to a catastrophic failure.

### 5.3.1 Characterization of the Action of Effect the Shock

In the excavation process of coal rock body, coal rock body within the special geological structure of the ground stress field (including huge thick hard top plate subsidence induced high-pressure stress region) region of the formation of strain elastic potential energy, elastic energy in the form of elastic stress wave transfer release in the rock body, relying on the size of the wave energy and the impact and extrusion effect on the coal rock body medium point of mass to produce the impact effect the role of

5.3 Twenty-first Century Impact Ground Pressure Gestation Mechanism … 113

the of effect is often a combination of erratic, and its The main characteristics of the performance:

① Stored the form of elastic energy in the coal rock body in the local area $\Omega$, it has to the coal rock body excavation space vulnerable surface direction transfer tendency potential (a tendency potential), there is a critical state, where the elastic potential energy reaches a kind of extreme value point.
② The ground stress field and the secondary stresses generated by the extraction process are superimposable, reinforcing the ground stresses and the destructive nature the of the occurrence of impact ground pressure in the form of elastic stress wave energy probability;
③ Ground stress field on the stored in the coal rock body size of the elastic energy, as well as the transferability of the coal rock body damage with the transfer path and the transfer distance to produce a direct control role, elastic stress wave energy transfer direction directly refers to the enclosure of the loose circle and mining space, that is, the enclosure of the vulnerable spatial surface;
④ The elastic deformation energy stored in the coal rock body damages and weakens the does work in a way that properties of rock body and releases the propagation the coal, and the impact direction of the wave energy causes the weakening of the broken strength of the coal rock body organization and structure, i.e., the crushing energy-decreasing effect, which is also a combination of the impact effect of the elastic potential energy and the crushing energy-decreasing effect. When approaching the weak in space surrounding the rock level, the coal rock body produces sudden destruction like an avalanche;
⑤ The hardness or denseness of the coal rock body determines the degree of elastic deformation and energy storage of the coal rock body caused by the ground stress field, and a larger elastic deformation means that the coal rock body is able to store or generate higher energy, thus promoting the occurrence of impact ground pressure.

Therefore, excessive gestures are a necessary condition, i.e., inherent for the occurrence of impact ground pressure a condition, i.e., in the geological tectonic zones, especially the extrusion-type tectonic zones and the stress mining the sudden release of elastic energy is extremely likely to form an impact dynamic disaster. There is no obvious direct correlation between the propensity of impact and depth, decompression zones of even the coal rock body with little mining depth may be a high stress zone, storing very high elastic strain energy, and there is also the possibility of impact ground pressure. Impact ground pressure is different from earthquake, its occurrence of energy level is small, need to external induced conditions, that is, need to excavation activities to prompt the coal rock body stress state produced by the static rotation of the change, and induced elastic energy along the easy to pass the path to the excavation of the space surrounding the rock vulnerable to the area of the polymerization of the transfer (Table 5.3).

**Table 5.3** Top and bottom plate conditions that produce impact-type dynamic instability

| Country/coalfield | Top and bottom plate lithology | Thickness of rock layers (m) | Rock compressive strength (MPa) | Rock slug (MPa) |
|---|---|---|---|---|
| United states of America | Limestone | >10 | ~ 100 | 25,000 |
| Georgia | – | >8 | >70 | – |
| India | Limestone | >8 | 50–75 | – |
| Sino | Limestone | 10–40 | 130 | – |
| France/Provence | Limestone | >10 | 200 | – |
| Germany/Ruhr | Limestone | 5–6 | – | – |
| Poland/Upper Silesia | Limestone | 15 | 90–130 | – |

## 5.3.2 Mechanism of Impact Ground Pressure and Impact Effect Equation

(1) Intrinsic spontaneous impact ground pressure

The author believes that most of the impact ground pressure formation is directly related to the geological structure, and the impact tendency or impact ground pressure formed in this case is called the inherent spontaneous type. Coal seam is in folds and other geological structures, similar to the inherent geological stress field, mining or other external forces continue to disturb shown in Figure 5.1 will the folded zone appears more intense geological stress field decompression equilibrium changes, which in turn prompted the original stress elements (concentration points) to produce changes in the formation of very high elastic potential energy. This change is due to the of mining spaciotemporal and spatial effects, induced by the deep stress field of the geological discontinuity decompression and slippage of the combined effect of the original stress field changes in the direction of the role of the original stress field and the role of the applied power load, the original stress field equilibrium is destroyed, and then produce or form the activated stress–strain elastic potential energy of the influence of the region $\Omega$, at this time it is very likely to lead to the energy release of from the impact sustained, resulting in the organization of coal rock bodies destruction and strength weakening, and ultimately cause impact ground pressure disaster. In this case, the induced elastic potential energy influence area scope of is very large, and it is very easy to form a very high energy $\Phi$ stored in the $\Omega$ area, and when the $\Omega$ area is subjected to the caused by rapid perturbation influence redistribution and concentration of the direction of the energy impact the, the resulting impact intensity is often also relatively large. The elastic energy of the coal seam with inherent spontaneous impact tendency $\Phi$ may sometimes include: (1) from the coal seam the energy the region of $\Omega$ $\Phi_0$; (2) the superimposed the superimposed stress of the auxiliary stress field of the neighboring coal rock strata energy formed by, and

## 5.3 Twenty-first Century Impact Ground Pressure Gestation Mechanism ...

total elastic potential energy the coal rock body $P_z$ the mathematical expression of the energy equation of the of is $\Phi$:

$$\Phi = KP_z U + \Phi_0 \qquad (5.1)$$

where, $\Phi_0$—the coal seems own inherent elastic potential energy, unit, J;

K–Transfer (conversion) coefficient (or elasticity coefficient), $K \leq 1$; auxiliary superimposed stress $P_z$.

$P_z$ Superimposed stress in MPa;

U–$P_z$ induced displacement deformation of the coal rock body (or coal seam), unit, m.

(2) Intrinsically induced shock ground pressure

Another more common type of impact phenomenon is due to the mining activities (such as the impact of overburden rock movement in the adjacent mining area, etc.) caused by the thick or hard rock layer overlying the quarry and mining area roof of the to generate a large area of compressive stress $P_z$ acting on the coal rock body. As shown in Fig. 5.2, in the area the elastic energy formed is superimposed and transformed to act on the coal seam $\Phi_z$, $P_z$, and make the $\Omega$ area form a very high elastic potential energy $\Phi$, resulting in overloading and overloading and destabilization of the supporting structure of the working area of the quarry and the supporting pressure of the coal seam. In this case, the roadway will be subjected to a great impact of the dynamic load of the overlying rock layer, causing a wide range of drastic changes in the elastic strain energy of the coal seam, often resulting in the mining face and the roadway to produce a shocking impact damage phenomenon,

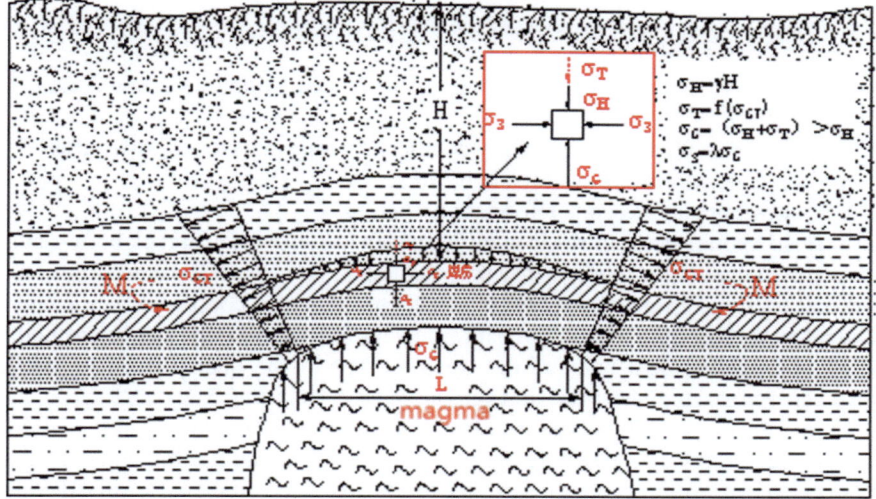

**Fig. 5.1** Schematic diagram of stress distribution in typical geological formations

extrusion of the coal body crushed and sprayed debris, dust, gas and other anomalies, in particular, causing instantaneous twisting and deformation of the roadway damage phenomenon, such as the Datong mining area, the Jining Mining Area. This kind of impact phenomenon is a kind of impact phenomenon, with the increase of mining depth, there are some medium-thick (or thick) coal seams with elastic strain performance and good performance of saving elastic energy, the coal quality is both hard and elastic–plastic high characteristic properties, and the action of compressive stress presents elastic–plastic and crispy (or plastic) brittle destructive characteristics. The cause of this impact phenomenon is also related to the mining method and process design, such as long-wall mining height, small coal pillars (or no coal pillars) and the space effect of the mining airspace and other factors, which seriously common in recent years affects the production safety, and has become an impact phenomenon that cannot be ignored. Therefore, this kind of large-area deformation movement induced by the formation of thick and hard top plate, very elastic–plastic characteristics of the coal seam of the impact power damage is called inherently induced impact ground pressure. As shown in Fig. 5.2, the energy of impact tendency or impact ground pressure induced by large-area compressive stress formed by the thick and hard roof plate (the overlying rock layer of the coal seam) in the quarry originates from the roof plate, while the elastic energy of the coal seam itself is extremely small; however, the coal seam has the physical property of storing elastic potential energy, and the coal seams elastic potential energy is rapidly increased under the inducement of the compressive stress from the thick and hard roof plate and obtains the energy roof plate, which $\Phi$. Therefore, it is proposed that a representative of the thick and hard roof plate form impact damage is the thick and hard is the most important factor in the impact damage. The mathematical equation representing the impact damage energy formed by the thick and hard top plate is expressed as:

$$\Phi = KP_z U \tag{5.2}$$

where, $\Phi$–Elastic potential energy generated by the coal seam under action, unit, J; $P_z$

K–Transfer coefficient (or elasticity coefficient); top plate compressive stress $P_z$.

$P_z$–Top plate compressive stress, unit, MPa;

U–Coal seam compression deformation, unit, m.

Intrinsic spontaneous, intrinsic induced impact tendency of coal rock body genus has the elastic potential wave energy propagation and transmission, work consumption and energy conversion release are subject to the law of conservation of energy guidelines, the energy propagation path is always chosen from the savings of high-energy $\Omega$ regional coal body to the weak surface of the quarry perimeter rock low-energy area (mining space) transfer and release, as in Fig. 5.3.

Elastic energy in the transfer process, will trigger the vibration of the medium point of coal rock body, resulting in the formation of impact friction and collision between the medium point of mass and the formation of regional coal structure damage and weakening of the strength of the impact friction of the touch action known as the elastic energy of the impact effect, the impact effect of the energy in

## 5.3 Twenty-first Century Impact Ground Pressure Gestation Mechanism ...

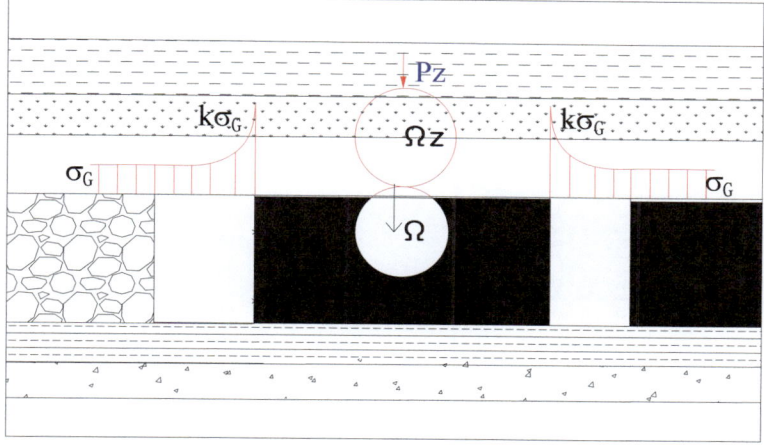

(a) Thick and hard roof compressive stresses acting on mined coal seams

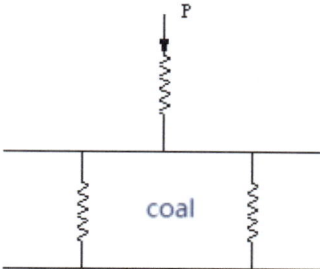

(b) Principles of mechanical action of thick and hard top plates

**Fig. 5.2** Principle of impact damage formation in a thick hard top plate

the coal rock body within the continued to occur unless the energy to consume all the energy, and its behavior is similar to the domino effect. The impact effect occurs when the internal coal rock body vibration and friction of the medium point to medium strength weakening and expansion process, this moment there is acoustic emission, micro-vibration, dust, temperature rise (thermal phenomenon), induced displacement and deformation of the surrounding rock and other physical phenomena, due to the internal coal rock body medium strength weakening and expansion, this is a process of work consuming the elastic energy. If the elastic energy is large enough, the elastic energy repeats its elastic impact effect, and the elastic energy continuously transfers and releases physical signals in the direction of the surrounding rock (i.e., the mining space).

The process of impact ground pressure is an instantaneous release of energy. The occurrence of this process is the result of the action of the two necessary conditions of the stored energy in the coal rock body and the influence of the corresponding

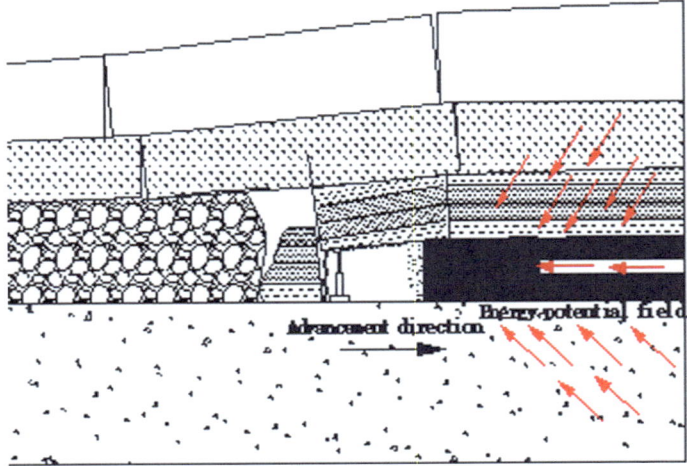

**Fig. 5.3** Distribution of elastic energy potential field transfer due to extraction

mining activities. Coal rock body Ω region of the stored energy by mining activities, induced by the rock body of elastic strain energy suddenly to the same direction of the transfer of the accumulation of energy, the instantaneous formation or the gradual formation of directional energy potential field, as shown in Fig. 5.3, this change in the process often results in the mechanical properties of the rock body of the damaging damage, so that the structure of the coal rock body and the traits of the organization of the weakening of the weak phenomenon triggered by the action of this elastic potential energy called the elastic potential energy of the Crushing effect. Formation of energy transfer accumulation and coal rock body traits of the structural organization of the weakening of the alternating, continuous energy along the easy transfer path to the mining space. This physical change process is called the impact effect of the elastic potential energy of the coal rock body. The size of the elastic strain energy that causes the impact effect of the coal rock body to occur is:

$$\phi = \sigma u \qquad (5.3)$$

where σ is the stress that directly causes strain damage to the rock mass, Pa; u is the amount of strain deformation that occurs rock mass under the action of σ, m.

It is not difficult to understand, the transfer of stored energy in the coal rock body depends on the continuous weakening of the medium or the existence of easy transfer paths for energy transfer, the process to consume energy to overcome the medium resistance and damage to the rock body trait organization, that is, the strain energy of the rock body medium damage weakening and consuming energy to do work, if we set the value of the work done to $W_0$, then when the transfer of energy is sufficient to overcome the damage weakened medium to do the work of the energy consumed by the damaged weakened coal rock will be suddenly thrown from the weak surface

## 5.3 Twenty-first Century Impact Ground Pressure Gestation Mechanism …

of the mining enclosure rock. If the value of work is to do more damage weakening medium, the coal rock will be suddenly thrown out from the vulnerable surface of the excavation surrounding rock, that is to say, if the coal rock body weakened by the damage stored $W_0$, when the transferred energy is enough to overcome the energy consumed by try $\Phi$ in the rock body $\Omega$ must be more than the energy consumed to complete the damage to the media trait is to be throat, thrown Ener structure to do the work, and the residual energy that exceeds the value of work is completely contributed to the kinetic energy required to throw out (or move) coal rock body. Therefore, according to the energy conservation theorem, it can be obtained:

$$\Phi = MU_0^2/2 + W_0 \qquad (5.4)$$

where M is the mass of coal rock ejected (or thrown up); $U_0$ is the initial ejection velocity of the ejected or thrown coal rock dust.

Equation (5.4) is called the coal-rock impact effect equation. The physical significance revealed by the can be summarized as follows: ① Whether the impact effect and impact ground pressure can occur or continue to occur depends on the coal rock body (5.3) and (5.4), the level of energy stored in the system. Energy along the medium advantage path to the mining space surrounding rock vulnerable surface direction transfer process need to damage the weakened medium (coal rock body) traits organization, and consume energy to do work $\Omega$ $W_0$, the size of the remaining energy directly affects the occurrence and continuation of the impact effect, if you reach the vulnerable surface of the space surrounding rock residual energy is greater than $W_0$, that is, to satisfy the Eq. (5.4), the elastic potential energy reaches a critical state, then the destructive impact pressure may occur. The elastic potential energy reaches a critical state, then destructive impact ground pressure may occur. It also reflects the result of work done by the elastic energy to produce the shock effect. The same stored energy $W_0$ $\Omega$, but also by the rock layer, media cohesion (or bonding) between the role of different, and produce different results of the impact effect, which means that the coal rock body occurs when the impact effect of a corresponding stored energy threshold value $W_T$. If the stored energy $\Phi$ is less than the energy consumed by the corresponding damage weakening medium to do its work, no matter how large the stored energy is, no destructive impact pressure will be generated; on the contrary, the stored energy if is higher than the energy consumed by the corresponding medium to do its work, even if the stored energy $\Phi$ is small, destructive impact pressure will be generated. ② When the storage energy in the coal rock body is certain, the duration of the impact effect is inversely proportional to the size of the work done on the medium, i.e., the larger the storage energy is to the medium, the shorter the duration of the impact effect is, and the more significant the crushing effect phenomenon is. (iii) Since the medium weakened by damage pushes down the weak surface of space surrounding rock with initial kinetic energy, if the weak surface of excavation is strengthened enough to absorb the residual energy ($\Phi$-$W_0$), it is possible to artificially control the magnitude of the residual energy to reduce it to zero, avoiding the occurrence of impact damage. ④ During the process of impact effect occurring in the coal rock body, i.e., the energy transfer process, a large amount of dust and fissures are generated in the transfer path. ⑤ the process of impact effect is a gradual change to sudden change, during which the coal rock medium undergoes the physical change from extrusion strain to weakened expansion displacement, and the change process is a process of absorbing energy, as well as the development process of displacement deformation of the space surrounding rock.

## 5.3.3 Coal Rock Body Energy Storage Relationship with Impact Ground Pressure

The difference in environmental conditions for the formation of coal seams makes the coal and rock masses in some areas in a relatively high in-situ stress field, so the coal and rock masses in this area store (or have) higher elastic potential energy. The energy storage characteristics in this coal rock mass is an inherent property of itself. From the perspective of underground coal-rock mass control, the elastic strain energy stored in coal-rock mass can be called elastic energy or elastic failure energy of coal-rock mass. The above definition can accurately characterize the impact tendency of coal and rock mass in the coal seam.

When the elastic energy stored in the underground coal and rock mass is not affected by the mining disturbance, the energy storage is generally in a relatively stable equilibrium state due to the inherent effect of the cohesion (cohesive force) between the longitudinal (transverse) structure of the coal and rock mass. If the adjacent coal-rock energy storage system is defined as a regional energy storage system, the energy in the energy storage system can be calculated by the following formula:

$$\Phi = \int_{u_x} \sigma_x u_x du_x + \int_{u_y} \sigma_y u_y du_y + \int_{u_z} \sigma_z u_z du_z$$

where $\sigma_x$, $\sigma_y$, $\sigma_z$ represent respectively the stresses x, y, z directions in the, in the three-dimensional spatial coordinates, $u_y$, $u_z$ represent the different directional in the above coordinates respectively strains, usually, the maximum value of $\sigma_x$, $\sigma_y$, $\sigma_z$ can be regarded as the value of principal stresses of the stress field.

The above equation can be regarded as a generalized functional expression representing the energy principal criterion for $\Omega$-area system, and its meaning is expressed as the strain elastic energy. available in system the area energy storage When there is no disturbance such as extraction, influence the transfer direction of the energy released by the system energy storage $\Omega$ will be indeterminate. When the energy storage system suffers from the influence of perturbation, the transfer direction will be determined according to the direction of the perturbation influence, and then it can be by using simulated the three-dimensional function solver. If used the boundary condition criterion is to judge the stability state of the system energy storage, it can be considered that when the total energy storage the system is stable, and becomes a

## 5.3 Twenty-first Century Impact Ground Pressure Gestation Mechanism …

very small value will be destabilized, and then the energy will be released to trigger the shock phenomenon when it becomes a very large value.

The corresponding elastic strain of value of is the principal the stress field stress dominant factor that the instability changes of the system triggers energy storage, and this stress plays an important role in causing the instability damage of the coal rock body and the shock phenomenon. excited by this stress the size of the energy value $\Phi_t$ can be found by the following equation:

$$\Phi_t = \int_{u_t} \sigma_t u_t du_t$$

where $\sigma_t$ denotes after disturbance the principal stress of the stress field; in the coal-rock body $_t$ denotes the principal stress the strain produced in the coal-rock body caused by the action of, and the direction of deformation is consistent with the direction of stress. In the process of the elastic impact effect occurring, when the elastic energy is transferred to the area near the surrounding rock, it only when the elastic energy's called destructive impact ground pressure) is enough to throw the damaged weakened medium and the surrounding rock away from the vulnerable surface and rushes to the mining space, i.e., it satisfies the equation remaining energy after ($\Phi$-$W_0$ $\Phi$ damages the medium to do the work $W_0$ (5.4). Therefore, it can be summarized that the shock effect produced by wave energy in the coal rock body is a necessary condition for the occurrence of shock ground pressure, while satisfying the shock effect equation of elastic wave energy is a sufficient condition for the occurrence of shock ground pressure. The occurrence or intensity of shock ground pressure depends on the value of, i.e., the residual energy that reaches the vicinity of the vulnerable surface of the surrounding rock $\Phi - W_0 = \Delta\Phi$ the size of. $\Delta\Phi$ is proportional to the damage degree of the shock ground pressure and the magnitude of the earthquake. When the energy arrives at the near area of the surrounding rock $\Delta\Phi$ is relatively small, due to the solidifying force and resistance of the supporting structure of the surrounding rock, it can only cause displacement and deformation of the surrounding rock or damage phenomena such as flake gangs and toppling, etc.; when the elastic energy is transferred to the near area of the vulnerable surface of the surrounding rock by the depth of the rock body, and it can reach the value of which gathers very high elastic energy in a very short period of time $\Delta\Phi$, the in the surrounding rock and its weakened will be weakened instantaneous and rapid action of. Medium will be like an avalanche thrown into the space, resulting in destructive disasters, this instantaneous process is the so-called occurrence of impact pressure, which causes damage to the energy of $\Delta\Phi = MU_0^2/2$, all involved in the development of the completion of the impact pressure formation of the energy to meet the impact effect of the equation of the quantitative value of the relationship between the impact effect, i.e., the impact of the conservation of energy theorem.

## 5.3.4 Impact Pressure Triggers and Energy Conversion

### 5.3.4.1 Of Elastic Principles and Characterization Energy Release

(1) Rate of energy release

Coal mining face and roadway excavation face are gradually carried out with the formation of space (face), here there is a problem of propulsion speed, with obvious dynamic or random spatial–temporal scale change characteristics. Based on the principle of elastic energy transfer and release in the coal rock body it is inferred, that the volume increment and exposed surface area of the excavated coal rock body play the role of energy release source, in short, the elastic energy stored in the deep coal rock body is continuously released to the surface of the excavation in the unit spatial and temporal scale. The intensity of energy release can be by the energy release rate per unit area visualized. Figure 5.4 $d\Phi/dS$ or energy release rate per unit volume $d\Phi/dV$ shows a simplified model of the step-by-step extraction process of an arbitrary cross-section, let the extraction surface change from the S surface to the surface, the surface area increment is $\Delta S$, and its volume increment is $\Delta V$, $S\prime$ so it can be seen that the value of the energy released by this extraction process can be expressed as follows: the area energy release rate $d\Phi/dS$; the volume energy release rate $d\Phi/dV$, $d\Phi/dV$ is suitable for expressing the geometrical shape of the regular ore body extraction. This expression can also objectively reflect and explain per unit area in the mining process the intensity of strain energy release effect) inherent. It can accurately and effectively reflect and understand the theory of coal rock body catastrophic instability and establish a scientific engineering $dS$ (or per unit volume $dV$ theory of. catastrophe prevention and control.

The rate of energy release $d\Phi/dS$ or $d\Phi/dV$ reflects the intensity of energy release, which fully and superficially reveals coal rock bodies the temporal characteristics of propagation of work dissipation. Objectively reflect and describe the state of propagation of elastic energy and the degree of acoustic emission phenomenon of coal rock body in local high stress area and high energy storage area. energy storage and in Fig. 5.5 shows the corresponding relationship between the number of acoustic emissions, the state of coal rock body and the rate of energy release in mining.

**Fig. 5.4** Simplified model of stepwise advancement of extraction

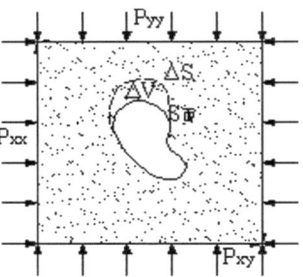

**Fig. 5.5** Curves of acoustic emission counts, coal rock mass state and energy release

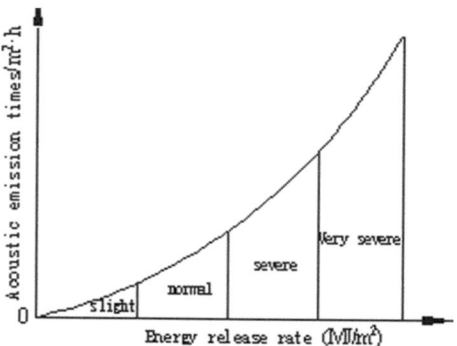

Figure 5.5 reveals that the number (or frequency) of acoustic emissions and coal-rock the degree of damage increase dramatically with increasing energy release rate. For coal seams with impact pressure and gas protrusion, the interaction between the energy release rate and the number of acoustic emissions is more obvious, which better reflects the strong impact and protrusion tendency media characterization. For coal seams without significant impact and protrusion tendency, the energy release rate and the number of acoustic emissions is relatively low, but if the internal energy of the coal rock body cannot be effectively released in real time, such as the mining advance speed is too fast, beyond the response time of the energy release and dissipation of the work done, the same strong impact or protruding phenomenon may occur. Therefore, the production cannot just pursue high speed, high efficiency mining progress, should be combined with the geological conditions of the mine, the development of reasonable mining and excavation intensity in time to control the spatial and temporal scales of effect on the impact pressure, prominent tendency of coal seams and thick $d\Phi/dS$ or $d\Phi/dV$ hard roof workings, to establish the this understanding concept of is particularly important.

(2) Timeliness of energy conversion

The research shows that acoustic emission is the characterization result of energy accumulation and release process. If the elastic energy released by coal and rock mass per unit time is greater than the energy required to destroy dv of coal and rock mass, the accumulation of residual energy becomes a possible opportunity condition, which leads to the weakening of damage strength of coal and rock mass skeleton, the expansion of original crack fracture of coal and rock mass, the deformation displacement of coal and rock, the instability and failure of coal and rock particles, and the generation of dust. Excessive residual energy continues to be released or propagated in the form of elastic waves or kinetic energy. This physical phenomenon is called a failure process, that is, the failure phenomenon of wave energy impact effect. During the failure process of coal and rock mass, there are two kinds of work failure phenomena, which are the transfer or release of residual energy in unit time and the conversion of elastic energy into kinetic energy in unit time. Here, the space–time scale is used to measure the energy conversion and work rate, which is

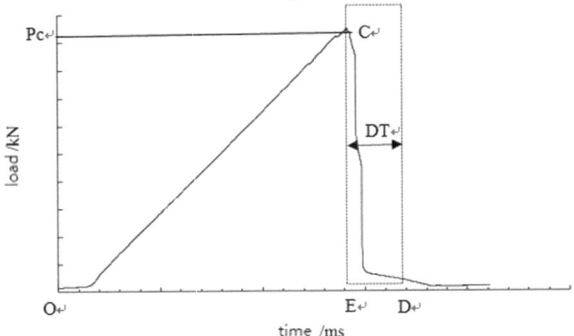

**Fig. 5.6** For coal rock bodies dynamic damage ageing principal curves

convenient for understanding and describing the degree of instability and failure of coal rock and the dynamic timeliness of work in the process of energy conversion. The timeliness of dynamic failure can be illustrated by the curve shown in Fig. 5.6, and the time trajectory from ultimate strength to complete failure of coal and rock specimens under uniaxial compression under experimental conditions is given.

Figure 5.6 shows the dynamic damage and spatial–temporal effect coal rock specimens the typical curves of. In the figure, Principle's for $_c$ denotes the maximum damage load experienced by the specimen, ED denotes the spatial–temporal scale of the damage experienced, CD denotes the process trajectory of the damage experienced, and OC denotes the loading process trajectory. The magnitude of the dynamic damage time DT reflects the inherent load resistance of the specimen and is used to illustrate and evaluate the impact propensity of the coal rock body. When tested in the laboratory in a certain range is not affected by the loading speed, the strength of the specimen and the capacity of the press, but can reflect the of the destruction experience of the coal rock spatial and temporal trajectory, and the application of the time to reflect the indexes to illustrate the impact propensity of the coal rock. The relationship between energy and time will still be analyzed in the following two aspects.

(1) Gradual formation of space. The use of elastic strain energy consumption to do work in time describes and illustrates the relationship between energy and time. here, the introduction of power indicators $P_\Phi$, the physical significance of which indicates the work done per unit of time to overcome the constraint force in the roadway excavation process. The nature of is a gradual unloading process. The progressive mining starting moment of mining is equivalent to Figure same as point C, and the completion moment is point D. So the $P_\Phi$ value is:

$$P_\Phi = \frac{W_1}{DT}$$

The value of is the work done to overcome the constraint force when digging progressively, and its value can be derived from the previous inference. Because the total work size and the size of the working face or tunneling scale corresponds to, so $W_1$ the larger the value of, that is, the more work done per unit time, but also indicates that the working face advancement or tunneling speed is faster, the corresponding energy release rate is also correspondingly $P_\Phi$ higher, corresponding to the probability of impact damage is also greater.

(2) Instantaneous and rapid formation of space. In this case, the binding force suddenly drops to zero, and the destruction of the coal rock body can occur after the end of mining activities. At this time, the residual energy value is used $\Phi_r$ to compare with DT to get the residual energy release rate index $V_\Phi$. Its physical significance characterizes the amount of energy released per unit time, and its magnitude can be expressed by the following equation:

$$V_\Phi = \frac{\Phi_r}{DT}$$

$V_\Phi$ The larger the value is, the larger the amount of energy transformed and released per unit of time after the formation of space in an instant, the higher the kinetic energy stored in the coal rock body, and the more serious the possibility of damage to the surrounding rock High stress areas are often encountered in mining, such as tectonic stress extrusion zone, containing high gas coal body and other areas of strain energy is very high, the project is often used to correspond to the high elastic energy storage area drilling or drilling water injection or gas pre-pumping or reduce the mining progress and other measures is the use of the principle of energy timeliness to make the high-energy areas to release the energy in due time.

### 5.3.4.2 Factors Inducing Impact Ground Pressure

Figure 5.7 illustrates several factors related to the occurrence of pressure bumping. Among these, natural factors include the fundamental geological stresses, which mainly consist of gravity and tectonic residual stresses. Secondary factors involve tectonic stresses, coal seam conditions, including fault and fold influences, coal seam occurrence, coal quality, and the coal seam's susceptibility to pressure bumping. In terms of technical factors, these mainly include local stress concentration, production rate control, selection of preventive measures, and excessive coal seam extraction. Management factors primarily involve issues such as stagnation in construction investment leading to inadequate supporting infrastructure, unreasonable operation procedure arrangements, and insufficient specialized training in pressure bumping prevention.

From a comprehensive analysis of all the factors, it can be concluded that human-induced factors account for a significant portion of pressure bumping occurrences. Therefore, it is crucial to strengthen equipment management and personnel training, as well as enhance awareness and forecasting capabilities concerning pressure bumping to ultimately achieve prevention and mitigation goals.

Additionally, the author believes that coal mining is a dynamic process. For coal seams with no significant tendency for pressure bumping, the energy release rate and

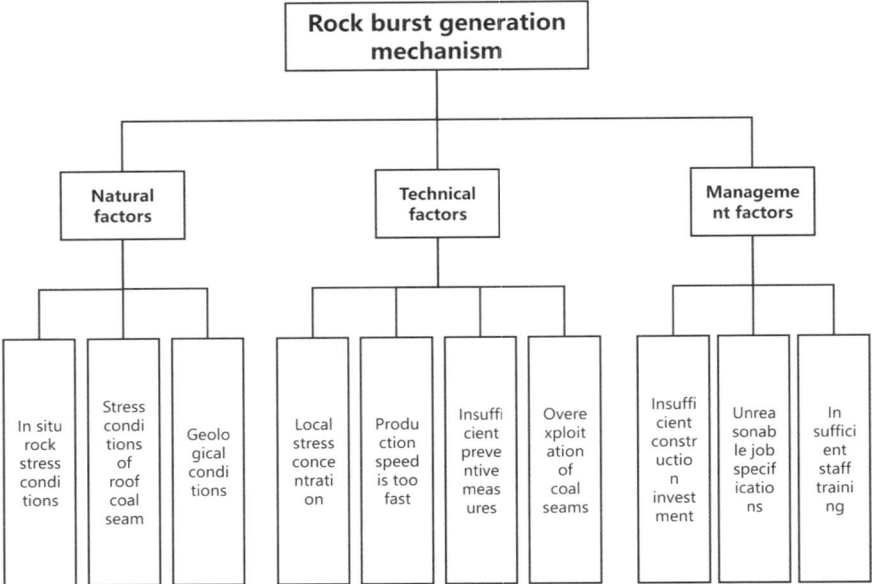

**Fig. 5.7** Impact ground pressure influences

acoustic emission frequency are relatively low. However, if the internal energy of the coal-rock mass is not effectively released in time—such as when the working face advances too rapidly, exceeding the energy release work and dissipation response time—pressure bumping phenomena may still occur. For example, in fast tunneling operations, miners may sometimes hear sound signals emanating from within the coal seam when advancing through the stone gate of a coal roadway. Therefore, production should not solely pursue high speed and efficiency in mining progress; instead, it is essential to combine the geological conditions of the mining area to develop reasonable extraction and tunneling schedules, and to improve continuous monitoring systems.

## 5.4 Predictive Forecasting Principles and Methods

Impact ground pressure is a coal-rock dynamic disaster in mining engineering practice, with strong engineering practice characteristics. Over the past 10 years, China has strengthened the theoretical research on impact ground pressure, while paying more attention to engineering practice, and various monitoring methods and prevention methods have been popularized and applied in actual impact ground pressure mines. Impact ground pressure prediction and forecasting methods must be based on

## 5.4 Predictive Forecasting Principles and Methods

the mechanism of impact ground pressure targeted selection of monitoring equipment, signal acquisition, monitoring methods, signal analysis. Impact ground pressure prediction can be according to categorized and managed the of the operating environment of the quarry different principles.

In the process of construction and mining, as long as there has been a coal seam impact ground pressure or rock formation impact ground pressure (rock explosion), the mine is designated as an impact ground pressure mine, such mines are required to improve the impact ground pressure prediction and forecasting system. For scholars from all over the world impact ground pressure prediction and forecasting has been an academic problem, the its suddenness of will also affect the prediction effect, scientific and reasonable impact ground pressure prediction and forecasting will be one of the coal mine safety production management personnel and scientific researchers urgently need to solve the problem. In this paper, we will introduce several widely used forecasting techniques. Occurrence.

### 5.4.1 Impact Propensity of Coal Rock Bodies

The stored deformation energy in coal and its propensity to generate impact-type failures can be quantified using one or more indices. According to the coal industry standard of the People's Republic of China, Classification of Coal Seam Impact Susceptibility and Method for Index Measurement (GB/T 25217.2–2010), the coal seam impact susceptibility index includes the Dynamic Failure Time (DT), Elastic Energy Index (WET), Impact Capacity Index (KE), and Uniaxial Compressive Strength (Rc). This standard applies to the classification of coal seam impact susceptibility and the determination of impact susceptibility indices for coal that can be processed into standard specimens under laboratory conditions. These indices are specifically divided into three categories: strong impact susceptibility, weak impact susceptibility, and no impact susceptibility.

Dynamic Failure Time (DT): Refers to the time interval, in milliseconds, from the limit strength to complete failure of the coal specimen under uniaxial compression. It is represented by DT.

Elastic Energy Index (WET): Refers to the ratio of the elastic deformation energy to the plastic deformation energy (dissipative deformation energy) of the coal specimen under uniaxial compression, when the applied force reaches a certain value (prior to failure). It is represented by WET.

Impact Energy Index (KE): Refers to the ratio of the stored deformation energy before the peak value to the dissipated deformation energy after the peak in the stress–strain curve of the coal specimen under uniaxial compression. It is represented by KE.

Uniaxial Compressive Strength (Rc): Under laboratory conditions, it refers to the ratio of the failure load of the standard coal specimen under uniaxial compression to the area of its bearing surface.

The impact susceptibility of rocks refers to their tendency to store deformation energy and generate impact-type failures. According to the Classification of Rock Impact Susceptibility and Method for Index Measurement (GB/T 25217.1–2010), the rock impact susceptibility index is represented by the Bending Energy Index. The Bending Energy Index refers to the bending energy stored by a unit-width cantilever rock beam under uniformly distributed load when it reaches the ultimate span, expressed in kilojoules (kJ), and represented by UWQS. This index classifies rock impact susceptibility into three categories: strong impact susceptibility, weak impact susceptibility, and no impact susceptibility.

Firstly, the impact susceptibility indices of coal and rock layers are determined through laboratory tests, as specified in the standards. These indices are then applied to formulate management measures and prediction guidelines for the impact hazards of coal and rock layers in mining areas. During field construction, based on specific engineering geological conditions and using engineering mechanics testing methods, simultaneous monitoring is implemented to make preliminary predictions and evaluations of the impact hazards (i.e., impact susceptibility) of coal and rock layers during the mining process.

## 5.4.2 Drill Chip Method

Drill chip method monitoring is a method of identifying the danger of impact ground pressure by utilizing wind drills or electric drills to implement drilling holes with a diameter of 42–50 mm in coal seams, based on the amount of discharged coal dust and its change pattern and related dynamic effects. It is also a real-time dynamic monitoring method. The theoretical basis of the drilling chip method monitoring is that the amount of drilled coal dust has a quantitative relationship with the stress state of the coal body, i.e., in the other conditions of the same coal body, when the stress state is different, the amount of coal dust in its drilled holes is also different. When the discharge rate per unit length increases or exceeds the calibrated value, it indicates an increase in the degree of stress concentration and an increase in the impact hazard. The indices for evaluating the impact hazard of the drilling chip method mainly include the drilling powder rate index and the dynamic effect. Among them, the drilling powder rate index refers to the ratio between the actual amount of coal powder per meter and the normal amount of coal powder per meter, which should be determined through the analysis of the actual measurement; the power effect of the drilling chip method refers to the phenomena such as jamming, suction drilling, top drilling, abnormal sound and impact in the hole produced in the drilling process. When one or more of these power phenomena exist in the drilling process, it can be determined that the area has an impact hazard.

The effectiveness of the application of the drill chip method requires specialized knowledge and experience (uniformity and comparability of operating methods and drilling conditions are important), and the method needs to take into full consideration the depth of the drilling hole, the orientation angle, the hazardous area (area of very

## 5.4 Predictive Forecasting Principles and Methods

high elastic energy storage), the effective time of monitoring, and the duration of continuous monitoring. It is extremely important and meaningful to analyze and summarize the changes in dust volume along the depth of the drill hole, especially at the bottom of the hole. At present, the amount of dust per unit length of the drill hole is used to assess the degree of danger (this assessment method is easy to grasp and operate). The theory of the drill chip method is consistent with the direction of the action of deep stress, elastic stress wave impact on the direction of dust generation, the deep or bottom of the borehole dust amount is greater than the shallow edge dust amount. As shown in Fig. 5.8, $\triangle L_1$ indicates the dust generated by the surrounding rock in the shallow part of the borehole, and $\triangle L_2$ indicates the dust generated by the bottom of the borehole in the deep part of the borehole. If the amount of dust in the shallow part of the borehole is more than that in the bottom of the borehole, the change of dust may be triggered by the pressure of the surrounding rock and the roof plate. From the occurrence mechanism of impact ground pressure, it is known that the coal body impact tendency and the thick and hard roof plate induced coal seam anomalies (pressure size, distribution pattern, and the range of pressure wave excitation and the degree of drastic change) in the distribution characteristics of the support pressure are the for predicting the impact ground pressure important elements. Coal body impact propensity is the inherent property of coal rock body to produce impact damage; supporting pressure distribution characteristics, i.e., supporting pressure peak size and wave excitation range and its distance from the coal wall, etc., are suitable to be detected by drilling chip method.

Drill chip method is the main method used at present, and its measuring point is generally arranged in the middle of the coal seam and other suspicious places, although is widely used, but the arrangement of its measuring point has certain defects, especially in the section with more complicated geological conditions, which can't guarantee to measure the most accurate this method range and direction of the impact wave of the elastic wave in the deep part of the impact action. According to the known mechanism of impact ground pressure, when impact ground pressure occurs, the stress of coal rock body exceeds the strength, but the stress exceeds the strength, the coal rock damage, that is, the maximum amount of drilling chips exceeds the limiting value does not necessarily occur in the impact ground pressure, so the

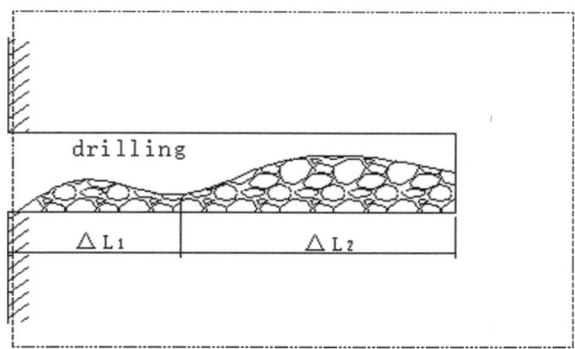

**Fig. 5.8** Borehole principle of the for predicting impact propensity drill chip method

drilling chip method is often used in conjunction with other monitoring methods, to establish a comprehensive evaluation system.

On the other hand, when the drill chip method is used for prediction, sometimes anomalous does not occur information, i.e., the strength phenomenon is not reached or exceeded. However, due to the high elastic energy or high stress in the dust and elastic energy coal seam has rheological properties, so that has high elastic energy in the region of a moment may produce static fatigue phenomenon, so that the ultimate strength of the coal rock body to reduce the stress is no longer increased, but also may occur destabilization damage, i.e., impact pressure, this situation is less common, but it does happen in practice. On the other hand, through several on-site construction, in the process of digging and drilling, no matter technicians or drilling workers should pay special attention to the dynamic effect produced in the process of drilling, this phenomenon just shows that the location of the region stores very high elastic energy, and then we should try to utilize more holes to relieve pressure and reduce energy.

### 5.4.3 Shock Ground Pressure Acoustic Emission (Micro Seismic) Monitoring Method

During the mining process, physical phenomena such as deformation, failure, and energy release occur in coal-rock mass, with seismic waves like sound waves propagating outward in the form of elastic waves, including microseismic events. The vibration frequency of microseismic events typically ranges from 0 to 150 Hz, with energies often greater than 100 J. Based on the relationship between microseismic phenomena and the stress of coal-rock mass fracture, monitoring of regional coal-rock mass fracture conditions can be conducted. By analyzing the distribution pattern of microseismic events, the impact hazard in the mining area can be assessed, thus guiding the monitoring, early warning, and prevention of dynamic pressure.

Acoustic emission methods must employ real-time dynamic monitoring and should not rely on random sampling methods. The correlation between acoustic emission phenomena in the mine and the dynamic behavior of the rock mass, as well as the impact effects of elastic waves, is highly significant. Experimental studies have shown that the frequency of acoustic emissions from coal-rock specimens generally increases with rising pressure, where there is a one-to-one correspondence between stress and acoustic emission signals. The emission signals cease when the force or elastic energy disappears, indicating a clear relationship. This is due to the non-elastic deformation or unstable structural states in the coal-rock mass under the action of the stress field. The sources of acoustic emissions are related to the type of rock mass, the emission signal levels, elastic–plastic deformation, the formation and growth of microcracks, as well as the development of pre-existing cracks and microcracks.

For the evaluation of dynamic pressure hazards, the characteristic signal parameters of coal-rock acoustic emissions are primarily recorded and analyzed in combination with the geological conditions of the stress field. The entry of coal-rock mass into an unstable damage stage results from the violent expansion of fractures within the coal-rock mass, while acoustic emission phenomena represent micro-expansion (cracks and microcracks in the coal-rock mass) exceeding the compressive strength limit. The further development of this phenomenon indicates the final damage to the coal-rock mass and medium weakening. According to the principles of dynamic pressure occurrence, final fractures will trigger high elastic energy shock vibrations, leading to dynamic pressure. The acoustic emission phenomenon is a physical manifestation induced by the impact effects of elastic stress wave energy during propagation, which leads to weakening and failure of the coal-rock mass. Dynamic pressure occurs when the surrounding rock of the roadway suffers sudden and intense impacts from deep elastic wave energy, resulting in destruction. Therefore, monitoring the acoustic emission activity in the coal-rock mass near the roadways can provide insight into the incubation and development process of dynamic pressure. In other words, in mines with dynamic pressure hazards, acoustic emission activities contain precursor information about dynamic pressure events.

(1) Principles of sensor (probe) installation and elastic wave characteristics

Applying the principle of acoustic emission from the coal rock body to determine the appropriate sensor installation location to monitor the acoustic signals, also known as vibration signals, excited from deep within the coal rock body. The choice of sensor installation location is very important, first of all, according to professional knowledge and experience to determine the possible hidden sources of sound and dangerous areas. Often because of some reasons, forcing the signal sensor cannot be installed in the sound source, can only choose to install farther away. At this time we should know, elastic wave energy propagation process to have energy consumption, loose rock consumption of energy is greater than the dense (high strength) rock, which are to reduce the propagation distance of the wave energy; elastic wave through different media rock layer, the level of the wave energy caused by refraction and reflection to reduce the wave propagation distance; fissure cracks and a variety of structures are not conducive to or even cannot pass elastic wave, then the wave will be bypassed, will reduce the wave energy to lengthen the propagation time, reduce the propagation distance. Propagation time, reduce the propagation distance. The larger the elastic wave energy is, the longer the propagation distance is, and the smaller the elastic wave energy is, the shorter the propagation distance is. However, the purpose of acoustic emission monitoring is to monitor the acoustic signal as soon as possible, and the low-energy signal at the moment of creep is the early signal, which cannot wait until the wave energy is very high to monitor the signal, which can realize the purpose of prediction and forecasting.

(2) Predictive forecasting information and sensor installation methods for impact ground pressure prediction.

At present, the common way of sensor installation is probably divided into: ① ground to establish real-time monitoring station multi-probe partition monitoring signal, the probe into the ground borehole, the way shown in Fig. 5.9. The advantage of this method is that the monitoring equipment is easy to maintain a wide range of control, the disadvantage is that even if you play a relatively deep borehole to install sensors, it is difficult to monitor the coal bed rock in the creep process of acoustic emission signals, because the elastic wave energy of the acoustic signals or vibration signals are very easy to propagate the same coal rock seams, and the direction of the propagation path is the direction of the excavation space, i.e., the vulnerable surface of the surrounding rock, and the ground direction of the propagation of the signals is very limited, it is difficult to really realize and very fast. It is difficult to realize and do early and fast monitoring to the impact pressure or mine earthquake precursor signals, cannot or extremely difficult to monitor the easy to recognize and meaningful information to achieve prediction and forecasting, even if the monitoring signals are difficult to say that it must be related to the impact pressure signals, or perhaps monitoring the signals are only ground subsidence process released acoustic emission signals, non-impact pressure precursor signals, generally can only be monitored after the occurrence of the mine quake disaster signals. The signals after the occurrence of mine earthquake disaster can generally only be monitored. ② mine quarry mining area to establish acoustic emission signals (AE signals) guarded real-time monitoring station monitoring mode as shown in Fig. 5.10 numbered labeled monitoring area, the advantage of this method is that due to the proximity of the possible occurrence of impact ground pressure hazardous sources, so even relatively low elastic wave energy signals may be monitored, in line with the idea of early prediction forecasting advocated concepts and requirements. The disadvantage is that the anti-interference ability of the monitoring equipment is relatively high, especially the software programming to fully consider the impact of other vibration signals and interference, to choose a reasonable signal threshold to remove the useless noise signals, the authors suggest that in the environmental conditions of the monitoring location to first monitor the threshold value of the noise signal as a basis for setting the threshold value, so that the first to remove the environmental conditions of the noise impact to reduce the nonevent monitoring information content.

**Fig. 5.9** Ground-guarded acoustic emission (micro seismic) monitoring model

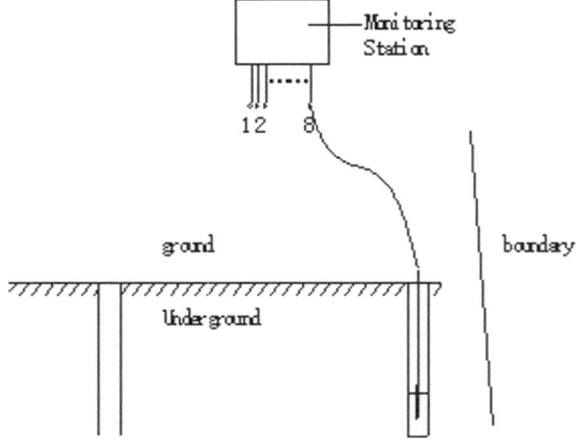

**Fig. 5.10** Quarry guarded real-time continuous signal acquisition monitoring model

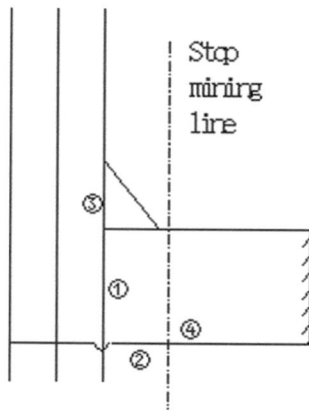

### 5.4.4 Mineral Pressure Monitoring Method

Mine pressure monitoring is the basic work of impact ground pressure monitoring, which can comprehensively reflect the interaction relationship between the support (pillar or support body) and the surrounding rock in the mining face or roadway. Mine pressure monitoring generally monitors the working resistance and downward shrinkage of the support, the opening rate of the safety valve, the specific pressure of the bottom plate, the distance between the ends, the top plate off the layer and the top and bottom plates, the distribution of the support pressure (coal seam stress), the force of the anchor rods (ropes), the horizontal deformation of the roadway, the initial top plate pressure step and the cycle of pressure steps, stress concentration coefficient, and so on.

Based on the research results of the mechanism of impact ground pressure, in the process of forming elastic strain energy within the coal rock body, the propagation of energy triggers the destruction of the character structure of the coal rock body, and the weakening and expansion of the medium occurs, and the coal rock body undergoes a stage-by-stage violent deep displacement in the direction of the mining space, causing the roadway wall surface to be violently shaken and deformed. In other words, the elastic wave energy propagation transfer by the impact of wave energy, coal rock body trait structure damage, so the impact of the region of coal rock body will occur decompression displacement changes, the frequency and amplitude of this signal change is different from the change of mining pressure. Through the analysis of this principle, we can use multi-point displacement meter or off-layer meter to monitor and record the displacement of the deep base point of the coal rock body and the deformation of the surrounding rock in real time, which can be used as the precursor signals of the propagation and destruction of the elastic wave energy. At the same time, we also need to monitor and record the axial force changes of the anchor rods in real time, especially focusing on analyzing the frequency and amplitude of the signal changes and the intensity of the changes. Apply displacement deformation and stress changes to analyze and judge the changes of deep stress or elastic energy. If a large area of thick and hard top plate (overlying rock layer) comes under pressure, i.e., a large range of compressive stress acts on the coal rock body, its compressive stress and deformation signals change drastically. This situation can also be considered through the monitoring of coal body pressure, displacement and deformation as a means to carry out the analysis and research of the creep signal of the coal rock body. This is actually through the mine pressure monitoring method means to realize the monitoring and forecasting of the early signals of the impact ground pressure, just monitoring the density of the collected signals to be greater than the general mine pressure monitoring information, and therefore requires the use of real-time monitoring records and computer software analysis of the prediction and forecasting methods.

## 5.4.5 *Extraction Stress Monitoring Method*

Stress is one of the necessary conditions for the occurrence of impact ground pressure, and the stress in the coal rock body includes 2 parts, the original rock stress and mining stress, the original rock stress is the stress that exists in the coal rock body in the process of formation or through the geological tectonic movement, and the mining stress is the stress that is produced by the change of bearing of the coal rock body caused by the mining activity. The problem of impact ground pressure is essentially the problem of mining stress. It is of great significance to monitor the mining stress of coal seam and learn the distribution law of mining stress field for early warning and control of impact ground pressure disaster. Mining stress is a dynamic field that changes in time and space with mining disturbance, and the mining stress formed due to mining activities is extremely complex. Nevertheless, with the progress and development of science and technology, it is now possible to realize the continuous monitoring of mining stress, thus providing a basis for predicting impact ground pressure.

In actual monitoring engineering practice, mining stress monitoring is also called stress monitoring or stress online monitoring. For the mining stress monitoring in the coal pillar area, it is usually pointing monitoring, i.e., 1 to 3 points are monitored to evaluate the impact hazard. For the mining stress monitoring in the face, it is usually facing monitoring, i.e., every 10 m along the direction of the roadway to set up a drill hole, through the stress monitoring of different drill holes to form the stress distribution in different locations at the same time, and then evaluate the impact hazard of a certain area or range.

## 5.4.6 Electromagnetic Radiation Monitoring Method

As a phenomenon that radiates electromagnetic energy outward during the deformation and rupture process of coal rock body under load, electromagnetic radiation is closely related to the deformation and rupture process of coal rock body. Electromagnetic radiation is enhanced with the increase of load, and with the increase of loading and deformation rate; the intensity of electromagnetic radiation mainly reflects the loading degree and deformation and rupture intensity of coal rock body, and the number of pulses mainly reflects the frequency of deformation and micro rupture of coal rock body. Through the monitoring of electromagnetic radiation in the mine roadway and working face, it can assess the force and rupture situation within a certain range of the surrounding rock in the working face and roadway, and predict the purpose of impact danger. In China, the electromagnetic radiation monitoring method was mainly used for coal and gas protrusion monitoring in the past, and was only used for coal mine impact ground pressure monitoring after the twenty-first century. According to incomplete statistics, from the beginning of 2002 to now, about 90 impact ground pressure mines in China have used EMR monitoring technology, of which about 70 mines use portable EMR instruments and about 20 mines use online EMR systems. The electromagnetic radiation method can be used for impact ground pressure monitoring and early warning work as a method for localized impact hazard evaluation.

As a rule, in the area with shock hazard, the electromagnetic radiation signal strength (E) is stronger and the number of pulses (N) is higher, exceeding the set critical value and increasing upwards. In the case of a coal rock body without impact risk, the electromagnetic radiation signal is weaker and the number of pulses is lower than the critical value.

## 5.5 Outlook for Impact Pressure Prediction Research

The process of coal-rock mass experiencing creep to catastrophic instability releases a variety of physical signals, and thus the monitoring methods that can be utilized are equally diverse. Therefore, the application of mine pressure monitoring methods and the analysis of monitoring data cannot be overlooked, and the application of a research-based monitoring and management mechanism should be emphasized. Regardless of the monitoring principle or method applied, the reliability of the signals, as well as factors such as signal acquisition quantity, continuity, intermittency, and singularity, must be considered. Multiple single signals can be integrated for comprehensive analysis and used as precursory warning indicators. Monitoring the resistivity changes in deep coal-rock masses can be used as early warning signals, as resistivity signals are well-suited for monitoring. The design technology of such monitoring instruments is not demanding and has a low level of difficulty.

At the present stage, the author leans towards developing acoustic emission (micro-seismic) monitoring technology, strain pressure sensor monitoring technology with high-frequency impact resistance, and high-resolution roof and surrounding rock displacement and deformation monitoring technology.

In conclusion, to better address the prediction and forecasting of dynamic pressure (rock bursts), it is essential to comprehensively utilize monitoring information, enhance the research on analytical methods with functional capabilities, and develop computer software to establish a computer-based information analysis and early warning platform. A comprehensive evaluation system, with acoustic emission monitoring signals as the primary data and other signals (specifically drill cuttings, displacement and deformation, and stress) as supplementary, should serve as the early precursory signal for the prediction and forecasting of dynamic pressure. In terms of the current development status of integrated technologies, this approach represents the most reasonable and simplest monitoring method.

**Open Access** This chapter is licensed under the terms of the Creative Commons Attribution-NonCommercial-NoDerivatives 4.0 International License (http://creativecommons.org/licenses/by-nc-nd/4.0/), which permits any noncommercial use, sharing, distribution and reproduction in any medium or format, as long as you give appropriate credit to the original author(s) and the source, provide a link to the Creative Commons license and indicate if you modified the licensed material. You do not have permission under this license to share adapted material derived from this chapter or parts of it.

The images or other third party material in this chapter are included in the chapter's Creative Commons license, unless indicated otherwise in a credit line to the material. If material is not included in the chapter's Creative Commons license and your intended use is not permitted by statutory regulation or exceeds the permitted use, you will need to obtain permission directly from the copyright holder.

# Chapter 6
# Principles and Prediction of Coal Seam Gas Outcrop Methods

## 6.1 Overview

Gas protrusion is a creeping deduction process with gas dynamics as the dominant role and coal rock body stress as the auxiliary, and compared with the disaster mechanism of impact ground pressure, it is the same shock geological disaster induced by elastic energy with dynamic characteristics presented in coal mine underground mining. However, because the gas in the coal seam has more significant rhea dynamic characteristics after being affected by mining disturbance, the elastic wave energy formed by the gas rheology in the coal rock body is more intense in the fluctuating characteristics of its energy in terms of appearance and characterization, i.e. the fluctuating characteristics of the creep process are very obvious. It can accumulate very high elastic wave energy $\Phi$ in the $\Omega$ region of the coal seam in a very short time, and with the help of coal and rock media as a bridge, it can be transmitted along the route that is easy to propagate and has the shortest path toward the vulnerable face of the mining surrounding rock. The elastic wave energy continues to impact the coal and rock media, resulting in damage to the coal and rock media damage impact effect phenomenon, the formation of the direction of the mining space directly after overcoming the medium resistance to do work $W_0$, the remaining energy continues to pass along the selected path direction, the impact range of wave energy continues to expand, and constantly have the gas influx into the transmission path, and then continue to expand the scope of the impact to form the main channel of the main stream of the gas transmission and the impact direction, repeat the process. Due to the compact strength of the surrounding rock layer, a large amount of gas accumulation is formed in the area close to the vulnerable surface of the surrounding rock in the mining space, and the elastic potential energy rises to reach enough energy to instantly form a gas-powered jet and air impact and roll up a large amount of coal and dust thrown into the mining space, the process is very gas-powered impact characteristics, and a large amount of gas continues to pour out of the transmission path in the area of the damage caused by the accumulation of energy, and the vibration caused by the gas-powered damage sometimes results in a large number of gas surges, the vibration triggered by the gas dynamic damage sometimes causes a very huge impact dynamic effect. Due to the suddenness and strong destructive nature of its occurrence, it not only jeopardizes the safe production and personal safety of coal mines, but also restricts the production capacity of mines, and affects the production efficiency and economic benefits of mines.

The world's first recorded herniation occurred on March 22, 1834, while digging a flat tunnel at the Issac Mine in the Rou are Coalfield, France. At present, coal and gas protrusion has occurred in more than 20 countries in the world. Among them, China is one of the countries with more serious protrusions, and the first recorded coal and gas protrusion in China occurred on May 2, 1950, at the Fuguo No.2 Mine of Liaoyuan Mining Bureau in Jilin Province. China's largest protrusion(the world's second)occurred on August 8, 1975, at Tianfu Mining Bureau Sanhui a mine + 280 m level, flat mine uncovering K1 coal seam, protruding coal rock 12780t, gas 1.4 million m 3. Cannon shot 3–4 s after the protrusion, a gust of wind in the hole, the small debris rolled up, blown to 50 m away, a large number of gas carrying coal dust gushed out, was black smoke, lasted for 40 min to gradually decline, from the protruding point of about 700 m away from the three dampers were destroyed, dampers outside the brazier and gangue equipped with the mine car rushed out more than 30 m away, a 1t more than the boulders were washed away more than 120 m, another piece of 3t heavy Another boulder weighing 3t washed out more than 60 m, turning two 90°sharp bends into the substation. With the development of China's coal industry, some mines have been protruding, especially with the development of new mines, the extension of old mines and the expansion of mining scale, the number of protrusions has gradually increased.

China has carried out a lot of research on the prevention and control of coal and gas outburst for a long time, but because coal and gas outburst is a kind of mine gas dynamic disaster phenomenon under the condition of complex geological environment. Due to the complexity and diversity of mining technology and geological conditions and the diversity of inducing factors, the law of outburst under various geological mining conditions has not been fully grasped so far, and it is still difficult to completely control and avoid the occurrence of coal and gas outburst. Therefore, some people once believed that rock burst and outburst disasters are invincible natural disasters. Based on years of research and study, the author believes that gas outburst is a kind of engineering disaster. For this kind of disaster prevention and control method of mining and geological symbiosis, there is still a lack of scientific and objective multi-professional knowledge to carry out comprehensive research and prevention and control evaluation system, prediction and evaluation system with monitoring equipment as the main body, and decompression and energy reduction management system with high stress elastic potential energy by means of control and prevention.

## 6.2 Principles and Phenomena of Coal Seam Gas Prolapse

### 6.2.1 Gas Highlighting Basic Factors and Energy

Using a microscopic approach, the factors affecting coal and gas outcrops are numerous and complex, including both geological and non-geological factors. Geological factors mainly include three aspects, i.e. the geological and structural conditions of the coal field or coal bed and the characteristics of the coal body structure; the gas parameters in the coal bed, including the gas pressure, gas content and closure conditions of the coal bed; and the state of the gestures in the mining area or coal bed. Among the above factors, the intrinsic characteristics of geological structure factors is the geological background elements affecting the coal and gas protrusion, the coal and gas is the material basis for the occurrence of coal and gas protrusion, and the ground stress state of the mining area or coal seam is the power or mechanical conditions for the occurrence of coal and gas protrusion. Obviously, coal and gas protrusion is the comprehensive effect of the above geological factors, but these factors do not necessarily play equal roles, and their roles or degrees of influence vary with specific situations.

The spatial effect and the influence of disturbing force formed by the roadway in mining cause the deep or distal coal seams to undergo expansion and displacement under the action of gas as the dominant power and the surface of the surrounding rock to undergo constant deformation, resulting in the local range of the coal and rock body appearing to varying degrees of fragmentation, and appearing to be uninterruptedly transformed from one energy(force)equilibrium state to another new energy(force)equilibrium state. Under the impact of elastic potential energy of gas, in most cases, these transformations are first of all the extrusion of coal and rock body inside and the deformation of wall surface of surrounding rock, structural destruction of coal and rock organization, and fissure development, weakening and breaking of coal organization structure, and the emergence of dust, particles of varying sizes of crushed coal, and the formation of miniaturized coal structure of the separation of crushing and destruction of the elastic potential energy balance of transferring the development of the dynamic process of change. This process is a combination of the result of the impact effect of elastic energy and the development of the phenomenon of crushing effect, and it is also the starting process of the development of the pre-emergence of gas, a process that must not be missing. Sometimes in the case of extremely high gas content, the elastic wave energy of gas in the rheological state of gas outflow channel is attached to the extrusion and sudden eruption of coal and rock, which can trigger the collapse of a large number of coal and rock locally, which is manifested as the violent outburst or protrusion of coal and gas.

From a macroscopic point of view, the occurrence or otherwise of protrusion is entirely determined by the action elements (size and direction of action) of the energy within the coal seam. As shown in Fig. 6.1, for a mine to run through the known risk of protrusion of coal seam stone doorway when the gas protrusion, in the head of the excavation close to the upper plate of the coal seam at a distance of 2 m to play four drill holes at the same time gunning to unload the pressure, the results of gunning at the same time caused a major gas protrusion accidents in the direction of the head of the excavation of a wide range of gas protrusion in the coal seam occurred above the direction of the coal seam (protrusion of coal dust of about 2t, a distance of more than 200 m), the protruding material was dominated by gas and dusty coal lumps, and many scale-like coal fragments were found, which indicated that the coal seam was strongly extruded before the protrusion. At that time, it was analyzed that it was safe to use similar digging method in the past, but the difference was that this time, the amount of gunning charge was a bit more. Today based on the analysis of the principle of impact ground pressure and gas protrusion, it is found that there are two factors that are conducive to gas protrusion in such a state: ① When the left side of the steeply inclined coal seam is close to the hanging wall of the coal seam, the compressive stress of the hanging wall of the coal seam is increased and the compressive stress Pz in the large area is completely superimposed on the coal seam, which forms the extrusion impact effect on the coal seam, and the coal body structure is destroyed and broken. As shown in Fig. 6.1, in the $\Omega z$ region, the elastic energy in the $\Omega z$ region completely acts on the upper plate of the coal seam, which makes the gas in the $\Omega$ coal seam in the large area of the upper plate form a ' free ' rheological situation, and the elastic potential energy in the $\Omega$ region increases. The direction of action of the energy (also the only direction of action) mainly refers to the heading face, and the structural relationship between the horizontal heading direction and the angle $\beta$ formed by the coal seam is also conducive to the accumulation of energy. The increase develops to the critical state of instability. ②Suddenly relax the shot, instantaneously remove part of the rock formation that constitutes the angle $\beta$, and suddenly remove the damping control of the elastic potential energy in the coal seam in front of the tunneling head, so that there is an uncontrollable critical state situation, which creates a state of sudden release of the high elastic potential energy of the coal seam that has long been activated, that is, coal and gas outburst. The occurrence of scale coal flakes is due to the long-term impact and friction extrusion between rock strata and coal seams. The coal seam under the heading face does not participate in the outburst. The reason is that the lower coal seam forms (or is in) an effective and sufficient decompression and energy reduction zone $\Omega s$ for a long time, and it is normal that no outburst occurs. If attention is paid to slow tunneling, sufficient decompression time is given, or multi-head (point) decompression is adopted, the outburst accident can be completely avoided.

6.2 Principles and Phenomena of Coal Seam Gas Prolapse        141

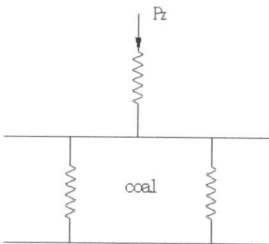

(a)Schematic diagram of the principle of high stress superimposed on a gas coal seam by a roof rock layer

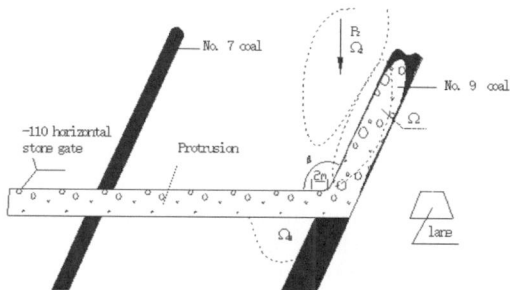

(b)damage in the area of gas protrusion

**Fig. 6.1** Principle of superimposed stress action on rock layers and gas protrusion characterization

## 6.2.2 Principles of Energy Transfer and Destruction in Coal Rock Bodies

### 6.2.2.1 Energy Transfer and Coal Seam Damage

Coal seams with a tendency to protrude form high elastic potential energy under the joint action of gravity(or compressive stress)and gas pressure of the overlying rock strata, and the structural strength of the coal body gradually decreases and crushes under the long-term impact of the elastic energy, which can be seen through the phenomenon of protruding materials(dust, gas, and small particles of coal briquettes). Dust and particles generated by the loosening and separation of the coal body, dust and particles falling from the coal body at the working face, or dust and particles thrown out during a protrusion, can cause the coal seam to lose its potential, gas internal energy and kinetic energy. From this it can be inferred that the fundamental difference in the breaking process of the coal body lies only in the amount of energy provoked by this process.

Under the action of rock compression stress and gas pressure (or under the induced action of rock pressure), when the coal body is gradually broken by impact extrusion and expansion, the potential energy of elasticity in the coal body in the region of

$\Omega$ will be elevated to reach the critical state, and then, under the triggering of the external force (the disturbing action of the mining propulsion), the elastic energy in the coal bed will gradually form the power impact energy that will move towards the direction of the roadway, the fissures will be increased, and the movement of the gas will be expanded. The paths and channels section and number and the surface area of gas desorption are all increased and enlarged. At this time, it provides favorable conditions for the existing equilibrium transformation of the potential energy of elastic energy in the coal body and the development of the rheological state of the gas, and makes it possible for the elastic rheological potential energy and kinetic energy to be continuously enlarged and strengthened.

When the internal structure of the coal body is suddenly broken, the flow change energy of the elastic potential energy dominated by gas appears instantaneously, forming the only directional impact direction, and a large amount of gas is concentrated in the local area of the coal seam without diffusion transfer. It has reached a very unstable critical state. At this time, it is very likely that gas and particulate dust will be ejected and thrown up. The latent (adsorbed) gas with extremely high static pressure energy in the extended cracks in the coal body is immediately converted into impact flow energy. At the moment when the coal body is broken into loose and separated granular coal blocks by the impact of kinetic energy, the activated gas produces impact flow change energy, which impacts on the area $\Omega$ of the granular coal block (i.e. broken coal body) that has formed a loose state of separation, resulting in a relatively high energy difference between the area $\Omega$ and the whole structure of the surrounding rock of the roadway. This elastic energy difference makes the crushing of the coal body more serious and the area $\Omega$ expanded. The essence of this physical change is the physical process of the flow change energy of the gas to overcome the resistance. At this time, the gas emission velocity becomes larger, the kinetic energy of the gas causes the movement of the coal dust, the flow change energy of the pulverized coal and the gas, and the suspension movement or rotation movement of the dust. At the same time, there are also phenomena of mine pressure, such as the sharp increase of monitoring displacement deformation signal and the fluctuation of monitoring pressure signal.

In addition to the above, rock pressure and gas pressure manifestation phenomena may occur during roadway excavation. For example, when the tunneling of the upper mountain roadway is close to exposing the coal seam, there is a loose coal dust collapse from the unpressurized coal body; when the coal collapses to tipping, there is an increase in the gas outflow, forming a gas flow, and in some cases, it can also bring out coal dust and dust. Therefore, the external characteristics of sudden dumping and protrusion may be similar.

At the time of protrusion, due to the fragmentation of the coal body in the localized area of the coal seam, the process of conversion of the variable energy of the gas flow begins to be generated, and the process is relatively complicated. In the unbroken coal body, the gas existing and acting on the pore crack wall surface is only in the static field of the adsorption state; at this time, the adsorbed gas does not directly participate in the crushing of coal or move to the direction of the roadway to do work. It only plays an energy reserve or latent role. However, with the continuous action of external forces, the expansion of the broken range of coal body and the adsorption of gas continue to turn into free rheological state, forming elastic potential energy, which can cause destructive elastic energy.

When the rock pressure stress gradient is quite large, if the coal body is broken in advance, even if the coal seam does not produce elastic potential energy or form flow change energy, this pressure can make the coal throw out, and even in the absence or with a small amount of adsorbed gas, the dust coal block will be ejected or thrown up (i.e.impact phenomenon). However, in this case, the distance thrown and the amount of dust are not large. Therefore, it should be pointed out and recognized that even in the coal body with very little adsorbed gas or weak adsorption capacity, outburst or instability may occur due to the combined action of a small amount of ' free ' gas and rock compressive stress. For example, the stress field condition of mining coal seam under the action of compressive stress of thick and hard roof strata will enhance and increase the total energy of elastic potential energy of coal seam, and provide the conditions conducive to the formation and development of impact and outburst.

Gas pressure (hydrostatic energy) is not only capable of ejecting gas through the fissure channels, dusting the broken pieces of coal by way of dust, but is also a cause of triggering the structural fragmentation of the coal body. When blasting is used to uncover coal seams with low permeability and containing very high gas, the role of the rebound force formed by the action of the drill bit by the gas flow when drilling holes is usually performed can be found and proved to be the causative factor of this feature. On the other hand, because of the inherent permeability of the coal, when preparing and boring along the seam, there is sometimes no gas pressure gradient or kinetic gas energy as large as in the case of a gas-broken coal body. Therefore, the role of gas flow and energy and broken coal body should be regarded as a method applicable to uncovering the coal seam to reduce the role of gas flow and energy, but also should be recognized that the coal body of the broken and dust increase is the pre-precursor phenomenon of the protrusion, impact and destabilization of the destruction of the pre-precursor phenomenon, some scholars will be protruding into the three periods known as the preparatory period, the initiation of the start and start-up period.

### 6.2.2.2 Elastic Energy and Salient General Characteristics

1. Top plate power impact triggers protruding phenomenon

For example, as shown in Table 6.1, it is the gas geological information data of a mine, the normal coal seam is primary structural coal, the damage type of coal is class II to III, locally class IV, and the joints of coal seam are relatively developed. The inclination angle of the coal seam is around 12°. It is known that the coal seam has a large gas content of about 22.5 $m^3/t$ and a gas pressure of 2. 6 MPa. The top and bottom plates of the coal seam are dark gray sandy mudstone, with poor permeability, which is unfavorable to the release of gas, and it belongs to the coal seam of protruding tendency.

In the mining area piece disk, stone door system has been formed after the start of the machine tunneling construction, in the digging face from the wind back to the mouth of the 21 m when ready to feed 2 m, suddenly occurred roof power impact damage phenomenon, accompanied by a loud noise triggered by a large vibration of the top and bottom of the plate slagging, after investigation, the occurrence of the phenomenon of power impact has the following characteristics:

(1) Length 5.6 m, the first 4.0 m is a full roadway pile, the last 1.6 m is a half roadway pile, the total amount of protrusion is about 46t.
(2) The upper part of the protrusion is a layer of broken coal lumps with no sorting properties, and there is a small amount of broken coal inside it.
(3) From the thrown broken coal rock and the accumulation state, there is no gas ejection channel (hole fissure).
(4) No accumulation of floating dust was seen on the surface or rear of the throw.
(5) The power source of on-site observation is mainly from the left side of the roadway. The coal wall on the left side of the roadway bulges 0.9 m; the roof of the roadway shows obvious deformation of the trapezoidal beam and the steel mesh caused by the influence of the large outward impact force. The anchor cable lock head is loose, the direct roof is obviously shaken and sinking, the front beam has a retraction phenomenon, and the tail column of the conveyor is broken. The deformation of the upper ladder beam; the floor scraper conveyor offsets to the right side of the roadway by 5°, and the maximum pushing distance reaches 0.9 m. After cleaning, it was found that the overall outward displacement deformation of the coal wall occurred in the lower side of the roadway facing the excavation face. The displacement deformation length of the lower side was 10 m, the thickness was 0.8 m, the floor heave was 0.5 m, and the excavation face was thrown out. The coal body is 3 m, and the amount of coal thrown is about 46 t.
(6) There is no obvious sorting phenomenon, the angle of coal body throwing is close to the natural angle of repose, and the gushing gas is $1280m^3$.
(7) The equivalent tons of coal gas outflow are $29.1m^3$/close to the coal seam gas content.

Abnormal phenomena occurring before the occurrence of destructive power impacts: continuous fracture of the roof plate of the roadway, obvious deformation of the top and bottom plates and the roadway gangs, failure of some of the anchor cables, serious deformation of the steel nets and ladder beams, and elevated changes in the amount of dust and gas after the power phenomena.

2. Characteristics of coal seam gas storage

The three-level return air downhill is arranged in the Ji 15 coal seam, and the exposed layer elevation is between − 590 m and − 820 m. The coal seam of the working face is relatively stable. The normal coal seam is the primary structure coal. The failure type of coal is II ~ III, and the local is IV. The coal seam joints are relatively

## 6.2 Principles and Phenomena of Coal Seam Gas Prolapse

**Table 6.1** Geologic information on coal seams

| Coal seam yield (°) | Trend | 120 | | Orientations | 30 | | Tilt (inclination of ship from vertical) | 12 |
|---|---|---|---|---|---|---|---|---|
| Coal seam thickness (structural) | 3.5 | Extraction level | | Coal seams 15 | | | | |
| Coal seam thickness and location | | | | | | | | |
| Jointed fracture of coal bed | More developed nodules | | | | | | | |
| seam stability | Coal seams are relatively stable | | | | | | | |
| Coal stone type | Semi-bright | | | | | | | |
| quality of coal | Coal grade | scorched fat | hydration | 3.7% | ash | 13.0% | volatility | 28.4% |
| Volumetric weight of coal | Coal firmness factor | Types of coal damage | | Coal Dust Explosion Index | | Spontaneous combustion ignition period of coal | | |
| 1.33 kg/m³ | f = 0.4 ~ 1.5 | II–III | | 31.7 | | 6–12 months | | |
| Coal seam gas content(m³/t) | Coal seam gas pressure (MPa) | | | Absolute gas outflow(m³/min) | | | | |
| 18 ~ 25 m³/t | 2.8 | | | 1.74 ~ | | | | |
| Prominent parameter test cases | | | | | | | | |
| Coal seam roof characteristics | The direct top is dark gray sandy mudstone, about 7thick ~ 12 m or so, and the composite top is about 1.3 m thick | | | | | | | |
| Coal seam floor characteristics | Substrate is grey-black ~ black sandy mudstone with fossilized plant fragments, approx.1.5 m thick | | | | | | | |

developed. The dip angle of coal seam is about 8 ~ 20°, and the dip angle of coal seam is about 11° at present. The roof of the coal seam is dark gray sandy mud-stone, the floor is black mud-stone, and the permeability is poor. The gas content of the coal seam is about 18 ~ 25m3, the gas pressure of the coal seam is 2.0 ~ 2.8 MPa, and the absolute gas emission during the tunneling process is about 1.7-4$m^3$ / min. When the excavation is 570 m away from the opening, the outburst strip is formed due to the influence of several cracks in the roof of the coal seam, resulting in the relative enrichment of gas and the increase of outburst risk. Therefore, the tunneling position of the three-level return air downhill roadway is in the dangerous area of gas outburst. From the information of gas before outburst and after outburst, such as outburst process, outburst site (no) airflow trace, gas emission before and after outburst, and the deformation and failure characteristics of roadway surrounding rock caused by dynamic phenomenon, it is shown that this dynamic phenomenon is not an ordinary coal and gas outburst.

3. Characteristics of Roof Power Impact

The depth of the three-level return downhill tunnel is about 1100 m, and the tunnel is in the lower part of the Li Kou big incline Ji, which is seriously affected by tectonic stress. According to the calculation, the vertical stress in the original rock stress of the tunnel is $Pz = 30$ MPa, and the maximum horizontal stress is more than $Ps = 40$ MPa, as shown in Fig. 6.2. Considering the secondary stress effect caused by the redistribution of surrounding rock stress after the roadway excavation(as well as the kinetic effect of the movement of the overlying rock layer in the neighboring air-mining area), the maximum stress of the surrounding rock of the roadway can exceed 80 MPa, which is in a high stress state. The overlying roof of the coal seam is 7.8 m thick of hard and dense sandy mudstone. At the same time, the coal seam is dry and brittle, by the strength theory of the impact ground pressure, the roadway excavation process of the surrounding rock may reach or exceed the strength limit of the coal rock body, the formation of the impact tendency, high stress and hard brittle surrounding rock to meet the conditions of the impact ground pressure of the elastic energy of the power of the impact action.

In addition, during the construction of the mine, hundreds of power impacts and impact ground pressures have occurred in the three levels of the belt downhill. According to the data analysis, the main stress direction of the three-level ground stress is in front of the left side of the digging face, which is consistent with the main stress direction of the ground stress in the tunnel where the impact ground pressure occurs, and most of the places where the impact ground pressure or power impact phenomenon occurs are in the direction of the left gang of the tunnel and the digging face.

4. Mechanism of Roof Power Impact and Gas Protrusion

If the spatial structural relationship composed of top plate, bottom plate and coal seam(gas)is regarded as the mechanical structural system of the excavation roadway perimeter rock system, the mechanical model of the structural system is shown in Fig. 6.3. In order to facilitate the exploration of the problem, the roadway excavated

## 6.2 Principles and Phenomena of Coal Seam Gas Prolapse

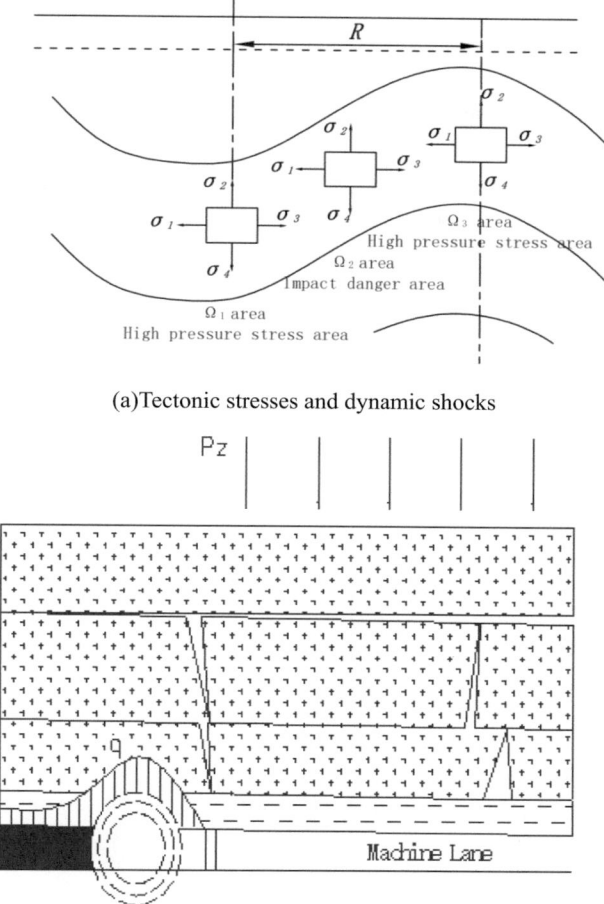

(a) Tectonic stresses and dynamic shocks

(b) Protrusion triggered by dynamic impacts on the excavated roof plate

**Fig. 6.2** Coal seam protrusion induced by high-pressure stress impact action on the roof plate

in the coal seam is regarded as an elastic body clamped by the top and bottom plates. In the sense of limit equilibrium, the coal body forms a high stress, high saving energy excavation headway due to the clamping effect of the top and bottom plates. Mechanical effects resulting from the action of the top and bottom plates clamping the coal body:

(1) Extremely high elastic potential energy is stored in the roof rock and coal seams under high stress;
(2) The roadway excavation headland area is in the location of very high elastic potential energy action or high stress concentration zone;

**Fig. 6.3** Principle of dynamic impact action of roof compressive stresses

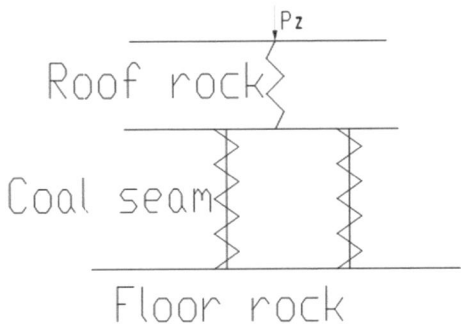

(3) The roof rock and coal seams are in a state of limiting equilibrium (or a state of systemic transient stability);
(4) The deformation and compressive stress effect of the roof rock layer is the dominant factor in lifting to increase the elastic potential energy of the coal seam and the coal seam destabilization.

The form of coal body destabilization is mainly related to the type and strength of energy release, analyzed from the perspective of impact ground pressure, if the size of the roof plate compressive stress is enough to brew an impact ground pressure, the compressed coal body will produce impact accelerated destruction. In the coal body of the primary structure of the weak surface and the new rupture surface of the strength of the sharp decline in the coal body internal rupture to a certain extent, in the roof plate power impact energy under the sustained action, the impact ground pressure occurs.

In addition to the clamping effect of high-pressure stress on the boring head, the coal body contains high gas pressure (2.85 MPa). The gas pressure is accompanied by the rupture of the coal body and releases the gas flow change energy. The superposition of the two energies will make the occurrence of impact and gas herniation possible and easier, with greater intensity and strength.

Comprehensive impact ground pressure and gas two aspects of the role of the analysis of factors, this type of power phenomenon is defined as: high gestures, to the roof type impact ground pressure as the dominant role in inducing the impact of the roof type impact ground pressure and coal and gas under the joint action of the power impact phenomenon. This kind of power impact phenomenon raises the risk of gas coal seam protrusion or called: impact ground pressure dominated coal and gas protrusion, referred to as roof impact protrusion.

## 6.2.3 Principles of Deformation(Elasticity)potential Measurement for Coal Rock Bodies

In order to approximate the potential that can be released when a coal seam passes from one state of stress to another, or when some portion of the seam is completely broken, the following two assumptions must be used:

(1) When gas power or energy action occurs inside the coal seam, the top and bottom plates give or allow very little deformation during the very short time of the change of the compressive stress in the coal seam, and the coal body only undergoes decompression displacement and deformation in the direction of the roadway space(the coal body is broken and swollen), and the direction of displacement and deformation is always the same as the direction of the propagation of the elastic wave energy, with the uniqueness of the direction.
(2) When a coal body undergoes directional compression for a short period of time and is then unloaded with a unidirectional load, the change in the unloading of the coal body proceeds in the form of the same principal mode of change as the process of unloading after prolonged extrusion and compression by stress or elastic wave energy in a coal seam.

According to the relevant data, the test for determining the movement speed of the top and bottom plates in gently sloping and sharply sloping coal seams proved that the speed was 0.5–2 mm/h or 0.14–0.56 dm/s at about 30 cm from the working face, and the data from the test of the Laboratory of Coal and Gas Outburst of the Institute of Mining of the USSR Academy of Sciences described that the test for the determination of the crushing speed of the coal and rock body triggered by the change of the state of stress or the impact and extrusion effect of the elastic wave energy showed that the crushing speed was the first element, and thus the first hypothesis was completely possible. Velocity, it was determined as 0.5–20 m/s, crushing is the first element, thus the first assumption is completely possible.

The second assumption, since the magnitude of the load compensates for the persistence of the loading in the test to determine the decompression curve, is also correct and objective to some extent.

As shown in Fig. 6.2, if the strain component is equal to zero along the cross-sectional direction y of the excavation roadway, the elastic potential energy of the coal seam along the direction x of the excavation roadway is:

$$w = \frac{1}{2}\varepsilon_x \sigma_x$$

where, $\varepsilon_x$ and $\sigma_x$ correspond to the strain and stress in the direction of the roadway.

If the residual modulus of elasticity is known $E_0$, then $\varepsilon_x$ can be expressed as:

$$\varepsilon_x = \frac{\sigma_x}{E_0}$$

When the stress state $\sigma_{x1}$ changes to stress state $\sigma_{x2}$, the elastic potential energy is:

$$w = \frac{\sigma_{x1}^2 - \sigma_{x2}^2}{2E_0}$$

Comprehensive energy power impact theory and academic point of view, such as the stone gate uncovering coal, roadway rapid excavation close to the gas coal seam (for example, gunning, including vibratory gunning), may and can instantly make the coal seam strength weakened coal body structure suddenly broken and formed a very unstable critical instability state, the elastic potential energy of the coal seam and coal body may be a sudden propagation and release.

In the two cases mentioned above, in addition to the role of the elastic potential energy of the coal seam, because it is related to the role of the compressive stress of the overburden layer above the coal seam, there is also a certain amount of elastic potential energy that has a positive effect on the release of gas rheology and kinetic energy of the coal seam. Because in most cases, the normal (or residual) modulus of elasticity of a protruding coal seam is one to two orders of magnitude smaller than that of its overburden, it can be assumed that the deformation of the coal seam is not significant and is not dominated by the coal body strain energy, but by the gas rheology and kinetic energy.

To summarize, the occurrence of coal and gas protrusion is the process of joint action of coal body compressive stress and gas rheological energy transfer, or the induced action of coal rock body compressive stress. Coal rock body is a kind of non-homogeneous object, which always has defects (micro-fractures, voids, inclusions with non-identical properties.). Therefore, when the coal rock body is subjected to external force, it will produce stress concentration in the defective parts and form damage development in the region of elastic potential energy accumulation. When the energy accumulation reaches a certain order of magnitude, the defective parts of the impact of elastic wave energy will first produce sudden microscopic rupture, due to the work to make the accumulated energy to be consumed while the remaining energy continues to do the work of propagation release. At the same time, this change also creates conditions for the re-transmission and accumulation of the remaining energy to expand the propagation, so that the change continues, gradually approaching the mining space, after a certain period of time the accumulation of energy accumulation may form the protrusion of coal and gas.

## 6.2.4 Types of Coal and Gas Outcrops

The following three types are categorized by prominent energy sources:

(1) Gas outburst. The basic energy source of outburst is the kinetic energy impact effect of a large amount of gas contained and accumulated in the coal seam

to destroy the coal body. The outburst location has obvious characteristics of coal and dust being transported by gas, and the angle of coal accumulation is smaller than the natural angle of repose of coal. There are a large number of very fine dust in the protrusion; the protruding holes are often pear-shaped, inverted bottle-shaped, with a small mouth and a large stomach; there is a large amount of gas gushing from the hole. The countercurrent of medium-sized outburst gas below 100 t can reach tens of meters, the countercurrent of large outburst below 1000 t can reach hundreds of meters, and the countercurrent of extra-large outburst gas can reach more than one thousand meters.

(2) Ground pressure protrusion. The basic energy source of protrusion is the coal seam elastic potential energy. Formed when the is subjected to(e.g. Compressive stress)extrusion induced by the gas rheology of the coal seam impacting the structure of the coal body Through the protruding phenomenon, the thrown solids in the elastic are piled up direction of the release of energy, and the angle of the coal heap is generally smaller than the natural angle of repose; the thrown coal and rock are in the form of different sizes of fragments, and sometimes the coal potential body is a large block of the whole extruded; there are often pocket, wedge and slit holes, and the external There are often pocket-shaped and wedge-shaped holes, large and small inside, wide outside and narrow inside; a large amount of gas gushes out, but from the holes there are very few gas backflow phenomena; the equipment can be pushed away and the brackets can be broken.

(3) Gravity protrusion. The basic energy source is dominated to enhance the rheological impact of gas to destroy the coal body structure by the gravitational potential energy of dumped coal, the dumped coal is piled up according to the direction of gravity, i.e. Piled up in the lower part of the original position, and the angle of coal piled up is the natural angle of rest; the dumped coal is in the form of different sizes of fragments; the dumped hole is in the form of tongue shape, pear shape, pocket shape, the depth of which extends along the inclined direction of the coal seam is from a few meters to dozens of meters, and the angle of axial inclination is greater than 45°, which is located in the original concentrated stress zone. The are located in the original concentrated stress zone; accompanied by a large amount of gas holes from the holes gushing out, but few there are gas backflow phenomena.

## 6.2.5 Prolapse of Coal and Gas at the Face

In the past dozens of years, gas protrusion occurred mostly in the excavation process of all kinds of roadways, and the probability of occurrence is relatively small due to the fact that the excavation process of the roadways in the preparation period has already given a spatial and temporal space for the release of coal seam gas and the reduction of energy. In terms of gas protrusion alone, today's mining technology for high-gas coal seams has become quite mature, and in the twenty-first century, the occurrence of gas protrusion accidents in coal seams has been greatly reduced, and it is not common to see more than two protrusion accidents in one year in the whole country. These are attributed to the pumping and depressurization of the gas coal bed, which greatly reduces the destructive effect of the elastic potential energy of the gas coal bed. As people pay more attention to environmental protection, the mining technology of full extraction and recovery of coal seam gas will be strengthened in the future, and the occurrence of gas protrusion disaster is expected to be solved completely.

At the present stage, the occurrence of coal seam gas protrusion or abnormal emergence accidents is mainly due to the following reasons:

(1) Changes in geologic conditions that cause an abnormal outpouring or protrusion of gas from a boring face or working face;
(2) Poorly structured and managed global and localized ventilation, resulting in abnormal outflows of gas that are not removed or diluted;
(3) The on-site monitoring and alarm mechanism or means are backward, failing to detect and remove potential accidents in a timely manner, and failing to issue alarms in a timely manner.
(4) The methods for evaluating the risk of gas protrusion in mines and coal seam gas protrusion are numerous and complex, which are not easy for engineers and technicians to grasp and utilize on site, thus creating difficulties in the application and management of engineering technology.

4. Mechanism of roof power impact and gas protrusion

If the spatial structural relationship composed of top plate, bottom plate and coal seam(gas) is regarded as the mechanical structural system of the excavation roadway perimeter rock system, the mechanical model of the structural system is shown in Fig. 6.3. In order to facilitate the exploration of the problem, the roadway excavated in the coal seam is regarded as an elastic body clamped by the top and bottom plates. In the sense of limit equilibrium, the coal body forms a high stress, high saving energy excavation headway due to the clamping effect of the top and bottom plates. Mechanical effects resulting from the action of the top and bottom plates clamping the coal body:

(1) Extremely high elastic potential energy is stored in the roof rock and coal seams under high stress;
(2) The roadway excavation headland area is in the location of very high elastic potential energy action or high stress concentration zone;
(3) The roof rock and coal seams are in a state of limiting equilibrium (or a state of systemic transient stability);
(4) The deformation and compressive stress effect of the roof rock layer is the dominant factor in lifting to increase the elastic potential energy of the coal seam and the coal seam destabilization.

The form of coal body destabilization is mainly related to the type and strength of energy release, analyzed from the perspective of impact ground pressure, if the size of the roof plate compressive stress is enough to brew an impact ground pressure, the compressed coal body will produce impact accelerated destruction. In the coal body the primary structure of the weak surface and the new rupture surface of the strength of the sharp decline in the coal body internal rupture to a certain extent, in the roof plate power impact energy under the sustained action, the impact ground pressure occurs.

In addition to the clamping effect of high-pressure stress on the boring head, the coal body contains high gas pressure (2.85 MPa). The gas pressure is accompanied by the rupture of the coal body and releases the gas flow change energy. The superposition of the two energies will make the occurrence of impact and gas herniation possible and easier, with greater intensity and strength.

Comprehensive impact ground pressure and gas are two aspects of the role of the analysis of factors, this type of power phenomenon is defined as: high gestures, to the roof type impact ground pressure as the dominant role in inducing the impact of the roof type impact ground pressure and coal and gas under the joint action of the power impact phenomenon. This kind of power impact phenomenon raises the risk of gas coal seam protrusion or called: impact ground pressure dominated coal and gas protrusion, referred to as roof impact protrusion.

## 6.3 Principles and Methods of Monitoring and Forecasting

### 6.3.1 Commonly Used Coal and Gas Outcrop Prediction Methods

The basis of coal and gas protrusion prediction is people's understanding of the protrusion mechanism, process, damage phenomenon and its influencing factors. During more than a century of research on the herniation mechanism, scholars in various countries have proposed various herniation prediction and forecasting methods. Due to the evolution of coal mining methods, these methods are characterized by the characteristics of the times and are related to the level of comprehensive technological development at that time. Among them, the following methods have been widely used:

(1) The single index method is used to predict the four single indexes: the failure type of coal, the initial velocity index of gas emission and diffusion in coal seam $\Delta P$, the firmness coefficient f of coal and the gas pressure P of coal seam.
(2) Drill cuttings method based on the maximum amount of drill cuttings per meter of drilled hole. The drill cuttings method is widely used because of its simplicity of operation, and the amount of drill cuttings is considered to be an effective indicator of the magnitude of the ground stress.
(3) Based on the gas geo statistics data, gas observation data and gas geologic maps, the gas geo statistics method that specifically gives whether the area is a protrusion danger area or a protrusion threat area after comprehensive analysis.

(4) According to the R value index method of the maximum amount of drilling cuttings $S_{max}$ (L/m) and the maximum initial velocity of gas emission $Q_{max}$ (L/min) in each borehole, the R value index method reflects the stress state of the working face, the physical and mechanical properties of the coal, the gas content and the permeability of the coal, which determine the main factors of the outburst risk.
(5) Combined reflection of ground stress, gas pressure and coal body strength D, K comprehensive index method.

This method is mainly used for the regional prediction of prominent coal seams and the prediction of protrusion danger of the coal working face exposed by Shimen. To divide the protruding area and protruding threat area.

(6) New monitoring methods highlight predictive methods

Nowadays, the prediction method of protrusion commonly used at home and abroad is the contact (such as drilling) prediction method. Although this method can predict the possibility of protrusion to a certain extent, the implementation of the operation is complicated, and it needs to take up the space of the coal tunnel, which affects the digging footage and the efficacy of the prediction project. Therefore, in recent years, domestic and foreign countries are explored the non-contact forecasting method without drilling, such as the use of coal body radiation temperature field, acoustic emission, gas outpouring kinetic phenomenon characteristics of the prediction of protrusion risk; there are also based on the electro-physical parameters of coal prediction of gas protrusion of the electrophysical method; based on the protrusion of the source of the regular relationship between the organic matter and the presence of markers of elements to predict the protrusion of the geochemical method; vibration acoustic resonance method to predict the impact of gas rheological energy in coal seam of roadway floor.

## 6.4 Principles for Predicting and Forecasting Impact Pressure and Salient Uniformity

### *6.4.1 Impact Ground Pressure, Prominent Characterization Appearances*

Coal rock body and gas (including methane gas and carbon dioxide) protrusion, according to the principle of mechanical analysis, and the impact ground pressure phenomenon are mining engineering in the energy power impact phenomenon, for the mining process of the objective existence of natural phenomena, is a serious in mining engineering disaster.

Based on analyzing a large amount of observation data and research results, as well as communicating with field technicians, it is found that impact ground pressure

## 6.4 Principles for Predicting and Forecasting Impact Pressure and Salient ...

and protrusion have a series of common characteristics: they both occur in the local destruction of; their destruction process is as rapid as an avalanche; they are brittle; they are both coal seams induced area to the coal seams by the high stress occur in by the dynamic impact destruction; they both adopt the methods of drilling debris in the production and test the emitted from the destruction of the coal rock acoustic pulse signals In production, and forecasting; mining methods that avoid high stress concentration are used in cases where there is a risk of impact pressure and protrusion; suitable roof management methods that reduce stress concentration at the edge of the ore body are used, and methods such as drilling chip method and testing acoustic emission pulse signals from coal rock damage and micro seismic methods are used for prediction mining of emancipated seams be used as a regional prevention and control measure; water injection, drilling and slotting, and unloading and blasting are used to reduce the stress in the adjacent coal rock body, so that the area of stress concentration $\Omega$ can be unloaded; and the area of stress concentration $\Omega$ can be unloaded. Stress concentration can area $\Omega$ can be decompression and energy transfer release, in order to prevent the occurrence of impact pressure and protrusion; can be utilized to reduce the elastic potential energy of many methods to induce the coal seam $\Omega$ regional decompression and lowering of energy to prevent the emergence of high-intensity impact pressure and protrusion.

Impact ground pressure and protrusion occurs before the coal rock body are occurring and to go through a coal rock body organization structure to withstand the elastic potential energy formed by the wave energy of the impact destructive effect, so that the coal rock body organization structure strength weakened or strain weakened, resulting in a large number of dust and particulate matter coal, that is, experiencing the impact effect of the role of the impact effect and the crushing effect of the destruction of the development of the initiation stage.

In the mid-1960s, Cook of South Africa and K hoot of the USSR proposed the energy theory of impact ground pressure and protrusion respectively. They both believe that the occurrence of impact ground pressure and protrusion is due to the destruction of the coal rock body and leads to the deformation of the ore body and the surrounding rock organization, the balance of its mechanical system is destroyed, the release of energy is greater than the energy consumed, the remaining energy is converted into the to make the coal rock kinetic energy thrown out, the surrounding rock vibration. The difference between the two is that the prominence has a gas effect. Therefore, it can be seen that the impact pressure just ignore the gas gush out and role of protruding or other gases(the authors agree that the process of impact pressure also has gas or other gases involved, but also therefore the impact of tendency to coal beds sometimes sound), so the impact pressure is also known as the coal rock or useful minerals protruding, and protruding to gas as the energy source of the elastic potential effect of the impact pressure, some countries in Europe, such as the United Kingdom, Poland will be without the phenomenon gas-containing protrusion is called coal body protrusion.

## 6.4.2 Unified the Mechanism of Impact Pressure and Protrusion Occurrence Understanding of

(1) Experimental method and the mechanism of uniformity occurrence

According to the theory of deformation and destruction mechanism of coal rock body, the process of deformation and destruction of coal rock body under the action of external force is a crack in the coal rock body destruction process dominated by the occurrence and development of and fissures. When approaching the peak strength, cracks in the local area develop rapidly and appear fissures. More than the peak strength, cracks and fissures are extremely developed and began to damage, including cracks including the emergence of and fissures, a broad strain concentration area or strength decline and disappearance, the nature of the coal and rock media in the has changed significantly, its ability to resist deformation with the deformation increases and in region decreases with the increase deformation may also increase the mutation damage, that is, it becomes " strain weakening" medium. According to the material stability criterion proposed by Drucker in plasticity theory, it is an unstable medium. Since the ability of strain-weakening medium to resist further deformation is reduced, further deformation will first occur near the strain-weakening region. There will be two different situations in the further deformation and failure development of coal and rock mass:

The first case corresponds to the rigidity test of a rock specimen, where the rock structure is gradually destroyed after the peak strength, and the deformation increases steadily until the specimen loses its strength completely. At this point the destruction process is stable and controllable, stop loading, deformation destruction process also stops;

The second situation is equivalent to the specimen for flexibility test, over the peak strength after the deformation damage process is uncontrollable, even if the loading is stopped, in the external perturbation, the specimen will automatically occur bursting damage, plastic brittle or brittle form of destruction, causing the specimen-test machine system balance damage, violent release of energy, is a non-stable power instability process with a dynamic response. Impact pressure, protrusion is the coal rock body deformation under the action of external forces, the local stress exceeds the peak strength, the formation of strain weakening strength reduction or damage to the region (or band), thus the occurrence of non-stable damage to the power instability process, the damage process and damage mechanism is the same.

To summarize, the occurrence of impact ground pressure and protrusion is in the process of deformation of coal rock medium affected by mining under the background of certain mining conditions and geological factors impact power destruction process. The deformation of coal and rock body before the occurrence of impact pressure and protrusion can be quasi-static, and after the cessation of impact pressure and protrusion, the coal and rock body obtain a great energy release and is in a new equilibrium state.

(2) Shock action of elastic wave energy and the mechanism of uniformity

## 6.4 Principles for Predicting and Forecasting Impact Pressure and Salient …

Impact pressure and protrusion are the that makes the coal and rock body change from one equilibrium state to another state development and evolution process. Therefore, we can study the characteristics of the equilibrium state of the coal-rock deformation system before impact pressure and protrusion, if the equilibrium maintains a stable state, then power will not occur impact damage; if it is a non-stable state, then the equilibrium instability will lead to power damage, and impact pressure and protrusion are likely to occur. Using this method to study the criterion of dynamic impact failure process of coal rock deformation system has more practical guiding significance for the prediction and prevention of rock burst and outburst.

The criterion for the stability of the equilibrium state of is generally adopted as the Dirichlet criterion, which holds that when regional deformation system within the coal rock body $\Omega$ is equilibrium stable, and unstable when it is extremely large. Let the total system of elastic potential energy of is extremely small regional $\Omega$ the general function of the total potential energy of the system be regional $\Phi(\sigma_x, \sigma_y, \sigma_z)$, according to the principle of differential extremum, when $\Phi' = 0$, the general function of the potential energy has an extreme value, i.e., it is a necessary condition for the system to be in an equilibrium state. $\Phi'' > 0$, the general function of the potential energy has an extreme small value, and the structural system equilibrium is stable; $\Phi'' < 0$, the general function of the potential energy has an extremely large value, and the equilibrium state of the system is non-stable; $\Phi'' = 0$, the system equilibrium state is regarded as a different equilibrium state of higher-order differentiation, and the structural system is in the stable to the unstable critical state of the transition from the. Therefore, the criterion of rock burst and outburst is the condition that the equilibrium system of coal and rock mass $\Omega$ is in critical state and unstable equilibrium state, that is:

$$\Phi' = 0 \quad \Phi'' \leq 0$$

If it is outburst, the total potential energy $\Phi$ ($\sigma_x$, $\sigma_y$, $\sigma_z$) includes the effect of gas dust rheological energy. The above Eq. is the necessary condition for rock burst and outburst.

Coal rock body equilibrium instability, the occurrence of power impact damage process must also meet the dynamic process criterion, generally using the energy criterion, that is, the energy released in the destruction process is greater than the energy consumed, then produce power impact damage, that is, the combination of the impact effect and crushing effect of the process.

If once the direction of action of the elastic potential energy forms a uniquely propagating action situation, the energy induced and involved in the propagation $\Omega$-system is $\Phi$, the energy consumed by the wave energy propagation process by overcoming the work done by the resistance of the medium is $W_0$, and the residual elastic potential energy is $\Delta\Phi$, then there is:

$$\Phi - \Delta W_0 > \Delta\Phi$$

Satisfy the above Eq. to have excess energy into continue to destroy the coal and rock body and flow change energy impact, as well as the formation of impact or protruding catastrophe, the above Eq. for the formation of impact or protruding sufficient conditions, but also impact pressure, protruding the occurrence of the energy theory of the important general function of the mathematical model.

### 6.4.3 Elastic Potential Energy in a Coal Rock Body Equation for the Impact Effect of

In the coal rock mass ( coal seam), the process of elastic potential energy impact effect damage is continuously carried out, and the information of coal structure damage, crack fracture development, acoustic emission, dust gas change, compressive stress and displacement deformation is generated. When the elastic energy is transmitted to the vicinity of the surrounding rock, only when the elastic energy $\Phi$ damage the medium to do work $W_0$, the residual energy $\Delta \Phi = ( \Phi - W_0)$ is sufficient to weaken the damage ( strength weakening). The medium and surrounding rock are separated from the weak surface and rushed into the mining space, which is called destructive rock burst or outburst. Therefore, it can be concluded that the occurrence of impact effect in coal and rock mass is a necessary condition for the occurrence of rock burst or outburst, and satisfying the impact effect equation is a sufficient condition for the occurrence of rock burst or outburst. The strength of rock burst depends on the value of $\Phi$ - $W_0$ = $\Delta \Phi$, that is, the residual energy $\Delta \Phi$ reaching the near zone of surrounding rock. $\Delta \Phi$ is proportional to the degree of rock burst and outburst damage and magnitude. When the energy $\Delta\Phi$ reaching the near area of the surrounding rock is relatively small, due to the consolidation resistance of the surrounding rock supporting structure, $\Delta\Phi$ can only cause displacement deformation or spalling, roof fall and other damage phenomena of the surrounding rock. When the elastic energy is transferred from the deep rock mass to the near area of the surrounding rock, and can reach a very high energy value in a very short period of time, the surrounding rock and its weakened medium will be thrown into the space like an avalanche under the action of $\Delta\Phi$, causing destructive engineering disasters. This instantaneous process is the so-called rock burst or outburst, and the energy causing the damage is $\Delta\Phi = MU_0^2/2$, all involved in the completion of the occurrence of rock burst or prominent energy to meet or comply with the quantitative relationship of the impact effect equation, that is, the principle of energy conservation.

## 6.4.4 Uniform Prediction and Forecasting Methods for Impact Pressure and Protrusion

The formation and development of are subject to the creep to drastic change impact pressure and protrusion objective law of the from, and from the stage of formation when the external force initiates the formation of creep by touching and extruding the coal body, many information such as acoustic emission signals that can be propagated along the medium of the media as well as displacement and deformation, and changes in compressive stress are released. Gas seams can release sound(ultrasonic signals)audible to the human ear, and impact-prone seams are sometimes able to produce the same sound audible to the human ear(the author: suggests that there may also be gas involved in the impact effect of the role of the impact effect),and a number of physical phenomena to prove that its occurrence and destabilizing and destructive process destruction of the creep development mechanism with a consistent characterization of the phenomenon. The power impact energy formed by the extremely high ground stress can induce the coal seam gas protrusion (including low-gas coal seam), both have the doctrine of disaster occurrence tendency and common technical concepts, that is, impact tendency, protruding tendency of technical principles, indicating that the two have the same or similar prediction and prevention and control measures.

### 6.4.4.1 Acoustic Emission (Microseismical)technology Prediction Methods

Research shows that in the process of deformation and destruction of coal and surrounding rock by force, rupture and acoustic emission signals, i.e., vibration signals, from are transmitted the source of vibration or sound waves, and when the intensity and frequency of the vibration or sound waves reach a certain value, sudden destruction of the coal-rock body will occur, and impact ground pressure and/or protrusion will occur. The vibration wave inside the coal rock body can be received by the detecting instrument (such as ground tone or pickup) installed inside the coal body, amplified and recorded. The acoustic emission source or.

Since the early 1970s, the United States Bureau of Mining has used standard microseismic technology to study the destruction of coal seam structure. At the same time, ultrasonic monitoring techniques have been used to monitor the energy of rock formation ringing, and researchers have utilized low-frequency (1–10 kHz) microseismical techniques to monitor and determine the most likely locations for impact ground pressure and protrusion, and high-frequency (40–110 kHz) microseismical techniques to determine when protrusion occurs.

Acoustic emission (microseismic) monitoring system is generally composed of several ground sound probes, amplifiers, filters and a multi-channel central recording device. The signal of the sound probe is recorded by the computer system of the central recording station. Each unit signal is regarded as an event, and each event has its energy value. The energy value and the number of events are recorded at a time interval of 0–1 min. Then, through data analysis and processing, the three-dimensional coordinates of each sound source are determined, and each sound source is drawn on the engineering drawing. According to the density contour, the location and range of the most frequent area of coal rock fracture activity can be determined. This method has the disadvantages of large engineering quantity and low efficiency, fuzzy information and low accuracy. All the vibration (vibration) information that occurs in the mining area is the monitoring target.

The results from the Huafeng coal mine in Xinwen, which uses microseismical technology to monitor coal mining activities, show that:

(1) The microseismical activity mainly occurs during coal mining and shelf removal, and the workface depressurization mainly occurs during coal fall. The maximum microseismical frequency is 1300 above times/hand under normal condition, the peak is 800 about times/hand about 50 ~ 100 in 1–2 d after the end of working facetimes/h.

(2) Microseismic (shock ground pressure) activity increases with production.

(3) During the monitoring period, there was no destructive impact ground pressure, and the several vibrations of less than 1.8 on the Richter scale were monitored, and the coal body structure of the working face was basically normal.

### 6.4.4.2 Coal Seam Temperature Prediction Methods

According to the data, the former Soviet Union N.A. Lerenko and other scholars in many coal mines, the coal body temperature of the key sections of the mining face was investigated and studied. The temperature of the working face is measured in a borehole up to 6 m deep, and the temperature of the heading face is measured in a borehole up to 2.2 m deep.

The results of temperature measurement concluded that: in the temperature gradient of the coal body in the near-working face section of the back-mining face$\Delta t2-0 \approx \Delta t3-1$($\Delta t3-1$ is the temperature gradient between the 3 m place and 1 m place of the coal drilling, and $\Delta t2-0$ is the temperature gradient between the 2 m place of the coal drilling hole and the mouth of the drilling hole), when $\Delta t3-1$ there is no danger of ground pressure and protrusion 3 °C. At the digging face $\Delta t2-1 \leq$ <2°C there are no signs of ground pressure and protrusion danger. $\Delta t2-1 2=\sim2.5$ °C there are signs of ground pressure and protrusion threat; $\Delta t2-1>2.5$ °C there are signs of protrusion or ground pressure danger. During the test period of the method, 620 m of roadway was excavated, 230 m of which were predicted on a daily basis, and all 6 herniations that occurred during the excavation were predicted, as well as 7 herniations in Kalinin and other mines. The method of forecasting using temperature gradient values was recommended by the Soviet Union Committee for Bulge Prevention.

Coal rock body by force or energy impact, triggering the coal body extrusion expansion friction broken a series of changes triggered by the temperature, including the temperature of the gas in the fissure changes. The increase in temperature of the prominent coal bed is favorable to the formation of rheological movement of the endowed gas. Our scholars have used the temperature condition as the theoretical basis for predicting the risk of protrusion, the principle is that gas desorption absorbs heat, leading to a reduction in the temperature of the coal seam, the more obvious the temperature reduction, indicating that the coal seam gas desorption ability is stronger, the greater the risk of impact ground pressure and protrusion. Practice shows that the higher the coal seam gas content, the more obvious this effect.

Practice has also proved that where the coal rock body there is a after continuous elevation of sudden and substantial decrease in temperature, it is a sign that there is a large geological structure near the working face (such as a sudden thickening or thinning of the coal seam, a sudden change in the angle of inclination) or a large fissure displacement, and there is the possibility of a large, impact, protrusion, or compressive stress.

### 6.4.4.3 Momentum Coal Rock Masses Prediction the Principle of Impact on Method for

Utilizing impact ground pressure and protrusion mostly occurring in the two gangs of the coal tunnel, i.e., the coal body around the quarry the characteristics of destruction. Assuming that the wave of elastic energy can induce the coal body mass point vibration and touching, the vibration speed of the mass point is v, the mass point propagation speed is V, it(the mass point)creep breeding process has the impact momentum characteristics of, so based on the characteristics of the breeding and development of the establishment of the elastic power impact equation of the coal body medium mass point, proposed with the view that the role of the elastic power impact of the mass point uniqueness characteristics of the direction of, that is, always pointing to mining space the role of the direction. Using the principle of contact mechanics-dynamic impact effect, predict the size of impact compressive stress or impact momentum and the number of occurrences (i.e., frequency) and damage phenomena. The (elastic potential energy) impact effect of energy and the basic principle of prediction are shown in Fig. 6.4. The expression of impact principle is also the technical principle of research and development of rock burst and outburst prediction monitoring instruments.

Based on the mechanism and thought of the transitivity and impact effect of the strain energy of coal and rock mass (rock) after being affected by the excavation activity, the elastic strain energy stored in the coal and rock mass is affected by the mining. After the strain energy in the $\Omega$ region of any coal rock orientation (direction) is far away from the mining space, after the formation of the directional energy action potential field, it is transmitted and accumulated along the weak surface of any radial mining space, during which the medium structure is subjected to severe impact extrusion damage, and the failure process has the characteristics of intermittent impact vibration. The momentum impact system model of energy transfer accumulation and impact process in coal and rock mass is established by using the

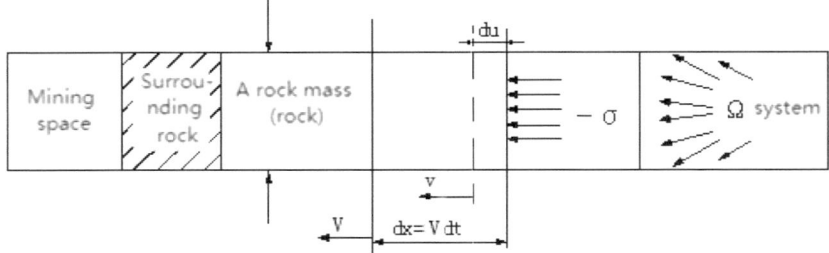

(a) stress wave at speed Model of transmission along an elastic rock body $V$

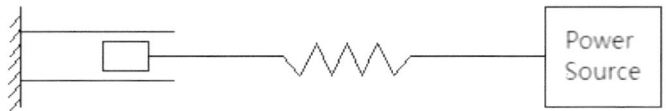

(b) Stress intensity transfer force source and elastomer composition spring-damper model of

**Fig. 6.4** Coal rock body energy impact system modeling

principle of momentum contact impact extrusion effect in contact mechanics, as shown in Fig. 6.4.

The physical change process of elastic stress–strain of coal and rock mass is regarded as the dynamic effect phenomenon in solid mechanics, that is, the impact effect phenomenon of elastic strain energy accumulation and transmission of rock mass and the stress wave phenomenon of solid after impact. Combined with the model shown in Fig. 6.4, the elastic strain energy in the rock mass is set to be a one-dimensional compression wave (also known as stress wave), and the stress pulse with strength of $-\sigma$ transmitted by the pressure wave from the $\Omega$ system along the coal and rock mass is considered. Within the time dt, the distance of waveform movement is $dx = Vdt$, and the mass of the medium involved in energy transfer is considered to be the contact vibration velocity v obtained by the unit of $\rho A dx$ under the action of pressure wave pulse, which is the inherent characteristic value of rock mass medium. $\rho$ is the medium density; a is the cross-sectional area of the stress wave. Then the momentum equation of the small unit cell is $-\sigma A dt = (\rho A dx) v = \rho A V dt$, that is:

$$\sigma = -\rho V v \tag{6.1}$$

At this point the cell is compressed by $du = vdt$, and the strain that occurs is:

$$-\frac{du}{dx} = -\frac{v}{V} = -\frac{\sigma}{E} \tag{6.2}$$

Combining Eqs. (6.1) and (6.2), the expression for the pulse velocity is obtained as $V = (E/\rho)^{1/2}$) or $V^2 = E/\rho$, where E is the modulus of elasticity of the coal rock medium.

From Eq. (6.2), it can be seen that the velocity v of the mass point in the elastic microelement body than the pulse velocity is much smaller. Through the analysis of this quantitative relationship, it can be obtained that: ① The energy impact stress pulse signal characteristic of $\Omega$ system reflects the dynamics of the system and the frequency response characteristics; ② When the stored elastic strain energy in the $\Omega$ system is small, then the stress pulse signal V, v is relatively weak, and the transmission characteristics and amplitude-frequency characteristics of the apparent stress wave will not be obvious; ③ The transmission velocity V of the stress wave and the transmission velocity v of the medium point are proportional to σ and E; ④ By monitoring the change of pulse frequency and amplitude of the rock stress wave, it is possible and easier to judge and recognize the intensity of the impact effect of the coal rock body, so it can be used to predict the early information of the impact ground pressure; ⑤ Since the eigenvalue of the stress wave the stress wave which V gets the size of is inversely proportional to the transferring distance and the density of the medium, the signal of the stress pulse in the process of the transferring process will be weakened gradually. Therefore, if the instrumentation equipment and analysis system are used to predict the early information of impact ground pressure by monitoring means, the sensing elements should be arranged as close as possible to the source of the earthquake or excavation space to ensure that more information, such as some relatively weak amplitude-frequency characteristic signals, can be captured to achieve the real sense of early prediction.

According to the energy impact system model of coal and rock mass and the stress pulse characteristics and transfer analysis of coal and rock mass shown in Fig. 6.4, it is easy to use energy-controlled sensors (such as passive magnetoelectric sensors that can convert impact energy into electrical energy) to transmit and accumulate energy from $\Omega$ system to mining space. The process, that is, the stress pulse signal generated by the impact extrusion during the impact effect of coal and rock mass, is monitored and collected. Based on such a kind of understanding of energy and rock burst, as well as the energy impact system model, the research and development platform of rock burst prediction and prediction equipment (sensing technology) is established, and the information is simulated and analyzed by computer, which is more in line with the energy dynamics and transmission characteristics of the source (energy) of rock burst. It can collect some pulse signals with relatively weak amplitude-frequency characteristics of stress waves in the early stage of rock burst and outburst (i.e., the stage of impact effect). Based on this mathematical model, a prediction system is formed through the simulation analysis of computer software. It can effectively solve the common problems of signal acquisition omission, lag and confusion in the current application of ground mine earthquake monitoring stations. It is often known and proved that the event signal is monitored after the occurrence of large vibration or rock burst damage, which can not capture and predict the early information of rock burst.

#### 6.4.4.4 Predictive Forecasting Method for Mining Pressure Monitoring Parameters

The method of monitoring mine pressure is applied to monitor the impact and protrusion precursor information, such as compressive stress, displacement and deformation, and other information that can reflect the impact effect characteristics of impact pressure and protrusion occurrence mechanism. Mine pressure monitoring instrumentation in the signal monitoring and acquisition of the signal acquisition time interval relative to the impact pressure and protrusion of the pre-signal aura, that is, the impact effect of the signal monitoring and acquisition of the time interval is large, therefore, due to the impact pressure and protrusion of the early signals generated by the characteristics of the more intense, fast change, the signal often comes suddenly and drastically, and sometimes disappears also fast. This requires the application of mine pressure monitoring instrumentation mode acquisition impact effect triggered by mine pressure information signal, ①signal acquisition density should be large, that is, the acquisition time interval as small as possible, up to the minute or second level signal acquisition, ②wait-and-see type continuous monitoring and acquisition of time is sufficiently adequate, ③and to regularly and timely analysis of monitoring data signals and data charts. The monitoring and acquisition signals are mainly compressive stress and displacement deformation.

### *6.4.5 Analysis of the Direction of Development of Predictive Forecasting*

(1) The development of the prediction technology of impact pressure and protrusion has many links with the coal mining process, so the monitoring should control the location of the signal monitoring and collecting points, and analyze and master the direction of the impact signal generating source and releasing channel.
(2) Carry out extensive research on the causes of the formation of elastic energy and the development of disaster mechanisms based on coal seams with a tendency to impact and protrude, and establish a monitoring application system for impact and protrusion energy and pressure with sound wave and mine pressure monitoring technology as the direction of development.
(3) Through typical cases, the results of China's geo stress test and the comprehensive analysis of many research results, it is shown that the size of the coal seam gas pressure Pw on the salient danger of prominence and contribution is very clear, and it is also convenient for the actual measurement and utilization of the production engineering operation, so it is proposed to classify the degree of prominence of gas-bearing coal seams as follows:

  I Gas pressure: Pw < 3 MPa, no danger of protrusion, pay attention to the superimposed effect of rock or roof compressive stress induced by power impact, and production of reasonable normal advancing speed of the working face. When the gas pressure is less than 3 MPa, it is the general ground stress order of magnitude.

II gas pressure: Pw ≥ 3 MPa, there is a danger of protrusion, pay attention to the superimposed effect of the pressure stress on the roof plate of the surrounding rock, take measures to decompress and reduce energy, and control the production of the working face should not be pushed forward too fast.

(4) Impact ground pressure and gas protrusion are all positive effects of elastic energy, so research on the effectiveness of pressure-reducing and energy-reducing technologies and methods for coal seams and surrounding rock strata(bodies) should be strengthened, as well as comprehensive energy-reducing and pressure-reducing measures with reasonable and moderate mining progress. It is fully recognized that energy and pressure reduction is the first element.

(5) Continuing to study in depth the existing indicators and methods for evaluating the risk of impact ground pressure and gas protrusion, and establishing indicators and evaluation methods and systems that are easy for engineers and technicians in the field to grasp and apply.

(6) Carrying out research on the change and mechanism of gas pressure in coal seams, developing applicable equipment and instruments for monitoring gas pressure in drilled holes with strong operability, and realizing a system of monitoring, evaluation and forecasting indexes for real-time on-line continuous monitoring and recording of gas pressure and data analysis and application.

**Open Access** This chapter is licensed under the terms of the Creative Commons Attribution-NonCommercial-NoDerivatives 4.0 International License (http://creativecommons.org/licenses/by-nc-nd/4.0/), which permits any noncommercial use, sharing, distribution and reproduction in any medium or format, as long as you give appropriate credit to the original author(s) and the source, provide a link to the Creative Commons license and indicate if you modified the licensed material. You do not have permission under this license to share adapted material derived from this chapter or parts of it.

The images or other third party material in this chapter are included in the chapter's Creative Commons license, unless indicated otherwise in a credit line to the material. If material is not included in the chapter's Creative Commons license and your intended use is not permitted by statutory regulation or exceeds the permitted use, you will need to obtain permission directly from the copyright holder.

# Chapter 7
# Principles of the Roof Slab Coming Under Pressure and Creeping Instability in the Integrated Mining Face

## 7.1 Overview

The development of mining support technology can be divided into two aspects: ① the lane support means from the early wooden shed beam support → I-beam support → swan wall → U-beam support → various forms of mixed support → anchor network cable support; ② working face roof support from the wooden shed beam → I-beam → various forms of mixed support → friction strut → various forms of mixed support → monolithic hydraulic strut → synthesis mining bracket development process, the development process. During this period, it has experienced 30–50 years of development. Due to the insufficient, unbalanced and unstable support strength of the monolithic pillar to the roof, the driving and control strength of the overall support of the working face to the roof rock layer is extremely unstable, inducing the vertical and horizontal range of the development of the roof high rock shaking and subsidence off-layer changes, the fluctuation is fast and high, and the height of the three bands of the overlying rock layer of the mining area subsidence off-layer is high (especially the height of the II and III bands), and the pressure of the mine shows the relatively intense characteristics, seriously affecting the stability of the surrounding rocks of the quarry. It seriously affects the stability of the peripheral rock of the quarry. Obviously, monolithic column and synthesis mining stent to maintain the stability of the working face roof and the ability to control the essential difference, we cannot treat and utilize the laws and characteristics of the two, and cannot be mixed into one.

However, the comprehensive promotion and application of anchor cable support and integrated mining support has greatly improved and enhanced the stability of roof and surrounding rock, the roadway surrounding rock and working face roof have been strongly supported and supported, and the support has more full driving control over the roof and overlying rock, limiting the deformation of the quarry roof and surrounding rock to the maximum extent. It is especially suitable for the control of the roof range of the working face. Especially in the twenty-first century, comprehensive mining support completely replaces the single hydraulic strut, comprehensive mining working face roof cycle to pressure law characteristics should be completely different from the single hydraulic strut working face roof to pressure show characteristics. Variety mining stent initial support (support force) to fully meet the various types of roof stability control requirements, due to the variety of stent initial support force is sufficient and powerful, so that some of the mine roof rock layer without obvious cycle to pressure step, research also found that there are many mines direct top (or direct top + old top) with the stent to promote the side of the push side collapsed, the actual situation with the calculation of the pressure to the step of the discrepancy between the gap (according to the current variety of stent to the roof of the top) the control effect is fundamentally different from that of the monolithic hydraulic strut, and the preferred monolithic hydraulic strut cannot guarantee the rated work resistance and support strength, and its support effect on the roof is usually passive. Therefore, the research and application of the characteristics and development degree of the mine pressure manifestation formed by the two support methods should be discussed differently. This is all related to the synthesized mining support initial support force and high support strength. At the same time, the full use of comprehensive mining bracket support technology development, promote such as small coal pillar, no coal pillar, filling and other new coal mining technology method of change and innovation and development, and its mine pressure manifestation law characteristics, phenomena must also have changed.

## 7.2 Recognition of the Subsidence of the Overlying Rock Formation and the Pattern of Pressure Coming from Its Top Plate

The control effect of the comprehensive mining support on the roof plate is fundamentally different from that of the monolithic hydraulic strut, and the preferred monolithic hydraulic strut simply cannot guarantee the rated working resistance and support strength, and its support effect on the roof plate is usually passive; there is a fundamental difference between the spatial and temporal structure and mechanical elements of the working face, the mining area, the support and the roof plate. Therefore, the research and application of the two types of support methods should be discussed differently with regard to the characteristics of the mine pressure and the appearance and degree of development of the two types of support methods.

## 7.2.1 Basic Characteristics of Subsidence Movement of the Overlying Rock Layer in the Mining Area

When we analyze and study the pressure (manifestation) and subsidence movement law of the roof plate of the working face, the roof plate of the mining hollow area and the overlying rock layer, we first set (assumed) that the initial bracing force of the synthesizing stent is sufficient to meet the design requirements, and that the distance between the working face and the stent's roof control is reasonable.

Today, a tendency of ideological concepts drives one of the coal mining methods we advocate and prefer to adopt: long-wall large-height coal mining method. As the coal mining face advances, the exposed roof slabs (direct roof or direct roof + old roof) in the open area will be subject to subsidence deformation, movement and damage under the mining pressure exerted by the overlying rock strata. The usual situation or general rule is that, according to the different degree of damage state of the rock layer, the deformation and destruction of the overlying rock layer can be divided into three zones, as shown in Fig. 7.1, where I: bubbling zone, II: fissure zone, and III: deformation and bending zone.

When all the roof collapsing method is used to manage the roof, after the coal mining face advances a certain distance, under the influence of the effect of the empty roof of the mining area, it firstly causes the direct roof to undergo subsidence creep and eventually forms a subsidence deformation movement damage range which is completely detached from the constraints of lateral cohesion of the rock layer (or bonding force), and the damage height range is commonly known as the bubble drop zone (Fig. 7.1-I). This part of the rock layer eventually forms the maximum self-gravity force W and collapses. This portion of the rock is temporarily supported by bracing in the coal mining face.

The rift zone refers to the rock strata above the caving zone and below the bending zone. The characteristics of this part of the rock strata are as follows: the front and rear of the rock strata are respectively supported by the support force of the support and the coal pillar, which can also be described as the shear force. The center of the rock strata (or the central area) is affected by the gravity W. Due to the combined action of the vertical subsidence movement force W and the horizontal (transverse) strata decompression expansion stress at the boundary of the subsidence movement, and because the decompression expansion extrusion effect precedes the subsidence movement, the movement of the rock strata produces the development of cracks or open cracks, but the rock strata can still be arranged in an orderly manner through extrusion (Fig. 7.1-II), which plays the role of the arch beam of the combined structure of the subsidence rock strata. If we use the reverse way to observe and consider the phenomenon of the results of the rock mass test using small specimens in the laboratory, the author believes that because the actual subsidence movement of the rock stratum is a large-scale structural rock mass mechanics change specimen and event, the composite structure arch beam has the characteristics of elastic deformation ability and anti-deformation ability, even if it is irreversible deformation. It is not possible to fully use the results of small specimens tested in the laboratory to form

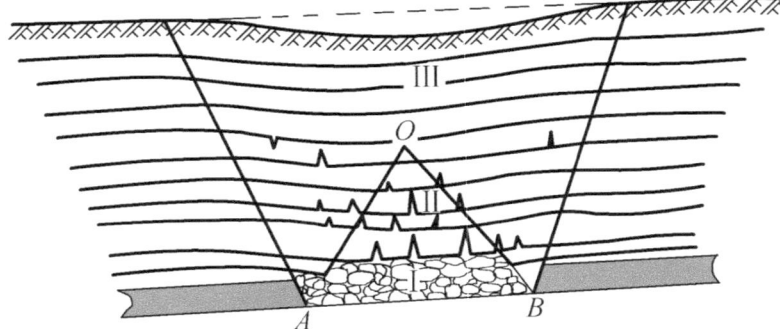

(a) Schematic depiction of the three zones of subsidence and collapse in the mining area

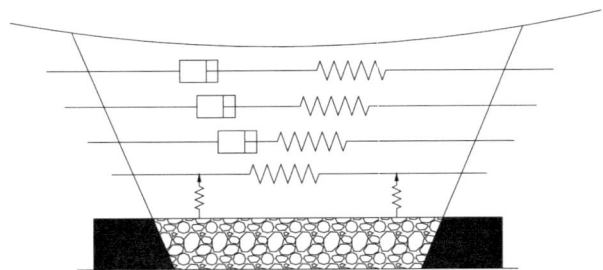

(b) Arch beams of decompression laminar assemblage in the rock layer overlying the mining area

**Fig. 7.1** Schematic diagram of the layered structure of the subsidence of the overlying rock layer on the roof of the mining area

ideas to understand this large-scale (by contrast) structural rock formation. The arch beam of this kind of rock stratum composite structure is elastic and has performance characterization. This method describes and expresses the subsidence deformation movement of goaf as a complete mechanical system, a stable equilibrium or unstable equilibrium composite structure system.

The deformation bending subsidence zone is generally defined as the subsidence bending of rock layers (mainly bedrock, or overlying rock layers) located above the fissure zone, and the influence of subsidence can be developed upward to the surface. The rock strata within this deformation zone will maintain their integrity and act as a laminated structural arch of sedimentary rock layers (Fig. 7.1-III), which is referred to as a structural arch.

Production practice and research have shown that the loading force on the support of the coal mining face is much smaller than the weight of the overlying rock strata on which the movement is generated, which is precisely the favorable effect of the laminated combined structural arch beams above the mining area. Only the movement of a part of the rock layer close to the coal seam can produce significant compressive stress effects on the coal body, roof slab and support of the working face. The so-called mine pressure control of the coal mining face is also the control of this part of the rock layer. This part of the rock layer is approximately equivalent to the total thickness of the fallout zone and fissure zone in the above three zones, which is generally 6–8 times of the mining height (estimation method).

The form of the movement and failure of the overlying overhanging strata determines the law of the appearance of the mine pressure and the requirements for control. There are basically two forms of movement from self-suspending to failure of overlying strata of roof, namely, bending failure and shear failure.

## 7.2.2 Academic Ideas and Current Developments

The coming pressure characteristics of the roof plate of the synthesized mining working face have a great influence on the stability of the stent and the service life and sealing performance of the stent. From the technical management point of view, the roof plate of the mining field can be divided into two areas: the roof plate of the coal mining face and the roof plate of the hollow area, the middle of which relies on the interactive connection of the synthesized mining stent to support the roof plate, as shown in Fig. 7.2 (I: the working face II: the hollow area), depending on the roof plate with the characteristics of the "beam or plate girder", the production process we hope that the roof plate of the face will be stabilized in order to stabilize the synthesized mining stent and the safety of the roof plate. The production process we hope that the roof of the working face is stable, in order to get the stability of the synthesized mining support and the safety of the roof plate. Usually, we pay more attention to the pressure law of the roof plate in the hollow zone and its manifestation characteristics and the pressure law of the roof plate in the hollow zone is the result of the combined effect of several mechanical elements. In short, the pressure fracture collapse law and characterization of the roof of the mining area and the mechanical characterization of the overlying rock layer's subsidence movement are inextricably linked, specifically analyzing the influence of the elements are: ① the mechanical properties of the rock layer, the modulus of elasticity of the rock layer, the shear strength of the rock layer, the rock layer horizontal stress (i.e., cohesion or adhesion); ② the roof of the quarry in the mining field spatial environment: the roof of the first face of the mine, the roof of neighboring mine, the roof of the mining area, the mining height of the roof of the thin coal mine, the roof of the mining area, and the roof of the mine is located in the mine space environment. The roof of the first mining face, the roof of the neighboring mining area, the roof of the large mining height, the roof of the thin coal seam and the roof of the medium-thick coal seam mining area. All of these elements will affect the stability of the production face roof and the pressure pattern, which also shows that the deformation and movement of the roof of the mining hollow area (the overlying rock layer) is very complicated. As a result, the theory of transferring rock beams centered on the movement of the overlying rock layer, i.e., the top plate of transferring rock beams, and the theory of masonry beams, i.e., the top plate of

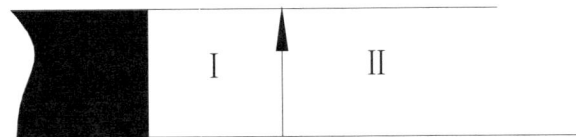

**Fig. 7.2** Principle of support and bracing of the roof plate by the integrated mining support

masonry beams, as well as the later theory of key layer theory, and there are even the theory of suspension beams masonry beams (Author: the suspension beams and the transferring rock beams are the same mechanical structural manifestation of the different statements).

The two different academic views are summarized and visible, and it is not difficult to find these two academic ideas and views through various books and magazines published by the coal industry. Combined with the two academic viewpoints summarized on behalf of the two viewpoints of the mining roof subsidence collapse principal description schematic diagram, detailed in Fig. 7.3 shown. The significance and purpose of researching and accurately describing the overlying rock layer subsidence movement on the mining surface is to accurately and effectively grasp and utilize the law of roof plate pressure to ensure the stability and support quality of the support, so as to satisfy and adapt to the change of geological conditions (such as coal bed inclination angle) and improve the mining rate. Therefore, the purpose of studying the movement of the overlying rock layer is to recognize the characterization of the pressure law of the roof plate and the appearance of fracture collapse, especially for the targeted research on the many changes brought about by different coal mining methods, such as the small coal pillar, no coal pillar filling method, the artificial roadside gangs along the open roadway, and the large mining height and long-wall method, etc. And to positively provide theoretical support for the maintenance of the stability of the bracket and the quality of the support.

## 7.3 Mechanics and Characterization of Subsidence Motion in Overlying Rock Formations

The focus of researching the law of roof plate pressure and the manifestation characteristics of the mining hollow area should be the comprehensive reflection and dynamic change characterization of the mechanics of the subsidence movement of the rock layer overlying the mining hollow area and the structure of the rock layer. The integrated force of the rock layer's subsidence kinematics has the most direct influence and result on the roof plate's pressure pattern and manifestation characteristics. So far, we have not been able to accurately monitor or calculate the magnitude and real-time change of the compressive stress or gravity when the rock strata are involved in the subsidence movement, but this is not important, as far as the current coal mining technology is concerned, we are concerned about the influence (or integrated influence) of the comprehensive force formed by the movement of

### 7.3 Mechanics and Characterization of Subsidence Motion in Overlying …

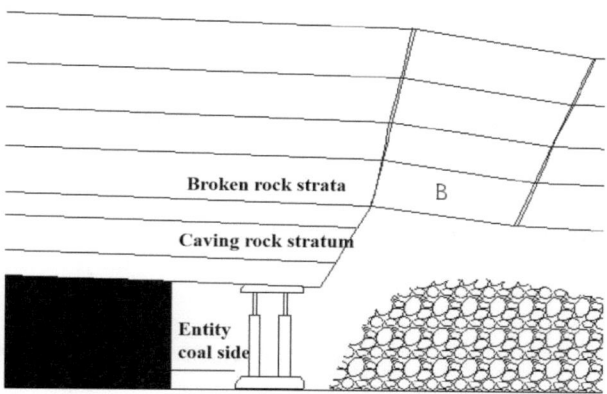

(a) Description of the fracture and collapse phenology of the top plate of the transfer rock girder

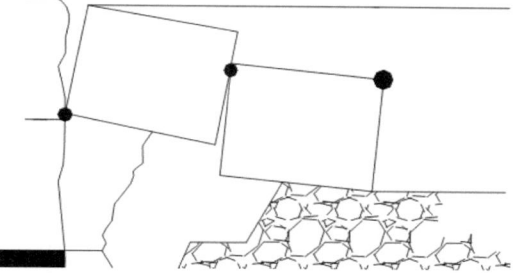

(b) Description of masonry beam roof fracture and collapse manifestations

**Fig. 7.3** Schematic diagram describing the principle of roof subsidence and collapse in a quarry

the overlying rock strata on the top plate's coming pressure law and manifestation characteristics; and how the spatial and temporal scales in the mining area affect We are concerned about how the temporal and spatial scales of the mining area affect the subsidence range of the overlying rock layer (or the surface), the development process of the movement and the final damage phenomenon or result.

### 7.3.1 Principles of Overlying Rock Movement and Changes in Subsidence Patterns and Phenology

The subsidence movement of the overlying rock layer in the mining airspace is divided into two stages for analysis, and most of the coal mining face roofs obey such a stage of the development process, except for the thin and very soft rock roofs:

(1) The stage of initial pressure on the roof plate in the hollow area. Due to the advancing development process of the working face, the area of the hollow zone gradually increases. As Figure shown in Fig. 7.4, due to the support of support I in front of the hollow area and the support of coal body II at the back, gravity W concentration and off-layer firstly began to form in the center line of the shape of the hollow area, Fig. 7.4a shows the state of sinking and collapsing of the overburden rock layer, and Fig. 7.4b describes the principle of the development process of the development of the sinking movement of the overburden rock layer to the ground surface. Due to the joint supporting effect of support I and coal body II on I, II at the top plate to produce support and shear force gradually increased, in other words, it also indicates that the gravity W and the deformation of the off-seam gradually increased. The formation and occurrence of the first pressure of the roof plate in the quarry: ① direct top or direct top + old top firstly produce creep and bending away from the layer at the center O–O of the shape of the mining area to the first fracture and fall; ② direct top or direct top + old top in the bracket and the roof of the point of action of the face of the O on the side of the mining area began to produce creep to the old top to pressure to the first fracture and collapse; ③ in the back of the mining area, the roof of the coal body II (direct top or direct top + old top) the same Creep development to complete the first fracture collapse. The final end of the roof initial pressure stage, the formation of the overlying rock layer subsidence collapse state appearance of the three bands, the formation of the initial pressure at the end of the moment of the collapse of the overlying rock layer state, usually, the schematic diagram of the roof collapse appearance state shown in Fig. 7.1 is used.

(2) The roof plate in the mining area enters the stage of normal cycle pressure. Assuming that the roof plate has sufficient elastic deformation capacity, with the advancement of the support, the roof plate rock beam hanging wall length L gradually increases, due to the formation of the overlying rock layers of the integrated transverse bond (also known as cohesion) between the rock layers (except for the thin or very soft rock), that is, the role of the structural arch beam of the overlying rock layer. At this moment, the stent, roof plate (overlying rock layer), and the mining area of the three combinations of the dynamic characteristics of the spatial and temporal structural correlation as Figure shown in Fig. 7.5, this structural relationship objectively for us to reveal the transfer of rock beams between the roof plate and stent of the interaction between the coupling mechanical relationship model, that is, academician Song Zhen qi's transfer of rock beams of the roof of the doctrine. The scale parameters and mechanical properties of the transferring rock beam are related to the deformation of the overlying rock layer or the comprehensive movement mechanics elements given by the subsidence movement of the ground surface, and it has the mechanical property characteristics of "plate beam and beam" (e.g. Elastic deformation capacity, elasticity and plasticity.), and the mechanical properties of the top plate of the transferring rock beam and the magnitude of the bracing force (or the initial bracing force) of the bracket will The mechanical properties

## 7.3 Mechanics and Characterization of Subsidence Motion in Overlying …

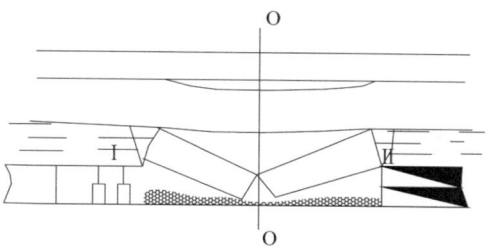

(a) Characterization of initial pressure

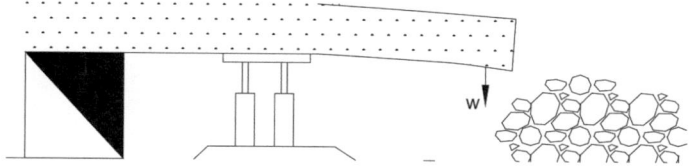

(b) Periodic pressure characterization

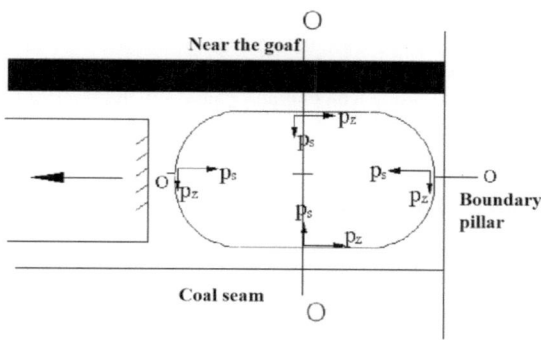

(c) Characterization of subsidence in mining areas

**Fig. 7.4** Development process of bending damage of the overlying rock layer in the extraction zone

of the roof plate of the transfer beam and the support force (or initial support force) of the support will affect the characteristics of the roof plate in terms of the period and step of the incoming pressure, the strength of the incoming pressure, and the incoming pressure pattern. However, the thin or very soft rock roof layer does not have the mechanical properties of the transfer rock beam, and there is no obvious mineral pressure manifestation characteristics and collapse pattern.

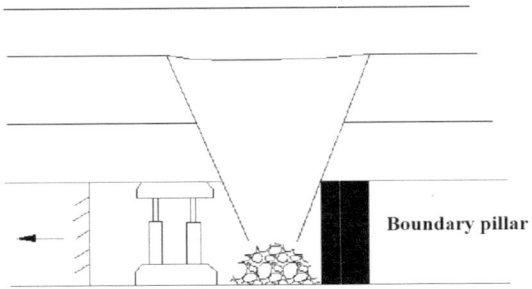

(a) Relationship between initial and cyclic pressure structure

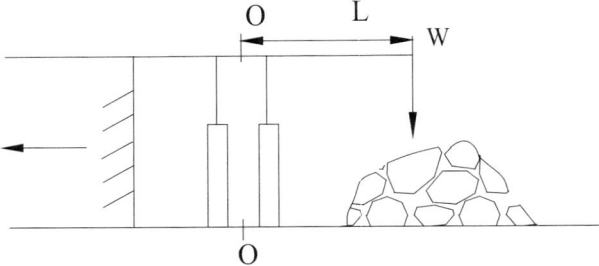

(b) Modeling of the interaction between the top plate of the transfer rock beam and the support

**Fig. 7.5** Schematic diagram of the temporal and spatial structural relationships of the stent roof mining zone

### 7.3.1.1 Top Slab Collapse Phenology

The development of the mining airspace to a certain space, the formation of a large area of exposed roof, due to the role of the space effect, the formation of the overlying rock layer sinking tendency of the movement trend, the influence of the movement trend and tend to develop the movement trajectory was inverted triangle shape, as Figure shown in Fig. 7.6. The height of the inverted triangle is significantly related to the thickness of the mined coal seam, and the openness of the inverted triangle is significantly related to the exposed area of the roof slab. The development of the inverted triangle shape (such as the surface direction of the degree of expansion or openness) is related to the area of the mining area, the strength of the overlying rock layer, whether it is the initial pressure or cycle pressure, this shape of the inverted triangle always exists, and from the bottom up the fracture of the overlying rock layer should be in the inverted triangle, unless extremely special circumstances, such as faults, encountered the influence of large pieces of gangue, the shape of this shape of the triangle morphology Size with the advancement of the working face and the development of changes. Overlying rock layer in the mining area from the center of the shape of the overhanging form layer by layer began to creep and develop to

7.3 Mechanics and Characterization of Subsidence Motion in Overlying ...

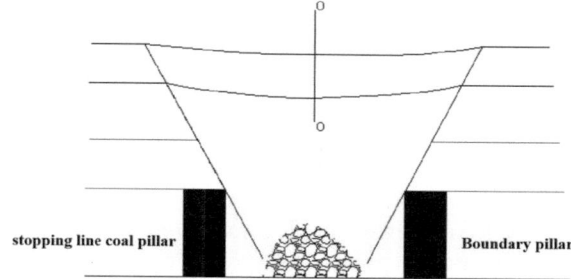

**Fig. 7.6** Inverted triangle of subsidence collapse in airfields

destruction, the basic movement mode of fracture collapse there are two kinds of fracture, namely, bending damage and shear damage.

### 7.3.1.2 Subsidence Motion

In short, the subsidence movement of the overlying rock layer firstly produces the development process of pulling displacement creep to bending damage to the rock layer, and in this process, the subsidence deformation always makes (or always has) a part of the rock layer to reach the critical instability state to collapse. Work face initial pressure to start, with the expansion of the mining area, the roof to the overlying rock layer first from the rock layer of the first layer of the roof (direct top) began to form the displacement movement and gravity concentration of the beginning of the subsidence tendency movement, and gradually evolve and develop the formation of the bottom-up layer by layer involved in the subsidence movement and collapse, as shown in Fig. 7.7. With the advancement of the working face, the overlying rock layer overhanging the exposed area is constantly expanding and began to move, the formation of horizontal movement force Ps and subsidence movement force Pz, under the action of this movement of the combined force P, from the bottom up the rock layer first occurred in the horizontal direction (or laminar direction) extrusion and expansion of the displacement and the next layer pulling on the first layer of the displacement to the centerline O–O, and then by the displacement into the subsidence movement, the formation of the action of P Rock layer bending deformation of subsidence movement, the movement direction for the centerline of the hollow area O–O, rock layer bending subsidence to a certain extent, gradually transformed into P → 0, each rock layer are formed a displacement crack development zone (or crack development area), from the bottom up to the beginning of the varying degrees of each rock layer by layer of the occurrence of the force of the transformation, that is, layer by layer decreasing loss of Ps (the end of the Ps → 0), and the role of the At the same time, each rock layer and layer by layer completely form the role of self-gravity W (at this time Pz → W) and the final formation of the collapse or bending deformation, such as Fig. 7.7b shows the structure of the role of the movement mechanics relationship. In this change process, the rock layer experiencing extrusion and expansion deformation produces a displacement fissure development zone and not affected by the displacement subsidence movement of the boundary between the region of the existence of a demarcation line, on both sides of the demarcation line of the same rock layer to form the performance intensity of the difference in the performance of the front of the line of the performance intensity of the side of the line of the big side (i.e., and the direction of the working surface has been the side of the side of the rock layer of the previous stratum of rock layer of the development of subsidence movement process to play a The author defines it as the stress intensity shear pivot point, as shown in Fig. 7.7c, the strength of the left side of the O–O line is higher than that of the right side, so the left side of the O–O line plays a supportive shear role for the upper rock layer during the process of rock layer movement. In this way, from the bottom up the rock layer development and formation of cracks and fissures, and ultimately in each layer of the stress intensity shear pivot point to form cracks and fissures or deformation or collapse. In the creep subsidence movement of the entire overlying rock layer each rock layer in turn to form the stress intensity shear pivot point, the cracks and fissures should occur in the inverted triangular morphology range line.

The deformation of the rock layer to the fracture collapse firstly occurs in the form center of the rock layer participating in the subsidence movement, that is, the middle part of the rock layer firstly fractures and collapses, and the degree of development of the fracture collapse of the rock layer or the development to the bubble fall is determined by the height of the space of the permissible movement of the lower part of the rock layer. Only when the space height of its lower permitted movement is greater than the sinkable value of the sinking rock layer, the movement of the rock layer will develop from bending and sinking to bubbling and falling. Otherwise, the rock layer will bend and sink until the bottom plate touches the bottom.

## 7.3.2 The Movement of the Overlying Rock Formation and Interaction Between the Working Resistance of the Support

The combined force W imparted to the top plate of the transfer rock beam by the overlying rock formation or the subsidence movement of the ground surface and the resulting moment action are carried entirely by the brace. At present, the bearing capacity of the brace or the force imparted to the top plate by the overlying rock formation is obtained in two ways:

(1) The method of obtaining the resistance value or support strength value of the stent through the calculation method, also known as the method of calculating the self-weight of the rock body, leaves a relatively large margin (unit MN/m$^2$ for the calculation value derived from the calculation method). The support strength value of the stent derived from the calculation method (unit MPa, KN/frame) is more suitable for use in stent selection. Commonly used method of

## 7.3 Mechanics and Characterization of Subsidence Motion in Overlying …

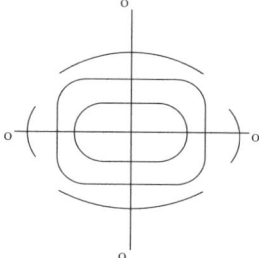

(a) Bottom-up creep modeling of overburden subsidence movement

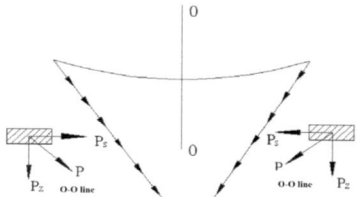

(b) Structure of the elements of the mechanical action of the overburden subsidence movement

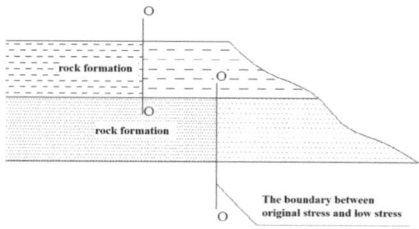

(c) Schematic diagram of the principle of formation of stress-strength shear pivot points in the upper and lower rock layers

**Fig. 7.7** Schematic diagram of the movement and collapse of the overlying rock layers in the extraction zone

calculating the method of multiples of the self-weight of the coal and rock body rated working resistance of the stent or stent support strength is the.

(2) The actual working pressure or bearing capacity of the support is obtained through the monitoring method, and we are accustomed to using the working resistance of the support (the monitoring value can be expressed in KN/frame, or in MPa). The monitoring method is more suitable for the technical management requirements of the support quality and roof safety involved during the production zperiod and analyzes the stability of the roof and stent through the evaluation of the change characteristics of the working resistance. At present, there are a lot of instruments and meters that can be used for stent pressure monitoring, as shown in Table 7.1.

**Table 7.1** Main technical specifications of several types of quarry support monitoring equipment

| Instrument type | | Display mode | Data utilization modalities | Data application | Stability | Working life/Month | Maintenance workload | Topicality | Price-performance ratio |
|---|---|---|---|---|---|---|---|---|---|
| Heddle (device form warp in weaving textiles) pluck classifier for rod-shaped objects, e.g. pens, guns; for army divisions; for songs racks supervisor conjecture | Mechanical Vibration Meter | Pointer | Manual periodic readings | General | Differ from | 2 to 5 | General | Differ from | General |
| | Digital pressure gauge | Liquid crystal | Manual periodic readings | General | Rather or relatively good | 6 to 12 | Few | Differ from | Your (honorific) |
| | Real-time monitoring of monolithic memory loggers | Liquid crystal | Micro-controller record storage, collector wireless transmission of collection data | (Of an unmarried couple) be close | (Of an unmarried couple) be close | 6 to 16 | General | (Of an unmarried couple) be close | Your (honorific) |
| | Computerized on-line monitoring system | Liquid crystal | Ground-based computerized on-line monitoring and data processing | General | General | 6 to 12 | Comparatively large | (Of an unmarried couple) be close | General |
| Alleys Dao (of Daoism) wear by wrapping around (scarf, shawl) rocky be subjected to an aerial bombing, hailstorm etc.) catch sight of in a doorway(old) supervisor conjecture | Mechanical monitoring instruments | Pointer, scale | Manual patrol readings | General | General | 2 to 6 | Few | Differ from | Your (honorific) |
| | Digital monitoring instruments | Liquid crystal | Manual patrol readings | General | Rather or relatively good | 6 to 12 | Few | Differ from | General |

(continued)

## 7.3 Mechanics and Characterization of Subsidence Motion in Overlying …

**Table 7.1** (continued)

| Instrument type | | Display mode | Data utilization modalities | Data application | Stability | Working life/Month | Maintenance workload | Topicality | Price-performance ratio |
|---|---|---|---|---|---|---|---|---|---|
| | Real-time monitoring of monolithic memory loggers | Liquid crystal | Micro-controller record storage, collector infrared wireless data collection | (Of an unmarried couple) be close | (Of an unmarried couple) be close | 6 to 16 | General | (Of an unmarried couple) be close | Your (honorific) |
| | Computerized on-line monitoring system | Liquid crystal | Ground-based computerized on-line monitoring and data processing | (Of an unmarried couple) be close | General | 6 to 12 | Comparatively large | (Of an unmarried couple) be close | General |

Such as YHY60 type pressure monitoring and recording instrument, applying the principle of single-chip computer storage and recording pressure monitoring and recording instrument (commonly known as infrared type pressure monitoring extension), as well as KJ12 type on-line monitoring and computer communication system for comprehensive mining stent, these two types of monitoring instrumentation collection density is large, 5 min collection, storage and recording of data once, continuous monitoring cycle is long, can realize the purpose of real-time on-line monitoring. It can complete the real-time on-line the initial bracing force of and the working resistance monitoring of, and the hydraulic stent analyze the working resistance curve to understand the top plate pressure step by step, the cycle of the pressure to show the law and master to understand the top plate of the mining area information about the collapse of. The development of higher level is to achieve the use of monitoring data information for early warning of roof pressure disaster, for engineering management to provide such as large area of roof pressure anomaly information or working face close to the roof fault and other early warning information instructions.

#### 7.3.2.1 Bracket Pressure Curve-Top Plate Coming Compression Strength Classification

For a long time, the working face top plate to pressure characteristics of the law is through the monitoring of the hydraulic bracket work resistance to reflect and realize, the top plate to give the bracket pressure strength or size of the mining area by the overlying rock layer movement and deformation of to the top plate (direct top) pressure, such as thick and hard top plate shows the physical and mechanical properties of the characterization of significant, such as elasticity properties, easy or not easy to collapse, and so on. Currently used top plate to pressure monitoring curve coordinate mode shown in Fig. 7.8a–c, through the curve analysis to understand the top plate at a certain moment of the pressure value or a stage of the process of the pressure change rule and change characteristics, to evaluate whether the data information is in a reasonable range of intervals. In the project, through continuous or a stage process of real-time monitoring of the stent the bearing capacity to obtain the top plate to pressure strength information and the top plate cycle to pressure manifestation (including the initial pressure step) law, the stent work resistance, the initial bracing force and other parameters, in order to meet the needs of the engineering and technical management (but also general routine production requirements).

Through the systematic analysis and summary of the accumulated tens of thousands of on-site monitoring data, it is found that in the production practice, the variation amplitude and frequency characteristics of the working resistance (pressure) curve of the fully mechanized mining support can be roughly divided into three types as shown in Fig. 7.8: three types of curves: dramatic change, moderate dramatic change and no dramatic change. Based on the 'transfer rock beam' theory of academician Song Zhenqi (overlying strata or roof of goaf) and the summary and analysis of the characterization of physical and mechanical properties of coal and rock mass (properties, stiffness, hardness and thickness, etc.), it is proposed that the

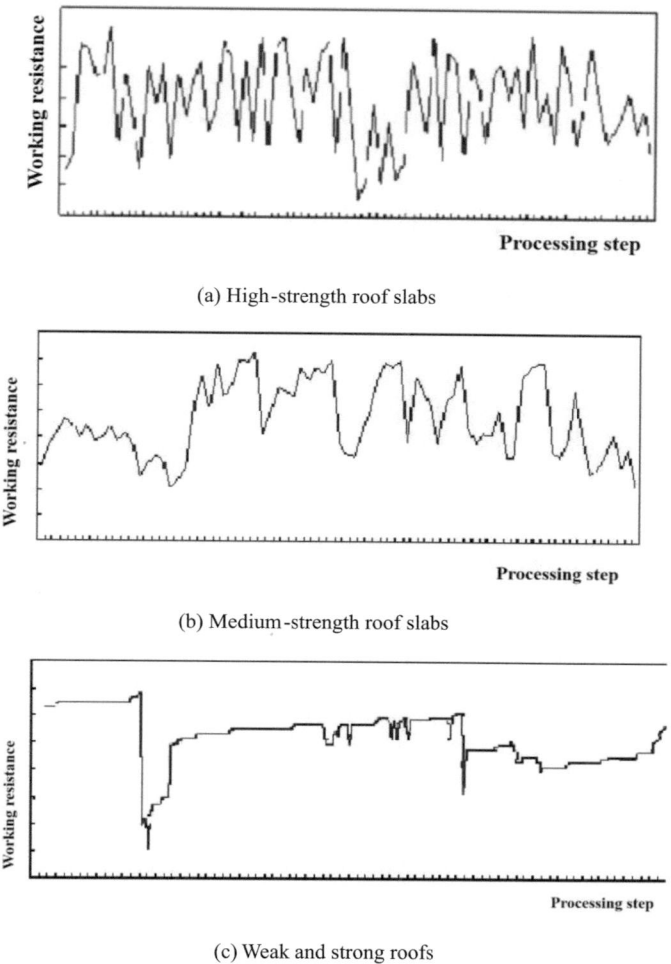

**Fig. 7.8** Pressure monitoring curves of types of roofs supports3

working resistance of the support and the roof pressure curve are the overall characterization of the movement mechanics of the overlying strata on the roof of the stope. It is the result of comprehensive reflection under the mechanical state of complex rock strata movement. The working resistance (pressure) curve law and development trend of the support are the comprehensive reflection of the mechanical characteristics of the overlying strata movement in the goaf or the mapping of the signs of the mechanical action characteristic model. Therefore, the support resistance curve characterization fully reflects the characteristics and laws of the roof pressure behavior during the support propulsion process. If the goaf, support and roof are regarded as the same space–time structure coordinates, and the support pressure monitoring curve is classified according to the roof type (mechanical and physical performance characterization), the roof strength of the overlying strata can be divided into three categories: high strength transfer rock beam roof (referred to as high strength roof), medium strength transfer rock beam roof (referred to as medium strength roof), weak strength transfer rock beam roof (referred to as weak strong roof), as shown in Fig. 7.8.

① High-strength roof plate and bracket are rigid (elastic) coupled to touch the roof, presenting elasticity characteristics of touching the roof is very obvious, easy to appear to come to the pressure cycle and step distance is too long, to come to the pressure of intense and difficult to collapse the roof plate phenomenon, often can hear the roof plate coming to the pressure of the sound of the fracture, the bracket is easy to be placed in an unstable movement situation, the emergence of abnormal coming to the pressure of phenomena, the engineering practice of corresponding to the thick and hard top of the roof plate coming to the pressure of the curve characteristics of the roof plate (or the top of the direct top) collapse in the mining area. The roof (or direct roof) of the mining area collapses showing the phenomenon of accumulation of large rock layers,

② The medium-strength roof plate and the stent are medium coupled to touch the roof, and the stent operates stably with satisfactory support effect. The collapse of the roof plate in the mining area shows the phenomenon of accumulation of large lumps of rock layers. Weak and strong type roof plate and stent are coupled to the roof the roof plate has almost no periodic pressure, the roof plate collapses with the advancement of the stent, the stent is little affected by the pressure of the roof plate, and the operation is stable, but the sometimes for the loose and easily broken roof plate support effect is unsatisfactory, and the working face is prone to the phenomenon of roof falling into the gangue. The hollow area The roof collapsing in is a phenomenon of accumulation of small pieces of rock layers, and the distance of roof control in the working face should be small rather than large.

### 7.3.2.2 Mechanical Structural Modeling of Bracket and Roof Plate-Leverage Principle

If the thick and hard roof plate commonly seen in engineering is regarded as an elastic plate and beam structure with sufficient shear strength, and the support, roof plate and mining area are regarded as same variables in the coordinate system, then they can be by Theas spatial–temporal mechanical structure of the roof plate, support expressed and described the roof plate and the support, mechanical model in and mining area relationship between. After ignoring the restraining effect of the coal pillar on the top plate in the roadway, using the three elements of force in physics shown in Fig. 7.9 law, we know that the support force given to the top plate by the stent point of or surface Ph (which is also the touching point between the stent and the top plate or surface) is Figure the O point in Fig. 7.9. Based on the academician Song Zhen qi on the study and research of theory of "transferring rock beam "centered on

## 7.3 Mechanics and Characterization of Subsidence Motion in Overlying …

the movement of the rock layer overlying the mining field, the authors believe that the core of the theory of transferring rock beam is the theory of "movement and change "of the roof plate, which reveals that the "movement and change "of the roof plate is reflected in the process of "movement and change". And change "of and change "of the roof slab The author, which reveals the characteristics of the rock beam reflected in the, and believes that the core of the theory is the theory of "movement slab we can understand that the comprehensive mechanical elements of the movement of the overlying rock layer are reflected in the performance characteristics and results through the transfer of the rock beam, therefore, from the mechanical point of view, this rock beam characteristic also represents the performance characteristics of the overlying rock layer, and therefore it can be the roof process of "movement defined as the characteristics and performance of the plate and beam of the roof slab of the transfer of the rock beam type. So far, the top plate of the mining face is regarded as a large-scale top plate rock beam structure set in the coal rock body at one end (such as point B in the figure) and has the rock beam feature, and the zigzag arrangement of brackets supports the large-scale plate beam structure, which gives a kind of structural-mechanical relationship shown in Fig. 7.9 and better reflects the in the dynamic in the process of the top plate coming to be pressurized extremely subtle classical mechanical principle-lever principle-.roof plate is revealed by the lever principle embedded structure spatial and temporal scale The structural-mechanical model of the, and the model is dynamic and has the time limit (effect) of "movement and change "in space and time. The support's, the support and the mining area roof touching pivot point (or face) is at point O; end A is the free end of the plate and beam; end B is the non-free end, which is in the deep part of the coal body above the front; then the gravity moment effect generated by end A is subject to the constraints of the pivot point O and end B. Moment of gravity generated by end A is $Q = W \times AO$, and moment of counter-gravity generated by end B is $q = P \times OB$ (here, W represents the vertical combined force, i.e. gravity, mining hollow area induced surface subsidence exerted to the roof plate in the process of the (Here W represents the vertical force exerted on the roof plate by the process of induced surface subsidence (movement of the overlying rock layer) in the mining hollow zone, i.e. gravity, and P represents the support pressure of the coal wall at the mining face. The moment effects of moments Q and q are dynamic (movement of the overburden rock layer) and interactive, varying with the cyclic pressure pattern of the roof plate.

Using "the academic point of view the theory of practical mine pressure", the roof slab pressure fracture collapse is divided into normal creep incremental pressure area (step) OA and abnormal leap change pressure area OB. The supporting force of the support for the roof slab is at point O, and the supporting force of the support at point O Ph produces the shear damage to the roof slab, so point O is the critical state demarcation point of the substantial roof slab pressure fracture collapse. Therefore, point O is the substantial roof plate to pressure fracture collapse demarcation point. The area to the left of point O of the critical state of the OB is defined as the danger zone of abnormal roof fracture and the area to the right OA is defined as the area of normal roof fracture. Once the abnormal pressure area is presented, the roof plate or the direct roof and the old roof lose the role of lateral binding force, and the

**Fig. 7.9** Principal model of brace and roof leverage

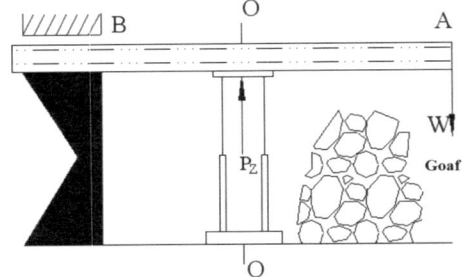

load force borne by the stent is several times or dozens of times more than that of the normal load coming under pressure. At this moment, the stent is in a fleeting unstable state, and if effective cannot be taken in tempered-control methods, large-area roof pressure may occur damage. During the advancing process of the working face, the roof plate of the overlying rock layer interprets a "movement and change "spatial structural movement mechanics mode, and the which is a fusion of process and phenomenon of revealing and applying this structural mechanics mode's creep law and leaping change characteristics of spatial and temporal scales have important engineering application value.

### 7.3.2.3 Bracket and Roof Space Structure and Dynamic Characteristics

Based on a large number of monitoring the analytical research and theoretical accounting of curve law, the following views and understandings are put forward: working resistance the working face bracket

(1) From the pressure step and engineering management considerations, the bracket and the roof (direct top or direct top + old top) can be used in Fig. 7.10 shows the relationship between the, Figure parameters of the spatial structural mechanics Fig. 7.10a indicates that the roof collapsed before, that is, before the pressure before the maximum step (or the roof plate-transfer of rock beams scale) $L_{max}$ and bracket of the maximum support force $Pz_{max}$, the bracket is subjected to the maximum pressure at this moment, to reach the critical state (or critical point) of roof instability and collapse. Figure 7.10b shows that the collapse of the roof plate after, that is, just after the completion of a pressure cycle of the roof plate rock beam scale Lmin and bracket support force Pzmin, at this moment, after the collapse of the roof plate after the bracket to withstand the pressure of the minimum.

Based on the monitoring summary analysis, it is found that where the pressure given to the brace by the roof slab (direct roof or direct roof + old roof) and the length L of the roof rock beams are in the normal creep incremental condition, the fracture zone (point) of the roof slab (direct roof) and the step L of the incoming pressure are the data disposed in the OA interval. When L increases gradually (in length) with

## 7.3 Mechanics and Characterization of Subsidence Motion in Overlying …

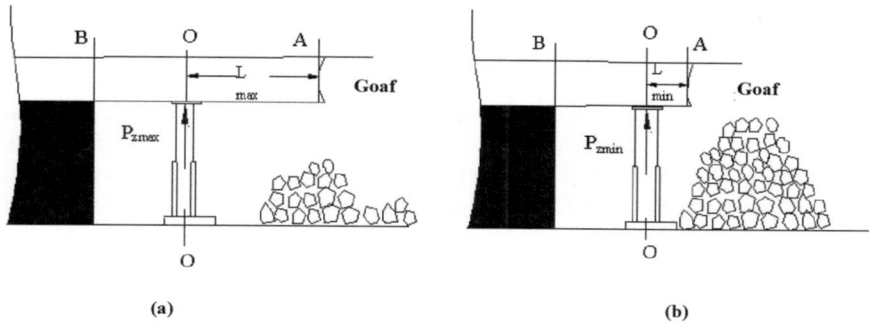

**Fig. 7.10** Schematic diagram of the relationship between the bracket and the top plate coming under pressure

the advancement of the brace, i.e., when L → Lmax, Pz → Pzmax, and when the overlying rock layer imparts gravity to the roof slab, W → Wmax, the creep of each structural-mechanical parameter develops to a state of extremity, i.e., the critical state of instability and collapse. At O–O, when, W → Wmaxthe bracing support force forms a great shear effect, and in most (normal incoming pressure) cases, the roof slab fracture and collapse occur at O–O. The creep of L accumulates and eventually reaches the critical state of instability, leading to the maximum deflection value of the roof slab at O–O or the roof slab subjected to the Maximum shear damage. Normal or most of the roof slab (direct roof or direct roof + old roof) pressure fracture damage and collapse occurs on the right side of the O–O line, i.e., L → when Lmin.

(2) If an in OB area to the left of the O–O center line (i.e., longitudinal, which is close to or greater than the heights of the two zones of the mining area rift or fracture tectonic situation, the author will call it an abnormal area or OB area. Based on occurs the upper rock layer of the roof plate in the area of) the working face's roof-control distance safety considerations, the author calls it an abnormal area or OB area, and the area of roof pressure is expanded from OA area to AB area, which is said to be the area of. Induced large-area roof pressure At this time, the stent and the roof plate are in a critical extreme instability state of, and the load borne by the stent (the coming pressure curve) has obvious characteristics of dynamic load impact, and the jittery the coming pressure curve change of continues to be elevated in a period of time (e.g., 1 day or 2 days), and it is very likely that the stent will fall over and be damaged over. A large area if it continues to develop Usually, there are the following three objective or subjective factors that can cause (or appear) abnormal situation: ① the initial support force of multi-bracket is insufficient (or multi-bracket leakage, fluid tampering, or unreasonable bracket selection or coal bed inclination angle), resulting in insufficient effect of the bracket touching the top, inducing the imbalance of the top plate to pressure; ② the overlying rock layer of is affected by man-made damage (such as water seepage belt in the upper mining area,

affected by man-made (such as water seepage zone of upper mining area, O-B area man-made damage to damage man-made damage to the top plate the roof plate, etc.); ③ O-B area is, (etc. the roof of overlying rock layer (along the direction of the direct top and the old top) Once occur in the deep cracks (gaps), including various tectonic structures and forward and reverse faults, there be a stent will steep increase in the curve of the or a continuous rise in the phenomenon work resistance. It is possible that the stent will be subjected to the impact load pressure, and of several times or more than the normal condition the roof plate the change of the situation will be so fast that the stent pressure safety unloading valve will not be able to be opened in time before the stent will be damaged or collapsed, and so on.

(3) In the O-A area on the right side of the O–O middle line, the roof pressure creep law is obvious, which belongs to the expected pressure law (pressure step) of engineering technology management. The size of the L value is related to the physical properties of the roof (i.e., the transfer rock beam or the overlying strata). The hard roof shows the obvious rigidity and toughness characteristics of the transfer rock beam. From the curve of Fig. 7.8a, it can be seen that the support and the roof show obvious elastic contact characteristics. In the O-A region, when $L \to L_{max}$, then $Q \to Q_{max}$, $G \to G_{max}$ (G represents the shear force generated by the support to the roof, Q represents the gravity moment), with Fig. 7.11 shows and describes its mechanical structure situation at this time. The size of Q value directly affects the law of roof caving, and its torque effect and influencing factors on roof caving have multiple characteristics. The G value is affected by the support force of the support, and the reasonable and sufficient G value is also one of the factors to ensure the timely and stable pressure step. Reasonable and sufficient G value is the support quality management requirements in the support must have sufficient initial support force.

(4) When $L \to L_{min}$, the top plate or direct top (or direct top + old top) collapses, i.e., when the top plate completes a pressure cycle (after the top plate collapses), coal body the maximum value of the pressure is the closest to the coal wall of the working face, and the maximum value is mostly in the range of 2–3 m from the coal wall supporting, which indicates that the top plate comes to the pressure to affect the change of the coal body supporting pressure of the working face and changes in a regular manner. When $L \to L_{max}$, when the direct top (or direct top or old top) is about to collapse, the maximum value of the support

**Fig. 7.11** Schematic diagram of the mechanical structure of the bracket and the top plate

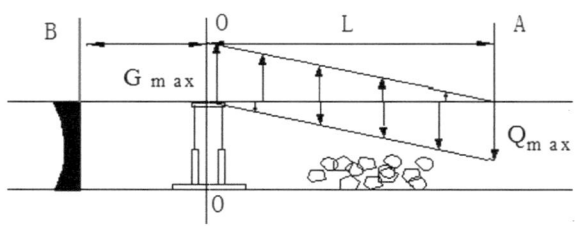

pressure of the coal body in front of the working face is mostly in the range of 3-6m from the coal wall of the working face, or even in a larger range, which is mostly presented in the high-strength or middle-strength type of the roof rock layer.

### 7.3.3 Integrated Mining Face Roof Plate to Pressure Law and Hollow Area Subsidence Collapse Characterization Appearance

Based on the analysis and study of the mechanical structure and mechanical action elements of the support and roof plate of the synthesized mining face, especially the analysis and discussion of the analysis of the action elements of the initial support force of the synthesized mining support (i.e., the support force), the extrusion and shear action on the roof plate of the support force Pz is determined, combined with the analysis of the subsidence movement and subsidence deformation of the overlying rock layer of the hollow area with the characterization of the phenological appearance of the incomplete fracture and the complete fracture and the analysis of the shear support of the stress intensity. Proposed with the analysis of the principle of shear action. The following conclusions or opinions are proposed and given:

(1) Under normal conditions, since the strength of the support force of the integrated mining support is sufficiently strong, the fracture or incomplete fracture of the top rock strata below the height of the three belts is unlikely to occur within the range of the top plate of the working face or within the range of the front of the support, as well as the rock strata in the upper part of the coal body (seam).
(2) In the working face of single hydraulic pillar, it is extremely difficult to ensure that the overall support strength of the working face is sufficient, and the support strength of the peripheral rock roof is insufficient, especially the low and insufficient control and control ability of the roof of the working face, which is extremely unsuitable for the development of the high-yield and large-scale quarry. Therefore, it often happens that the overall support force is insufficient, which triggers the fracture or fissure of the upper roof rock layer within the working face, and abnormal pressure phenomenon occurs.
(3) The descriptions and calculation methods of the amount of deformation of the rock layer subsidence given by the stent to the roof in some books on mining pressure are not suitable to be applied to the working face of generalized mining. The Eq. for calculating the amount of rock layer departure S overlying the mining airspace, or the amount of bracket-given departure subsidence S, is more suitable for the calculation of the roof slab of the working face of a monolithic column, but not for the calculation of the amount of bracket-given departure S of the generalized mining working face.
(4) Under normal conditions, the range of fracture of subsidence deformation of the rock layer overlying the mining area should be on the two waist side lines of

the inverted triangle, as shown in Figs. 7.6 and 7.7 to characterize the phenotype and the principle of subsidence movement.

## 7.4 Characteristics and Technical Management of Shallow Buried Mining Roof to Pressure Manifestation

The so-called shallow buried layer refers to the mining of coal seam outcrops from the surface height (distance) is very small, due to the role of the gullies buried layer thickness varies, usually less than 300 m, most of the coal seam outcrops in the range of 100 m to 250 m. The main refers to the last decade or so, represented by Shenhua's modern mines with advanced equipment. In mining, completely using integrated mechanized coal mining methods; mine to the quarry full development using trackless car transport, long wall large mining height, large roadway and other large-scale spatial effects brought about by creep instability phenomenon is obvious. This development and change characteristics to China's northwest, inner Mongolia region of the shaft mining coal seam as a representative of the mining area. From the point of view of mine pressure manifestation and pressure law, the movement of the rock layer overlying the mining area is often traction surface of the unstable movement or vibration, the performance characteristics of the characteristics of mountainous hills and the top plate to the pressure law of the extremely unstable gully effect. Shallow buried seam mining roof pressure is very special, although the roof pressure has regularity, the overlying rock layer has the performance characteristics of transfer rock beam, but compared with the deep coal seam mining, even if it is a as Fig. 7.8 (medium-strength transfer rock beam type roof shown in, the shallow buried seam mining pressure is very intense, the roof pressure process has great dynamic load impact, the bedrock layer under the loose soil layer on the ground surface is easy to be affected by the mining impacts and The bedrock layer under the loose soil layer on the surface is very easy to be affected by the mining and sink and collapse (there is even no obvious bedrock layer in the local area). The technical requirements for support equipment and management are extremely challenging. After summarizing and analyzing the relevant technical data, we summarize the experience of mine pressure management and the characteristics of mine pressure manifestation in the following aspects.)

(1) In shallow buried seam mining, it is not suitable to directly adopt deep well calculation method to calculate the supporting strength when selecting the type of comprehensive mining support, to predict the working resistance of the support and the value of the mineral pressure of the overlying rock layer and the initial support force of the support, and it is better to design and apply the special support which has a large specific pressure of top and bottom plates, and to reasonably determine the relationship between the supporting strength and the specific pressure of the top and bottom plates;

## 7.4 Characteristics and Technical Management of Shallow Buried Mining …

(2) synthesized mining the monitoring bearing pressure curve of the stent, i.e., the work resistance curve is characterized by the same law as shown in Fig. 7.8a, with high amplitude and frequency of pressure change, large dynamic load coefficient and unstable phenomenon, and unsatisfactory coupling of the stent and the roof plate touching the roof;

(3) Poor stability of the bracket, even if the bracket selection leaves a considerable support strength calculation margin, there are often unstable phenomena, the top plate to pressure strength and show characteristics are not fully estimated;

(4) Since the corresponding mining pressure management experience is little, it is reasonable and effective way and means to predict and test the support strength of the support by strengthening the monitoring efforts, mastering and applying the method of monitoring the working resistance of the synthesized mining support to realize the prediction and forecasting of the law of the roof plate coming under pressure;

(5) Shallow buried coal seams are mostly thick coal seams, with large area of coal seams and relatively good geological conditions, and most of them are mined by large-scale structural working face, so the subsidence movement of overlying rock strata in the air space of shallow buried mining face is in essence the subsidence movement of all rock strata including loose soil layer on the ground surface, and the height of "three belts "in the influence range of rock strata collapse is very large, so it is very easy to produce or form abnormal situation of large area of roof plate coming under pressure. The height of the "three zones "in the affected area of rock fall is large, and it is very easy to generate or form an abnormal situation of large-area top plate pressure, so it is more difficult to control the stability of the stent, and the stent is prone to shaking or large-area collapsing and falling of the stent, which is especially significant in the process of top plate pressure.

(6) Shallow buried layer mining pressure law and pressure to show the characteristics of the surface is susceptible to changes in geomorphological changes, the working face to advance to the surface of the valley near the gully and subsidence zone should pay special attention to the top plate pressure changes, timely grasp and control of the operation of the support status, sometimes as deep wells to be used as the face of the work over the fault as the same technical measures.

In short, from the perspective of safety production management, we do not need to pay too much attention to analyze the complexity and diversity of the overlying rock layer on the roof plate is which kind of rock layer composition and scale. Research has shown that through real-time monitoring and mastering the change rule of the working resistance of the stent, mastering and recognizing the normal or abnormal development of the stent bearing pressure, we can completely realize the technical goal of predicting and warning of the disaster such as the large area of the roof plate in the working face coming to the pressure. Grasp and understand the performance characteristics reflected by the overlying rock layer, analyze and summarize the coming pressure law curve of the supporting body, rock layer and roof plate, and focus on analyzing and summarizing the occurrence phenomenon and information of the coming pressure law and abnormal data (normal and abnormal), which is effective in actively preventing the damage caused by abnormal coming pressure of the roof plate.

## 7.5 Influence of Filling of the Mining Airspace on the Pattern of Coming Pressure on the Roof Plate

The purpose of filling the mining area is to minimize the deformation and subsidence movement of the overlying rock layer, which leads to the formation of surface subsidence and fissures and cracks, as well as the consequent damage to the environment and ecology. Currently applied filling materials and techniques cannot guarantee a 100%subsidence reduction rate. Therefore, the role of filling can achieve to slow down and reduce the degree of subsidence, reduce the surface deformation and fissure cracks brought about by the destruction of the problem, as well as mining earthquake and other disasters. Therefore, the development and promotion of filling engineering and technology in the hollow area should be strengthened, especially the long-wall and large-height mining method and similar principles of mining.

Since the current filling method of the hollow area cannot guarantee the substantial sinking reduction rate to reach 100%sinking reduction effect, the existing filling technology has little effect on the top plate coming pressure law, and it cannot produce the effect of changing or favoring the operation and stability of the support. If the filling material with volume expansion characteristic can be developed, it will produce sufficient filling effect. This will have a strong supporting effect on the overlying rock layer of the hollow area and 100%subsidence reduction effect, which will completely solve the contradictory problems of mining and surface damage.

## 7.6 Determination of the Strength of the Top Plate to Pressure and the Selection of the Synthesized Mining Bracket

At present, the support selection is mainly based on the rock mass weight method to determine the support strength. The support strength can also be determined according to the roof and floor conditions, such as the compressive strength of the roof and floor, and the accumulated experience of mine pressure management technology. The following conclusions and understandings are put forward on the accumulated experience and research experience.

## 7.6.1 Strength of the Top Plate Coming Under Pressure and the Strength of the Support of the Mining Support

(1) China's synthesized mining support has been moving towards the stage of accumulating experience and perfecting and maturing development from production to application as a whole. However, the research results of mine pressure theory still follow some mine pressure research results and some evaluation standards of monolithic pillar working face to a different extent. The mine pressure law and manifestation characteristics of the synthesized mining face, as well as the top plate and the overlying rock layer subsidence movement law characteristics are different from the monolithic column support mine pressure phenomenon, for example, the roof of the synthesized mining face sometimes appeared to have no cycle to pressure phenomenon of the face, the top and bottom of the plate approaching the amount of the relationship between the strength of the support curve is not convenient to monitor and utilize (and the monolithic column face can be). When the initial bracing force or initial bracing support strength of the synthesized mining strut is sufficient, under normal circumstances, the rock layer on the roof of the working face, i.e., the rock layer on the roof of the strut support force touching the roof area to the range of coal seams, cannot undergo the phenomenon of off-layer fracture, and the bracing force of the strut touching the roof produces a great shear effect on the roof plate. It is extremely difficult for the average working resistance of monocoque pillar to reach 3000–5000KN.

(2) From a macroscopic point of view, the degree of stability of the roof plate of the working face, the economic indicators and the working condition of the stent are the evaluation indexes or good or bad signs of the reasonableness of the initial bracing force or working resistance of the stent. If applying simple, easy to measure, comparable and accurate is the basic principle of selecting the type of stent for comprehensive mining, then the change of pressure (working resistance) of the stent for comprehensive mining is the best to be used as the evaluation mark. Other information phenomena, such as the amount of top and bottom plate migration, the degree of top plate crushing, the deformation and subsidence characteristics, the stability of the stent, and other phenomena have the problem of inconvenient operation and utilization.

(3) There are many methods for determining the working resistance of the synthesized mining support, but the self-weight calculation method and monitoring method of the rock body are the best. The self-weight calculation method is more in-depth and mature, and the model of combining theoretical analysis and actual measurement data is more practical and reliable. The other methods are mostly based on the observation data of mining pressure, and have similar characteristic principles.

(4) The initial bracing force of the stent, the strength of the rock layer on the roof (often the strength reflected in its thickness) and the strength of the incoming pressure play an important role in the size of the stent's working resistance, sometimes the direct roof plays a role, and more often it is the old roof that

plays a key role. Therefore, when considering the expected initial bracing force and working resistance of the stent, it is necessary to take into account the possible incoming pressure pattern and manifestation characteristics of the roof rock layer. It is even necessary to consider some abnormal pressure situation (change of lithology and strength of roof rock layer).

The reasonableness of the synthesized mining stent selection has a direct impact on the stability of the roof, the stent selection conforms to the movement of the rock layer of the roof mechanical structure relationship, then the stent can adequately control and harness the dynamic characteristics of the roof plate and the law of change. If the stent selection is not reasonable, the support strength is small, and cannot produce sufficient support force to the top plate rock layer, it will cause unstable top plate of the working face, and there will be untimely off-layer, fracture, and difficulty in over-supporting the roadway. It will induce the top plate rock layer of the mining area: direct top, or direct top + old top, and the unstable coming pressure of the high-level rock layer, and the phenomenon of off-layer sinking, high pressure, and even the phenomenon of power impact, The top plate comes to the pressure fracture irregular characteristics, the appearance of untimely time and space comes to the pressure law and the appearance of characteristics, the bracket is not stable and other unsafe phenomena (early monolithic column working face common phenomenon). The following is a brief summary of the two basic methods.

### 7.6.1.1 Calculation Method of Self-weight of Rock Body to Determine the Strength of Support and Working Resistance of the Support

Our country and many coal mining countries in the world have widely used the rock body self-weight method to calculate the support strength, although the specific algorithm parameters are different, but the principle is similar. Comprehensive many algorithms at home and abroad, combined with the classification type of the strength of the roof rock layer, tendency to summarize the proposed use of the following Eq. to determine the rated support strength of the bracket:

$$P_Z = Q_n \times M \times \gamma \qquad (7.1)$$

where Pz-stent rated support strength, KN/m$^2$;

M-Mining height, m;

$\gamma$-Top rock layer rock mass, KN/m$^3$;

Qn-Multiple times of height.

Qn mining high multiplier, the use of simple, easy to operate. However, the stability of the rock layer is different under the condition of different physical properties (strength) of the rock layer in use. Therefore, the actual use of mining height multiplier, in determining the selection should also be considered to reflect the physical properties of the roof rock layer, in short, to consider the cohesion of the rock layer, strength, elastic deformation modulus and other factors, as well as the coal seam expansion coefficient (or relaxation coefficient), and at the same time to be determined through the combination of experience and considerations.

Internationally, different countries have different application principles to consider based on the different conditions of coal seams (e.g., plains vs. Hills, depth, general thickness of coal seams, etc.), such as: In the former Soviet Union, Qn = 6–8, Germany takes Qn = 12, India, Japan take Qn = 5, The United States takes Qn = 16.

In addition to the application of Eq. (7.1) for calculating the support strength of stent, the above countries have also researched and summarized some other calculation methods with characteristic features (such as the support strength of a special type of stent). At present, there is no unified standard specification in this field in China. However, it also adopts the method of calculating the self-weight of the rock body, and the project generally tends to adopt: the top plate rock layer without obvious pressure cycle, i.e., weak and strong type 3 types of top plate, take Qn = 6~8; for the cycle of obvious pressure, the top plate rock layer is stable, the middle and strong type 2 types of top plate take Qn = 9~11; for the obvious strength of the thick and hard top plate rock layer, i.e., the high strength type 1 type of top plate take Qn ≥ 11.

The support strength determined by the self-weight calculation method of the roof strata in Eq. (7.1) is the pressure value of the rock mass, which is the rated support strength relative to the support. If the Pz value is converted into the working resistance value of the support of the specific frame type, the support area of the support should be specified. Using the basic data such as Pz value, working resistance and coal seam thickness, the specific frame type can be selected and determined, as well as related supporting equipment.

In addition, in practice, sometimes through the summary and analysis of a large number of support pressure (working resistance) monitoring data, we have obtained the average working resistance value (representative) of the support generated by the roof strata in this area, or the critical working resistance value Pzj, then the rated working resistance value of the support can be calculated according to the stability coefficient Kn of the roof strength classification, or the stability and rationality of the support in the work can be evaluated and analyzed. The determination and selection principle of Kn is: three types of weak and strong roof strata, Kn = 1.05 ~ 1.15; 2 types of medium-strong roof strata, Kn = 1.15 ~ 1.25; 1 type of high-strength roof rock, Kn = 1.25 ~ 1.35.

### 7.6.1.2 Approved Calculation Method for Stent Parameters

The calculation process of applying the stent parameter approval calculation method is to approve the calculation of the support strength calculated by the method of Eq. (7.1) according to the parameters of the determined type and specification of the synthesized mining stent. Usually, the parameters of the stent have been determined: rated working resistance, KN; support strength of the stent type, MPa or KN/stent and geometric scale, height and so on. It is important to note that the value of the support strength calculated by Eqs. (7.7–7.1) represents the value of the pressure formed by the roof plate to the upper rock layer. Comparative analysis should be made such as the support area of the bracket and other numerical conversion, while paying attention to the rated value and the maximum value, as well as the unity of the unit.

Example illustration, basic information: a mine coal seam thickness of 5.5 m, the rated support strength of the whole frame of the existing bracket frame type is 110t/m², the working resistance is 900t, and the rock mass capacity of the roof rock layer is 2.5t/m³. Others are summarized.

Pz = (6 ~ 8) × 5.5 × 2.5 = (82.5 ~ 110) t/m², take the rated support strength Pz = 1100KN/m², or Pz = 1. 1 MPa. Considering that the top rock layer is mostly sandstone, the inclination angle of the coal seam is 2° ~ 5° (the local deflection zone is 12° ~ 15°), and the direct top is relatively stable. The analysis determines that the existing frame type can continue to be used.

### 7.6.1.3 Calculation Method Based on the Cyclic Pressure Principle

Former Soviet Union scholar A. E. Satzonov once thought that the fully mechanized mining face was divided into two cases: non-periodic weighting phenomenon and periodic weighting phenomenon with the participation of the main roof, as shown in Fig. 7.12, Figure (a) indicates the principle of the load pressure action of the stent when there is no cycle to pressure on the top plate under the influence of the old top, and Figure (b) indicates the principle of the calculation of the cycle to pressure under the influence of the old top. In recent years, the authors through analysis and exchange with field engineers and technicians also found that there are some mining area mines exist no cycle to pressure phenomenon, this situation is the stent advance process, the direct top that is, after the collapse, that is, the stent while advancing the roof collapsed. Combined with the summary of the Soviet scholars, the calculation method for the two cases is given.

(1) When there is no periodic pressure, the calculation formula of working resistance of fully mechanized mining support is:

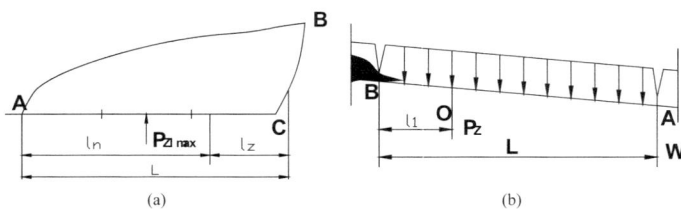

**Fig. 7.12** Principle of calculation of working resistance of stent without and with cyclic coming pressure

## 7.6 Determination of the Strength of the Top Plate to Pressure …

$$P_{Zlmax} = \gamma \cdot a \cdot S (\text{kN/frame}) \quad (7.2)$$

where $\gamma$-top rock layer capacity, kN/rack;

a-Width of bracket support, m;

S-Fig. 7.12a Area inside curve ABC, m².

$$S = \int_{l_2}^{l_0} l_n K(l+\lambda) d_r$$

$$l_0 = l_n + l_2$$

where $l_n$-Maximum support width of the support;

$l_2$-The length of the overhang above the top surface of the bracket.

What is supplemented here is the minimum value of the calculated value of the area S, that is, the lower limit cannot be less than the actual measured area of the support. Therefore, here S can be directly calculated by the length and width of the bracket.

(2) Under the condition of cyclic incoming pressure, the working resistance of the synthesized mining support is calculated as

$$P_z = P_{Z1\,max} + P_q$$

where $P_q$-The extreme value additional load produced by the main roof.

The total moment to point O will be equal to zero:

$$\frac{1}{2}l_1^2 \gamma h\alpha - \frac{1}{2}\alpha\gamma h(L-l_1)^2 + W(L-l_1) = 0$$

where $l_1$-Distance from the coal seam to the support to apply the equilibrium force, m;

h-Thickness of the old roof collapse rock layer, m;

L-step of old roof collapse under the given deformation condition, m;

$\alpha$-width of the support, m; -width of the bracket support, m;

W-Reaction force of the collapsed rock layer on the old top rock layer (or transferring rock beam) at the moment of the critical point (here the critical point represents a point of spatial–temporal extremes), kN;

$P_q$-Extreme force exerted on the brace by the old top rock layer at the moment of the critical point, KN.

For the equilibrium of the forces acting on the rock beam at the moment of the critical point:

$$W + P_q - \gamma h_0 \alpha L = 0$$

Solving the above equation gives

$$P_q = \gamma h_0 \alpha L \left[ 1 - \frac{L - 2l_1}{2(L - l_1)} \right]$$

At the time of the old top coming under pressure, the bracket working resistance is

$$P_Z = \gamma \alpha \xi \int_0^{l_0} l_n(1 + \lambda) + h_0 \cdot L \left[ 1 - \frac{L - 2l_1}{2(L - l_1)} \right] \quad (7.3)$$

The example of the Kuzbass mine in the USSR was used uniformly over a period of time:

$$P_{Z1\,max} = 0.30 \sim 0.40 \text{ MN/m}^2$$
$$P_Z = 0.60 \sim 0.65 \text{ MN/m}^2$$

It is used as a guideline for selecting the type of stent for integrated mining, which is beneficial to the production and technical management of the stent manufacturer.

### 7.6.1.4 Stent Resistance Monitoring Method

The basis of the bracket resistance monitoring method is to base on collapse zone the mining area calculates the rated working resistance of the bracket height of the (fallout zone).

$$P_Z = H_{caving} \cdot F \cdot \gamma (K_D + 1)$$

where $P_Z$-Design resistance of hydraulic support when the old top comes under pressure, KN;

$H_{caving}$-Height of collapse zone, m;

$F$-Area of support at the top of the bracket (usually measured using the actual rack type), m$^2$;

$\gamma$-Capacitance of the rock layer in the collapse zone, kN/m$^3$;

$K_D$-Dynamic load factor (usually taken as a maximum).

7.6 Determination of the Strength of the Top Plate to Pressure ...

According to the comprehensive mining surface stent pressure monitoring data and calculation analysis of the hydraulic stent selection design of comprehensive mining working resistance. Sometimes, the following Eq. can be used to calculate the rated working resistance of the stent:

$$P_z = \frac{M}{K_p - 1} \cdot F \cdot \gamma(K_D + 1)$$

where M-thickness of coal seam, m;

Kp-The coefficient of looseness of the coal seam (or the coefficient of fragmentation and expansion), and others are the same as above.

If the conditions are suitable, it is recommended to use the above four methods to calculate separately, and finally take the maximum value for selection or evaluation to analyze the reasonableness of bracket operation and selection.

### 7.6.2 Initial Support and Working Resistance

The initial support force of the stent is generated when all the columns of the stent rise quickly under the working pressure of the pumping station and touch the roof beam and the roof plate to give force, forming the support force of the stent for the roof plate. The initial support force is an important performance parameter for the stent. Its role is to slow down and reduce the natural subsidence of the roof plate, maintain and increase the stability of the roof plate, so that the stent can work under constant resistance as soon as possible. Therefore, it can be said that the ability of the support to control the roof plate depends entirely on the role of the initial support force. We should consider three basic factors to improve the control and management of the initial bracing force of the stent on the roof plate.

(1) Increasing the initial support force can increase the average support resistance of the stent during the working cycle. The initial bracing force determines the average support resistance of the stent when moving the stent, after moving the stent and within the working cycle, and the magnitude and speed of change of the load pressure of the roof plate. If the initial support force is increased to a certain degree, the support force of the stent is sufficient to balance with the pressure of the roof plate, and if the support force is continued to be increased, the effect is not obvious.

(2) Increasing the initial support force is conducive to reducing the crushing degree and stability of the roof plate at the end of the working face. Increasing the initial support can quickly crush the floating coal and crumbly layer, so that the support can quickly play the role of working resistance, and prevent the top plate from breaking prematurely.

(3) Increasing the initial bracing force is beneficial to preventing the amount of roof sinking during frame removal, albeit for a very short period of time, because the

localized range of support strength will be drastically reduced when the frame is removed, and it is possible that the top plate may sink rapidly. A high and sufficient initial bracing force can make the pressure of the hewing support after shifting the frame rise quickly, which is conducive to controlling the stability of the roof plate.

In the project, the unit of initial support force is MPa, and the initial support force of the stent is generated by the hydraulic power source of the pumping station, so the pumping station and the stent have an important and excellent matching relationship with each other. The initial support force is a key process parameter for the stent to effectively control and manage the roof plate, too small can't be done, and too high is easy to cause premature damage to the hydraulic piping system components. According to the long-term observation and the summary of practical experience, and combining with the information of relevant industry technical standards, it is proposed to use the following principles to grasp the ratio of initial support force and working resistance as a technical management reference. For the weak and strong top plate which is not obvious in the law of pressure from the top plate, the initial support force is required to reach 55–70% of the rated working resistance; for the medium and strong top plate rock layer which is obvious in the law of pressure from the top plate and the stability of the top plate, the initial support force is required to reach 70–85% of the rated working resistance; for the thick and hard top plate rock layer which is strong in the law of pressure from the top plate, i.e., the high and strong type of top plate, the initial support force is required to reach 80% of the rated working resistance. When the thick coal seam is mined in layers, the initial bracing force of the lower layer should be more than 2–3 times of the mining height and the weight of the roof rock layer. There is no standardized and unified requirement about the calculation relationship between initial support force and working resistance or management requirement, but only a guiding technical requirement.

**Open Access** This chapter is licensed under the terms of the Creative Commons Attribution-NonCommercial-NoDerivatives 4.0 International License (http://creativecommons.org/licenses/by-nc-nd/4.0/), which permits any noncommercial use, sharing, distribution and reproduction in any medium or format, as long as you give appropriate credit to the original author(s) and the source, provide a link to the Creative Commons license and indicate if you modified the licensed material. You do not have permission under this license to share adapted material derived from this chapter or parts of it.

The images or other third party material in this chapter are included in the chapter's Creative Commons license, unless indicated otherwise in a credit line to the material. If material is not included in the chapter's Creative Commons license and your intended use is not permitted by statutory regulation or exceeds the permitted use, you will need to obtain permission directly from the copyright holder.

# Chapter 8
# Coal Rock Body Energy Principal Analysis Method

Based on the principle and viewpoint of the conservation of elastic energy of the mechanical structural system or elastic system that carries the structural body and its generalized characteristic features, the establishment of the understanding and analysis of the conservation of energy system to maintain the equilibrium of the sufficient conditions for the structural system of the total potential energy to have an extreme value (or stationary value). Since the total potential energy of the structural system is a general function, the problem of analyzing the stability of the equilibrium is mathematically reduced to the problem of solving the extremum of the general function, that is, a differential problem, with the help of the differential method, it is possible to establish the energy discriminant between the stable equilibrium of the (mechanical structure) system and the unstable equilibrium. In solving the stability problem, if the same linear assumptions are used in both the energy differentiation method and the static differential calculus, then both methods will lead to the same differential equations, so it can be said that the energy method and the static differential calculus are equivalent, however, for many practical problems, it is still easy to establish the differential equations, but it is not easy to solve the equations, and the problem can be solved if the energy method is used in the approximation of the problem, which means that the energy method is the same as the static differential calculus. That is to say, compared with the static method, the energy method is more attractive because it is particularly suitable for analyzing and explaining the principles of force transfer, transmission, change, disappearance and control, as well as approximate calculations. Therefore, the discussion herein focuses on introducing the energy method and utilizing the energy criterion to analytically investigate some of the basic concepts of nonlinear theory.

## 8.1 Energy Balance Guidelines

Using the condition that the first order differential of the total potential energy of the theory of elastic conserved systems is equal to zero, i.e., the principle criterion of energy equilibrium, it is possible to determine whether or not an equilibrium state of the structural system exists, but it is not possible to decide the form of equilibrium, i.e., whether the system is in stable equilibrium, unstable equilibrium, or follow-on equilibrium, and in order to determine the mode of equilibrium of the system, the mode of stable equilibrium is discussed below with the principle criterion of energy.

### 8.1.1 Single-Degree-of-Freedom Structural Systems

A mechanical system if the total potential energy of the system is said to be a single-degree-of-freedom structural system a function of. When $\Pi$ can be expressed as a generalized displacement $\phi\phi = \phi_0$, the system is in initial equilibrium. In order to analyze the stability of this state. A small disturbance can be applied to the system to produce a small displacement change $\delta\phi$ (also known as displacement differential increment), then the system deviates from the initial state to a neighboring position $\phi_0 + \delta\phi_0$. At this point the total potential energy of the system will change as follows:

$$\Delta\Pi = \Pi(\phi_0 + \delta\phi_0) - \Pi(\phi_0)$$

Expanding into a Taylor series around, we have $\Delta\prod\phi_0$

$$\Delta\Pi = \frac{d\Pi(\phi_0)}{d\phi}\delta\phi + \frac{1}{2!}\frac{d^2\Pi(\phi_0)}{d\phi^2}(\delta\phi)^2 + \frac{1}{3!}\frac{d^3\Pi(\phi_0)}{d\phi^3}(\delta\phi)^3$$
$$+ \cdots = \delta\Pi + \frac{1}{2!}\delta^2\Pi + \frac{1}{3!}\delta^3\Pi + \cdots \quad (8.1)$$

Equation $\delta\Pi = \frac{d\Pi(\phi_0)}{d\phi}\delta\phi$

$$\delta^2\Pi = \frac{d^2\Pi(\phi_0)}{d\phi^2}(\delta\phi)^2$$

$$\delta^3\Pi = \frac{d^3\Pi(\phi_0)}{d\phi^3}(\delta\phi)^3$$

They are called-order differential, second-order differential, = -order differential,..., and so on of potential energy functional Ⅱ, respectively.

Considering the energy standing (extreme value) condition, at the first order derivative of the potential energy is zero or the first order differential is zero, i.e., we have:

## 8.1 Energy Balance Guidelines

$$\phi_0 \frac{d\Pi(\phi_0)}{d\phi} = 0 \text{ or } \delta\Pi = 0 \qquad (8.2)$$

Thus Eq. (8.1) becomes:

$$\Delta\Pi = \frac{1}{2!}\delta^2\Pi + \frac{1}{3!}\delta^3\Pi + \cdots \qquad (8.3)$$

The conditions representing the change in potential energy of the structural system expressed in the above equation can be interpreted as a measure of the work consumed by the external force when the system deviates from the equilibrium position, according to which the equilibrium form of the structural system can be judged. When the structural system is subjected to a small disturbance that produces a small displacement change away from the equilibrium position $\delta\phi$, if a disturbance energy needs to be added, or if the added second-order trace disturbance work (or energy) is positive, i.e., the structural system is in stable equilibrium. It should be noted here that to the second order differentiation $\Delta\Pi > 0$ the effect of is minimal and is therefore omitted. Accordingly, if the second order trace energy interference work described above is negative, the structural system is in unstable equilibrium. If the second-order trace interference work is zero, the system is in casual equilibrium or critical state, which implies that there may be an infinite number of neighboring equilibrium positions when both the magnitude and direction of the external force are constant. Higher order differentiation on compared.

The energy principal criterion for the three equilibrium forms is written as the following energy criterion discriminant.

(1) The system is in stable equilibrium ($\delta\prod = 0$)

$$\delta^2\Pi > 0, \quad \Delta\Pi > 0, \quad \Pi = \min \qquad (8.4)$$

This is the famous code of the principle of least potential energy as expounded by Dirichlet scholars.

(2) The system is in unstable equilibrium ($\delta\prod = 0$)

$$\delta^2\Pi < 0, \quad \Delta\Pi < 0, \quad \Pi = \max \qquad (8.5)$$

(3) The system is in casual equilibrium or critical state ($\delta\prod = 0$)

$$\delta^2\Pi = 0 \qquad (8.6)$$

For the contingent equilibrium, the positive and negative values of $\Delta\Pi$ depend on the positive and negative values of higher order differentials which are not zero. These high-order differentials are very important for determining the stability characteristics of a mechanical structure system in supercritical state, but they are not important for calculating the critical load.

## 8.1.2 Multi-degree-of-Freedom Structural Systems

The two-degree-of-freedom structural system is used in conjunction to illustrate how the above energy criterion for a single-degree-of-freedom system can be generalized to a multi-degree-of-freedom structural system or a generalized general structural system energy criterion. A mechanical structural system if the total potential energy of the system can be expressed as two generalized displacements is called a two-degree-of-freedom structural system a function. At this point, therefore, the corresponding expansion of Eq. (8.1) can be made as: $\phi$ and $\theta$

$$\begin{aligned}\Delta \Pi &= \Pi(\phi_0 + \delta\phi, \theta_0 + \delta\theta_0) - \Pi(\phi_0, \theta_0) \\ &= \frac{\partial \Pi}{\partial \phi}(\phi_0, \theta_0)\delta\phi + \frac{\partial \Pi}{\partial \theta}(\phi_0, \theta_0)\delta\theta + \frac{1}{2!}\left[\frac{\partial^2 \Pi}{\partial \phi^2}(\phi_0, \theta_0)\delta^2\phi\right. \\ &\quad + 2\frac{\partial^2 \Pi}{\partial \phi \partial \theta}(\phi_0, \theta_0)\delta\phi\delta\theta + \frac{\partial^2 \Pi}{\partial \theta^2}(\phi_0, \theta_0)\delta^2\theta + \cdots \\ &= \delta\Pi + \frac{1}{2!}\delta^2\Pi + \cdots \end{aligned} \quad (8.7)$$

Equation

$$\begin{aligned}\delta\Pi &= \frac{\partial \Pi}{\partial \phi}(\phi_0, \theta_0)\delta\phi + \frac{\partial \Pi}{\partial \theta}(\phi_0, \theta_0)\delta\theta| \\ \sigma^2\Pi &= \frac{\partial^2 \Pi}{\partial \phi^2}(\phi_0, \theta_0)\delta^2\phi + 2\frac{\partial^2 \Pi}{\partial \phi \partial \theta}(\phi_0, \theta_0)\delta\phi\delta\theta + \frac{\partial^2 \Pi}{\partial \theta^2}(\phi_0, \theta_0)\delta^2\theta \end{aligned} \quad (8.8)$$

Still referred to as first order differentiation, second order differentiation.

The energy criterion for the equilibrium of a structural system: i.e., its extremal condition is such that the first-order differential is zero, since $\delta\theta$, $\delta\theta$ are relatively independent and arbitrarily small quantities. Thus, the following two equilibrium equations can be obtained:

$$\begin{aligned}\frac{\partial \Pi}{\partial \phi}(\phi_0, \theta_0)\delta\phi &= 0 \\ \frac{\partial \Pi}{\partial \theta}(\phi_0, \theta_0)\delta\theta &= 0 \end{aligned} \quad (8.9)$$

That is, at $(\varphi_0, \theta_0)$, both are zero, and the condition for the system to be in stable equilibrium, i.e., $\frac{\partial \Pi}{\partial \theta}$, $\frac{\partial \Pi}{\partial \phi}$ for, is that $\Pi$ to be minimal the second-order differentiation of for all $\Pi$ must be positive. Incidentally, in algebraic theory, the quadratic homogeneous polynomials with respect to variables $\delta\phi$, $\delta\theta$ given by the second order differential of Eq. (8.9) are called positive definite quadratic forms.

To summarize, for a two-degree-of-freedom system, the equilibrium criterion is that the first-order differential of the potential energy is zero, i.e. $\delta \prod = 0$. if The

## 8.1 Energy Balance Guidelines

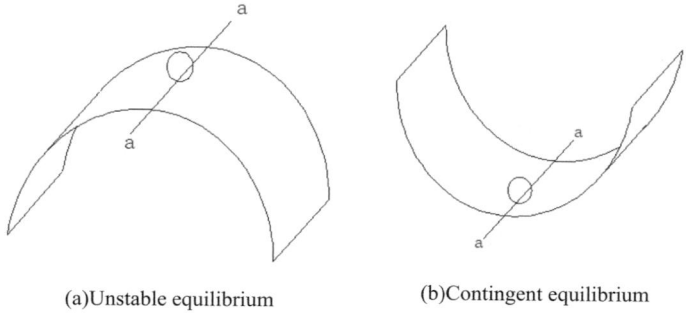

(a) Unstable equilibrium      (b) Contingent equilibrium

**Fig. 8.1** Principle of balance and stabilization of system energy

equilibrium is stable; if there is at least one specific second-order variant the second-order differential of is positive definite, i.e., $\delta^2 \prod > 0$ it is unstable of, and it if there is at least one specific second-order differential of $\delta^2 \prod < 0$ is asymptotic of. $\delta^2 \prod = 0$ and the rest of the second-order variants $\delta^2 \prod > 0$.

In order to more clearly understand the above-mentioned energy discriminant of unstable equilibrium and contingent equilibrium, we observe and consider a small ball that always contacts with a given surface, as shown in Fig. 8.1. In Fig. 8.1a, the ball is balanced on the convex surface. Since the second-order differential is only equal to zero in a specific direction a-a, and is less than zero in other directions, the equilibrium is unstable. In Fig. 8.1b. The ball is balanced on the concave surface. Since the second order differential is equal to zero in a specific direction 'a–a', and is greater than zero in other directions, the equilibrium is met.

Now the system instability criterion of the structure is discussed. It is known that the system will lose stability under critical load. According to the energy criterion of the equilibrium form, the critical load refers to the minimum load when the load increases from zero to $\delta^2 \Pi$. In other words, when the load is $P = P_{cr}$, the condition that the second-order differential has a relative extreme value can be expressed as

$$\frac{\partial(\sigma^2\Pi)}{\partial\sigma\phi} = \frac{\partial(\sigma^2\Pi)}{\partial\sigma\theta} = 0$$

According to this condition and using Eq. (8.8), it is obtained:

$$\frac{\partial^2\Pi}{\partial\phi^2}(\phi_0, \theta_0)\delta\phi + \frac{\partial^2\Pi}{\partial\phi\partial\theta}(\phi_0, \theta_0)\delta\theta = 0$$
$$\frac{\partial^2\Pi}{\partial\phi\partial\theta}(\phi_0, \theta_0)\delta\phi + \frac{\partial^2\Pi}{\partial\theta^2}(\phi_0, \theta_0)\delta\theta = 0 \qquad (8.10)$$

This equation is about a linear chi-square equation which has nonzero solutions coefficients determinant of zero. That is: $\delta\phi, \delta\theta$

$$\begin{vmatrix} \frac{\partial^2 \Pi}{\partial \phi^2}(\phi_0, \theta_0) & \frac{\partial^2 \Pi}{\partial \phi \partial \theta}(\phi_0, \theta_0) \\ \frac{\partial^2 \Pi}{\partial \phi \partial \theta}(\phi_0, \theta_0) & \frac{\partial^2 \Pi}{\partial \theta^2}(\phi_0, \theta_0) \end{vmatrix} = 0 \qquad (8.11)$$

It is known from the algebraic theory of quadratic forms. The limiting condition for a quadratic form to be positive definite is that the determinant of its coefficients is zero. From this and considering Eq. (8.8) it is straightforward to obtain the expression (8.11) for the destabilization condition of the system.

The above analysis of two-degree-of-freedom systems is also applicable to multi-degree-of-freedom systems. In short, for multi-degree-of-freedom systems, the energy criterion of equilibrium is that the first-order change of potential energy is divided into zero, that is, $\delta \Pi = 0$; if the second-order differential of the potential energy is positive definite, that is, $\delta^2 \Pi > 0$, the equilibrium is stable, and the critical load is the minimum load value when the load increases from zero to $\delta^2 \Pi$.

### 8.1.3 Combined Structural Systems

It will be discussed here how the above energy criterion for equilibrium and equilibrium forms of finite-degree-of-freedom systems can be generalized to the description of composite functions of combination structural systems characterized by continuity.

(1) Energy Equilibrium Criterion

First a brief introduction to the concept of generalized functions. If the total potential energy a of the structural system can be expressed as $\Pi$ function $w(x)$ of displacement, i.e., there:

$$\Pi = \int_{x_0}^{x_1} F[w(x)] dx \qquad (8.12)$$

where

w-unknown factor of $x$;

F-the known function w.

The value of $\Pi$ does not depend on a finite number of discrete variables, but on one or more continuous functions, the potential energy $\Pi$ is said to be a generalized function of the displacement $w(x)$. It should be emphasized that if the structural system is in stable equilibrium, then $w(x)$ should be such that the generalized function $\Pi$ is minimal.

In conjunction with the cantilever beam structural system subjected to a uniform load q shown in Fig. 8.2, the discussion illustrates how the structural equilibrium and energy criterion can be used to derive its equilibrium differential equations and natural boundary conditions.

# 8.1 Energy Balance Guidelines

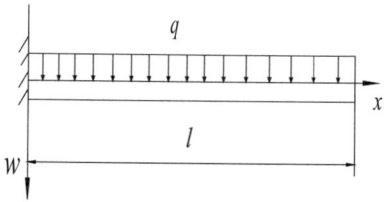

**Fig. 8.2** Uniform load Structural system of cantilever beams acting with $q$

If the bending equilibrium state of a cantilever beam is denoted by $w_0(x)$, the equilibrium differential equation is:

$$EIw_0^{IV} - q = 0 \qquad (8.13)$$

The geometric boundary conditions are:

$$x = 0, \quad w_0 = w_0' = 0 \qquad (8.14a)$$

The force boundary conditions or natural boundary conditions are:

$$x = l, \quad M = Q = 0 \qquad (8.14b)$$

Since the equilibrium of this beam (system) is stable, the total potential energy of the system, $\Pi$, is relatively minimal and must satisfy the extreme value condition, $\delta \Pi = 0$.

For this equilibrium problem, the total potential energy of the system is:

$$\Pi = \Phi + \phi_z \qquad (8.15)$$

where: $\Phi$–bending strain energy and there is $\Phi = \frac{EI}{2} \int_0^l w''^2 dx$

The external potential energy with respect to the unbent linear state is:

$$\phi_z = -\int_0^l qw\, dx \qquad (8.16)$$

Thus:

$$\Pi = \int_0^l \left( \frac{1}{2} EIw''^2 - qw \right) dx \qquad (8.17)$$

Comparison of Eqs. (8.17) and (8.12) shows that:

$$F = \frac{1}{2}EIw''^2 - qw \tag{8.18}$$

In order to examine the small change in the characteristic function of the potential energy $\Pi$ of the system when $w = w_0$, i.e., let:

$$w = w_0 + w_1$$

where $w_0$ and $w_1$ are both continuously differentiable functions that satisfy geometric boundary conditions and are called permit functions or allowable functions. Also called displacement differentiation or virtual displacement. For analytical purposes take:

$$w_1(x) = \varepsilon\eta(x) \tag{8.19}$$

where

$\varepsilon$–an optional tiny parametric variable whose value is independent of $x$;

$\eta(x)$–the given licensing function. Where: $\varepsilon$-an optional minor parameter whose value is independent of;

The total potential energy of the system at this time is:

$$\Pi + \Delta\Pi = \int_0^l \left[ \left( \frac{1}{2}EI(w_0'' + \varepsilon\eta'') \right) - q(w_0 + \varepsilon\eta) \right] dx \tag{8.20}$$

From the above equation and Eq. (9.17) we have:

$$\Delta\Pi = \varepsilon \int_0^l (EIw_0''\eta'' - q\eta) dx + \varepsilon^2 \frac{EI}{2} \int_0^l \eta''^2 dx$$

$$= \delta\Pi + \frac{1}{2}\delta^2\Pi \tag{8.21}$$

where

$$\delta\Pi = \varepsilon \int_0^l (EIw_0''\eta'' - q\eta) dx \tag{8.22}$$

$$\delta^2\Pi = \varepsilon^2 EI \int_0^l \eta''^2 dx \tag{8.23}$$

## 8.1 Energy Balance Guidelines

If $\varepsilon$ is made sufficiently small, there will be $|\delta \prod| > |\delta^2 \prod|$. According to the condition $\delta \prod = 0$ of the extreme value of the potential energy; and examined that $\varepsilon$ can take any value that is non-zero, from Eq. (8.22), we can get:

$$\int_0^l (EIw_0'' \eta'' - q\eta) dx = 0 \tag{8.24}$$

Through two partial credits, there is:

$$(EIw_0'' \eta')\big|_0^l - (EIw_0''' \eta)\big|_0^l + \int_0^l (EIw_0^{IV} - q)\eta \, dx = 0 \tag{8.25}$$

Noting that $\eta(x)$ is an arbitrary function and satisfies the geometric boundary conditions, i.e., $x = 0$, $\eta = \eta' = o$, it follows from the above equation:
Equilibrium differential equations:

$$EIw_0^{IV} - q = 0, \quad 0 \le x \le l \tag{8.26}$$

Natural boundary conditions:

$$x = l, \quad EIw_0'' = 0, \quad EIw_0''' = 0 \tag{8.27}$$

Solving the differential Eq. (8.26) yields a solution satisfying all the boundary conditions as:

$$w_0 = \frac{q}{EI}\left(\frac{x^4}{24} - \frac{lx^3}{6} + \frac{l^2 x^2}{4}\right) \tag{8.28}$$

In addition, it can be seen from Eq. (8.23) that there is $\eta(x)$ for any $\delta^2 \prod > 0$, which indicates that the equilibrium state is stable. Therefore, the above energy balance criterion is called the principle of minimum potential energy.

From the above analysis, it can be seen that the equilibrium differential equation and the natural boundary conditions can be derived from the expression of the total potential energy $\prod$ of the structural system by using the energy balance criterion and through the differential operation. Therefore, if the permit function $w_0$ can satisfy the conditions of $\delta \prod = 0$, it must satisfy the equilibrium differential equation and the natural boundary conditions.

(2) Euler's equations

Generalizing the above analysis to the product function $F$ as a function of $x$, $w$, $w'$, $w''$ then we have:

$$\Pi = \int_{x_0}^{x_1} F(x, w, w', w'') dx \qquad (8.29)$$

Still taking $w = w_0 + w_1$.
both (…and…) $w_1 = \varepsilon \eta = (x)$.
Where: $\varepsilon$ and $\eta$ have the meanings described earlier. So there is:

$$\Delta \Pi = \int_{x_0}^{x_1} [F(x, w_0 + \varepsilon \eta, w'_0 + \varepsilon \eta', w''_0 + \varepsilon \eta'') - F(x, w_0, w'_0, w''_0)] dx$$

Expanding the expression for F in Eq. into a Taylor expansion yields a first-order differential:

$$\delta \Pi = \varepsilon \int_{x_0}^{x_1} \left( \frac{\partial F}{\partial w_0} \eta + \frac{\partial F}{\partial w'_0} \eta' + \frac{\partial F}{\partial w''_0} \eta'' \right) dx$$

where: $\frac{\partial F}{\partial w_0} = \frac{\partial F}{\partial w}\big|_{w=w_0}$, and so on.

Considering the necessary condition $\Pi$ for the generalized function $\delta \Pi = 0$ to have an extreme value, and since it is a non-zero arbitrary value, we have from the above equation:

$$\int_{x_0}^{x_1} \left( \frac{\partial F}{\partial w_0} \eta + \frac{\partial F}{\partial w'_0} \eta' + \frac{\partial F}{\partial w''_0} \eta'' \right) dx = 0$$

Repeated partial integration of the above equation yields:

$$\left( \frac{\partial F}{\partial w'_0} \eta \right) \Big|_{x_0}^{x_1} + \left( \frac{\partial F}{\partial w''_0} \eta' \right) \Big|_{x_0}^{x_1} - \left( \frac{d}{dx} \cdot \frac{\partial F}{\partial w''_0} \eta \right) \Big|_{x_0}^{x_1}$$

$$+ \int_{x_0}^{x_1} \left( \frac{\partial F}{\partial w_0} - \frac{d}{dx} \cdot \frac{\partial F}{\partial w'_0} + \frac{d^2}{dx^2} \cdot \frac{\partial F}{\partial w''_0} \right) \eta \, dx = 0$$

Using the conditions satisfied by the licensing function $\eta$, i.e., $x = x_0$, and when $x = x_1$, $\eta = \eta' = 0$, the above equation becomes:

$$\int_{x_0}^{x_1} \left( \frac{\partial F}{\partial w_0} - \frac{d}{dx} \frac{\partial F}{\partial w'_0} + \frac{d^2}{dx^2} \frac{\partial F}{\partial w''_0} \right) \eta \, dx = 0$$

## 8.1 Energy Balance Guidelines

Since $\eta(x)$ is an arbitrary function, the above equation holds if:

$$\frac{\partial F}{\partial w_0} - \frac{d}{dx}\frac{\partial F}{\partial w_0'} + \frac{d^2}{dx^2}\frac{\partial F}{\partial w_0''} = 0 \quad x_0 \leq x \leq x_1 \tag{8.30a}$$

If the initial bending equilibrium state is denoted by $w(x)$, i.e., by removing the subscript in Eq. (8.30a) above, we have:

$$\frac{\partial F}{\partial w} - \frac{d}{dx}\frac{\partial F}{\partial w'} + \frac{d^2}{dx^2}\frac{\partial F}{\partial w''} = 0 \tag{8.30b}$$

The solution $w(x)$ of this equation ensures that the potential energy $\Pi$ is standing. Equation (8.30b) is called Euler's equation and it is similar to Eqs. (8.2) and (8.9).

$$F = \frac{1}{2}EIw''^2 - qw$$

If the function given in Eq. (8.18) is substituted into Euler's equation, we have:
$\frac{\partial F}{\partial w} = -q$, $\frac{\partial F}{\partial w'} = 0$. $\frac{\partial F}{\partial w''} = EIw''$
Thus, it can be obtained: $EIw_0^{IV} - q = 0$.

From the above analysis, the equilibrium criterion for a continuous system can be expressed as the Euler Eq. (8.30b) for the product function F in the total potential energy expression of the system.

Moreover, if the product function F is a function of $u(x)$, the two dependent variables $w(x)$ and $u$ and the first-order derivatives of E and the first- and second-order derivatives of $w$, viz:

$$\Pi = \int_l F(x, u, w, u', w', w'')dx$$

Then Euler's equation is:

$$\frac{\partial F}{\partial u} - \frac{d}{dx}\cdot\frac{\partial F}{\partial u'} = 0$$
$$\frac{\partial F}{\partial w} - \frac{d}{dx}\cdot\frac{\partial F}{\partial w'} + \frac{d^2}{dx^2}\cdot\frac{\partial F}{\partial w''} = 0 \tag{8.31}$$

Further, if the generalized function is expressed as:

$$\Pi = \int_A F(x, y, u, v, w, u_x, u_y, v_x, v_y, w_x, w_y, w_{xx}, w_{xy}, w_{yy})dxdy$$

where: $x$, $y$-independent dependent variables;
$u$, $x$, $y$-dependent variable.

Then Euler's equation is:

$$\frac{\partial F}{\partial u} - \frac{\partial}{\partial x} \cdot \frac{\partial F}{\partial u_x} - \frac{\partial}{\partial y} \cdot \frac{\partial F}{\partial u_y} = 0$$

$$\frac{\partial F}{\partial v} - \frac{\partial}{\partial x} \cdot \frac{\partial F}{\partial v_x} - \frac{\partial}{\partial y} \cdot \frac{\partial F}{\partial v_y} = 0$$

$$\frac{\partial F}{\partial w} - \frac{\partial}{\partial x} \cdot \frac{\partial F}{\partial w_x} - \frac{\partial}{\partial y} \cdot \frac{\partial F}{\partial w_y} + \frac{\partial^2}{\partial x^2} \cdot \frac{\partial F}{\partial w_{xx}}$$

$$+ \frac{\partial^2}{\partial x \partial y} \cdot \frac{\partial F}{\partial w_{xy}} + \frac{\partial^2}{\partial y^2} \cdot \frac{\partial F}{\partial w_{yy}} = 0 \quad (8.32)$$

Equations (8.31) and (8.32) are often applied in plate and shell theory.

(3) The Trefz Guidelines

The total potential energy of the system is now analyzed as an example of an equal straight rod subjected to axial pressure P at both ends, the conditions used to determine the critical load of its system:

$$\Pi = \Phi_a + \Phi_b + V \quad (8.33)$$

where: $\Phi_a$–axial shrinkage strain energy: $\Phi_a = \frac{EA}{2} \int_0^l u'^2 dx$
$\Phi_b$–bending strain energy. $\Phi_b = \frac{EI}{2} \int_0^l w''^2 dx$.
$\phi_z$–External potential energy,

$$\phi_z = -P \int_0^l u' dx - \frac{P}{2} \int_0^l w'^2 dx \quad (8.34)$$

So:

$$\Pi = \int_0^l \left( \frac{EA}{2} u'^2 + \frac{EI}{2} w''^2 - Pu' - \frac{P}{2} w'^2 \right) dx \quad (8.35)$$

In order to study the second-order differentiation of the potential energy, the displacement is made to change, i.e., it is taken:
$u = u_0 + u_1, w = w_0 + w_1$.
Where: $u_0, w_0$–the initial deformation state of the compression rod;
$u_1, w_1$–the differential or increment of displacement.
Since the initial state of the lever is unbent and straight, there is:
$u_0 = \frac{Px}{EA}$, $w_0 = 0$.
Thus $u = \frac{Px}{EA} + u_1, w = w_1$.

## 8.1 Energy Balance Guidelines

Substituting the above equation into Eq. (8.35) and organizing yields the second order differential expression:

$$\delta^2 \Pi = \int_0^l (EAu_1'^2 + EIw_1''^2 - Pw_1'^2)dx \tag{8.36}$$

The above second order differential equation is positive definite when the value of the load P is very small, while the critical load, is the minimum load $P_{cr}$ when the above definite integral is non-positive. in other words, the equilibrium of the system will change from stable to unstable when $p = p_{cr}$. The Trefz criterion for determining the critical load holds that the system is destabilized under the condition that the first order differential of the second order differential is zero, i.e.:

$$\delta(\delta^2 \Pi) = 0 \tag{8.37}$$

If $\delta^2 \prod$ is regarded as a displacement generalized function, write the following corresponding Euler equations based on the Trefz criterion:

$$\frac{\partial F}{\partial u_1} - \frac{d}{dx} \cdot \frac{\partial F}{\partial u_1'} = 0|$$

$$\frac{\partial F}{\partial w_1} - \frac{d}{dx} \cdot \frac{\partial F}{\partial w_1'} + \frac{d^2}{dx^2} \cdot \frac{\partial F}{\partial w_1''} = 0 \tag{8.38}$$

Style: $F = EAu_1'^2 + EIw_1''^2 - Pw_1'^2$.

Substituting the above equation into Euler's Eq. (8.38) yields:

$$u_1'' = 0 \tag{8.39a}$$

$$EIw_1^{IV} - Pw_1'' = 0 \tag{8.39b}$$

Equation (8.39b) is the differential equation of the compression rod when the load reaches the critical value. This shows that the system instability condition can be expressed as the Euler equation for the product function in the second-order differential, and the critical load can be obtained by solving this equation.

In general, the increment of the potential energy of the structural system can be expressed as:

$$\Delta \Pi = \delta \Pi + \frac{1}{2!}\delta^2 \Pi + \frac{1}{3!}\delta^3 \Pi + \frac{1}{4!}\delta^4 \Pi + \cdots \tag{8.40}$$

Each non-zero term in the terms at the right end of the above equation is greater than the sum of the subsequent terms. The first order differentiation $\delta \Pi = 0$ of the generalized function represents the equilibrium condition, i.e., the principle of potential energy extremum. If the second-order differential of the generalized function $\delta^2 \Pi > 0$, the equilibrium of the structural system is stable. If the second-order differential of the generalized function $\delta^2 \Pi < 0$, the equilibrium of the structural system is unstable. When the second-order differential of the generalized function is A = 0, the structural system will be in an intermediate state of transition from a stable to an unstable equilibrium, i.e., the critical load of the critical state is the minimum load corresponding to the second-order differential that is non-positive definite. The critical load equation derived from the Trevor's allowed rule $\delta^2 \Pi = 0$ can be obtained from the Euler equation for the product function in the second-order differential expression. Since the second-order differential expression is a quadratic chi-square generalized function, its differential result must be a linear chi-square differential equation. It should be noted that at that time $P = P_{cr}$, $\delta \Pi = \delta^2 \Pi = 0$ to study the initial post-buckling properties near the critical state, it is necessary to examine the higher order nonzero differentiation of the potential energy generalized function.

## 8.2 Determination of Critical Load by the Energy Method

As mentioned earlier, the application of the energy differentiation principle allows for the derivation of the equilibrium differential equations of the system. In addition, an approximation of the critical load can be determined based on the energy criterion. Combined with a typical mechanical problem of an isotropic compression bar subjected to axial pressure P, we analyze and explain how to apply the energy method to solve the critical load, and analyze and study the critical state and the state parameters which are very important for the stability analysis of the structural system and for the safety of the mechanical structure.

As you know, the structural system is in the critical state, that is, with the equilibrium state, $\delta \Pi = 0$ and $\delta^2 \Pi = 0$. For the compression rod system, now consider the linear equilibrium position of the critical state and the neighboring micro-bending equilibrium position: since $P = P_{cr}$ in the action of the compression rod is in the critical state, and so $\delta \Pi = 0$. Therefore, the potential energy of the compression rod can be regarded as an invariant constant, i.e., $\Pi$ = cost. If the straight-line equilibrium position of the compression rod is taken as a zero point of the potential energy, the then the potential energy at the equilibrium position of its micro-bending deformation remains zero, i.e., there is:

$$\Pi = \Phi + \phi_z = 0 \tag{8.41}$$

Of which:

$$\Phi = \frac{1}{2}\int_l EIw''^2 dx, \quad \phi_z = \frac{P}{2}\int_l w'^2 dx \qquad (8.42)$$

Substituting Eq. (8.42) into (8.41) has:

$$P_{cr} = \frac{\int_l EIw''^2 dx}{\int_l w'^2 dx} \qquad (8.43)$$

Understanding the principle of differentiation as presented in the previous section, the displacement differentiation $w_1$, given at the linear position $w_0 = 0$, gives:

$$\Pi + \Delta\Pi = \frac{1}{2}\int_l EI(w_0'' + w_1'')^2 dx - \frac{P}{2}\int_l (w_0' + w_1')^2 dx$$

Available:

$$\delta^2 \Pi = \int_l EIw_1''^2 dx - P\int_l w_1'^2 dx$$

It follows that the critical load given by Eq. (8.43) can still be obtained by $\delta^2 \Pi = 0$.

It should be noted that the use of Eq. (8.43) to determine the critical load, if the deflection curve equation $w(x)$ is the equation of the real deflection curve after the instability, the critical load and the exact solution is the same, if $w(x)$ is not the equation of the real deflection curve after the instability, but only to meet the geometric boundary conditions, the solution is only an approximation of the exact solution, and the approximation is greater than the true value of the critical load. The above conclusion can be physically explained as follows: since the deflection curve is not the real deflection curve after the bar is destabilized, the deformation of the bar is restricted, which is equivalent to adding constraints on the bar or enhancing the stiffness of the bar, which leads to a larger solution than the real one.

## 8.3 Nonlinear Stability Theory

Up to this point, some concepts of the basic classical stability theory have been introduced earlier and applied to analyze and solve the critical state and critical load of the compression rod. However, for the second equilibrium path beyond the branching point as well as the extremum point instability problem, the linearized theory mentioned above is powerless. Considering the value and development of the stability theory and its application areas, a few simple models will be analyzed using the energy principle criterion, by which some important concepts about nonlinear stability theory will be introduced. In addition, since actual engineering structures may have various imperfections in their geometry and loading methods, defects are bound to have a non-negligible effect on the yielding properties. In order to study the effect of initial defects on buckling, w-T-Koiter proposed the concept of "defect sensitivity", which is briefly analyzed and explained.

### 8.3.1 Symmetric Stable Branching Point Problems

(1) Improving the system

The rigid rod shown in Fig. 8.3 is hinged at the lower end but has a rotational spring constraint with a rotational stiffness factor of $\beta$. The spring is not stressed when the rod is in its initial lead state. Let us analyze the stability properties of the model of the above perfected system under load.

When the rigid rod is deflected by an angle $\theta(|\theta| < \pi)$ from its initial state, there is:

Spring force potential energy:

$$\Phi = \frac{1}{2}\beta\theta^2 \tag{8.44}$$

The upper end of the rod is discerningly displaced: $\lambda = l(1 - \cos\theta)$.
Load potential energy:

$$\phi_z = -P\lambda = -Pl(1 - \cos\theta) \tag{8.45}$$

Total potential energy of the system:

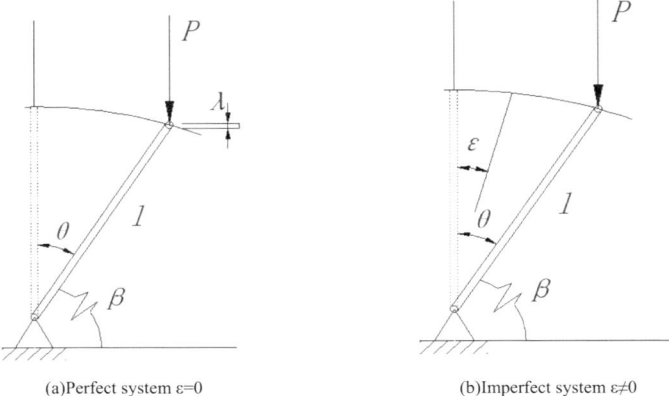

(a) Perfect system $\varepsilon=0$    (b) Imperfect system $\varepsilon\neq 0$

**Fig. 8.3** Non-stability cutoffs for rigid rod systems

## 8.3 Nonlinear Stability Theory

**Fig. 8.4** Critical branching points and symmetrically stabilized equilibrium paths

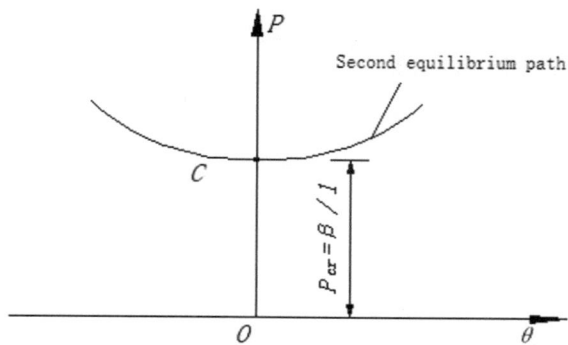

$$\Pi(\theta, P) = \Phi + \phi_z = \frac{1}{2}\beta\theta^2 - Pl(1 - \cos\theta) \tag{8.46}$$

Let the first-order differential or first-order partial derivative of $\Pi$ be equal to zero, then we have:

$$\frac{\partial \Pi}{\partial \theta} = \beta\theta - Pl\sin\theta = 0 \tag{8.47}$$

The above equation is the basic equilibrium equation for the structural system, which is plotted as a path-of-motion equilibrium curve as shown in Fig. 8.4.

It is easy to see that $\theta = 0$ is a possible equilibrium state for all values of P. In other words, the graphical vertical axis depicts the original equilibrium path. The curve plotted after the load exceeds the branch point C:

$$P(\theta) = \frac{\beta\theta}{l\sin\theta} \tag{8.48}$$

The second equilibrium path or post-buckling equilibrium path is depicted and given, and the value of P corresponding to the branching point C is the critical load, i.e.:

$$P_{cr} = \lim_{\theta \to 0} \frac{\beta\theta}{l\sin\theta} = \beta/l \tag{8.49}$$

In order to determine the stability of the equilibrium path, now consider the second-order differential or second-order partial derivatives of $\Pi$:

$$\frac{\partial^2 \Pi}{\partial \theta^2} = \beta - Pl\cos\theta \tag{8.50}$$

The stability of the original equilibrium path is discussed first:

$$P \left\langle P_{cr} = \frac{\beta}{l}, \frac{\partial^2 \Pi}{\partial \theta^2} \right\rangle 0, \text{ stable equilibrium}$$

$$P > P_{cr} = \frac{\beta}{l}, \frac{\partial^2 \Pi}{\partial \theta^2} < 0, \text{ unstable equilibrium}$$

$$P = P_{cr} = \frac{\beta}{l}, \frac{\partial^2 \Pi}{\partial \theta^2} = 0, \text{ critical equilibrium} \tag{8.50}$$

Then analyze the stability of the second equilibrium path and:
perceive $ctg\theta = \frac{1}{\theta} - \frac{\theta}{3} - \frac{\theta^3}{45} - \cdots$  $|\theta| < \pi$.
is then obtained from Eqs. (8.50) and (8.48):

$$\frac{\partial^2 \Pi}{\partial \theta^2} = \beta - \frac{\beta \theta}{\sin \theta} \cos \theta$$

$$= \beta(1 - \theta ctg\theta)$$

$$= \beta \left( \frac{\theta}{3} + \frac{\theta^3}{45} + \cdots \right) > 0$$

Thus, it can be seen that the second equilibrium path is stabilized everywhere.

Finally, the stability of the branch point $P = P_{cr}$, the critical equilibrium state, is studied when $\theta = 0$

$$\frac{\partial \Pi}{\partial \theta} = \frac{\partial^2 \Pi}{\partial \theta^2} = 0$$

$$\left. \frac{\partial^3 \Pi}{\partial \theta^3} \right|_{\theta=0} = (Pl \sin \theta)|_{\theta=0} = 0$$

$$\left. \frac{\partial^4 \Pi}{\partial \theta^4} \right|_{\theta=0} = (Pl \cos \theta)|_{\theta=0} = Pl$$

It follows that if $\Pi(\theta, P)$ is expanded by Taylor series in the neighborhood of the critical branching point C, the first nonzero term is obtained as:

$$\frac{1}{4!} \left( \frac{\partial^4 \Pi}{\partial \theta^4} \right) \bigg|_{\theta=0} \theta^4 = \frac{1}{24} P_{cr} l \theta^4 = \frac{1}{24} \beta \theta^4 > 0$$

This means that the potential energy $\Pi$ is locally minimal, so the critical state is stable.

To summarize, as the load is gradually increased from zero to time $P = P_{cr} = \beta/l$, the bar always remains straight and does not tilt. If the load continues to increase continuously, the bar will keep tilting along a stable post-buckling path.

## 8.3 Nonlinear Stability Theory

(2) Defective systems (imperfect systems)

A simplified model of the defective system is shown in Fig. 8.3b. When the rigid rod has a small rotation $\varepsilon$, the spring is not stressed. The above system is perfect if the initial defect parameter $\varepsilon = 0$ is taken.

Let the total angle of turn of the bar be $\theta$. At this time:
Spring force potential energy:

$$\Phi(\theta, \varepsilon) = \frac{1}{2}\beta(\theta - \varepsilon)^2 \tag{8.51a}$$

Load potential energy:

$$\phi_z(\theta, P) = -Pl(\cos\varepsilon - \cos\theta) \tag{8.51b}$$

Thus, the expression for the total potential energy of the defective system is:

$$\Pi(\theta, P, \varepsilon) = \frac{1}{2}\beta(\theta - \varepsilon)^2 - Pl(\cos\varepsilon - \cos\theta) \tag{8.52}$$

Thus obtained:

$$\frac{\partial \Pi}{\partial \theta} = \beta(\theta - \varepsilon) - Pl\sin\theta \tag{8.53}$$

$$\frac{\partial^2 \Pi}{\partial \theta^2} = \beta - Pl\cos\theta \tag{8.54}$$

The balanced path is obtained: from $\frac{\partial \Pi}{\partial \theta} = 0$ or $\delta \Pi = 0$

$$P(\theta, \varepsilon) = \frac{\beta(\theta - \varepsilon)}{l\sin\theta} \tag{8.55a}$$

$$\frac{Pl}{\beta} = \frac{\theta - \varepsilon}{l\sin\theta} \tag{8.55b}$$

Accordingly:

$$\frac{Pl}{\beta} \left\langle \sec\theta, \frac{\partial^2 \Pi}{\partial \theta^2} \right\rangle 0, \text{ table equilibrium}$$

$$\frac{Pl}{\beta} > \sec\theta, \frac{\partial^2 \Pi}{\partial \theta^2} < 0, \text{ unstable equilibrium}$$

$$\frac{Pl}{\beta} = \sec\theta, \frac{\partial^2 \Pi}{\partial \theta^2} = 0, \text{ critical equilibrium} \tag{8.56}$$

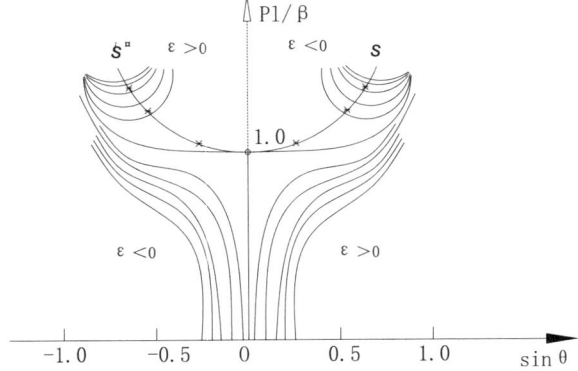

**Fig. 8.5** Symmetric nonlinear stabilization principle and variation curve

For different defect parameters $\varepsilon$, Eq. (8.55) can be plotted as a symmetric stabilizing, or nonlinear stabilizing, curve as shown in Fig. 8.5. In the figure, the solid line represents the stable equilibrium path, the dashed line represents the unstable equilibrium path, and the dotted line $ss$ represents the equation of the trajectory of motion in the critical equilibrium state:

$$P = \frac{\beta}{l \cos \theta}$$

Through the use of structural system example analysis of the defective system can be seen, when the load from zero constantly along the stability of the path increases, the system does not exist in the phenomenon of instability, only when the load exceeds a certain value, the displacement increases more rapidly and has been. Because the curve shown in Fig. 8.5 has the characteristics of symmetry, and the critical state is stable, so this type of stability problem is called symmetric stability critical branch point problem. For example, in the engineering structure, by the axial pressure of the four-sided simply supported rectangular plate stability problem belongs to this type.

## 8.3.2 Symmetric Unstable Branching Point Problems

A rigid rod of length $l$ is connected at its upper end to a horizontal spring with tensile stiffness coefficient $\alpha$, and at its lower end it is articulated to a rigid foundation as shown in Fig. 8.6. An initial defect parameter $\varepsilon$ is introduced, i.e., when the spring is not stressed, the upper end of the rigid bar has a small horizontal displacement $\varepsilon l$. After the application of load P, the system produces a total horizontal displacement $\theta l$.

## 8.3 Nonlinear Stability Theory

**Fig. 8.6** Symmetric unstable branching point principle

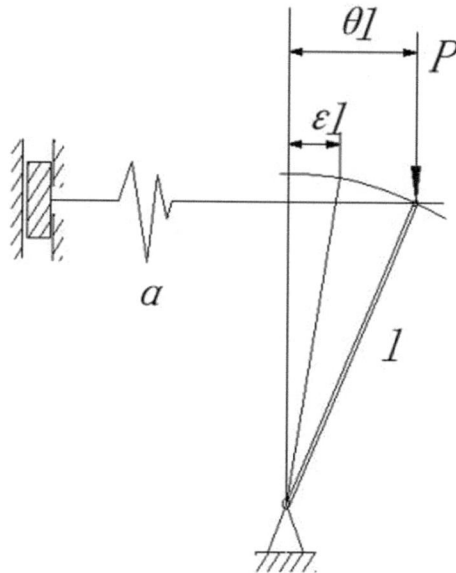

Spring force potential energy:

$$\Phi = \frac{1}{2}\alpha(\theta l - \varepsilon l)^2$$
$$= \frac{1}{2}\alpha l^2(\theta - \varepsilon)^2 \tag{8.57}$$

Load potential energy:

$$\phi_z = -P[\sqrt{l^2 - (\varepsilon l)^2} - \sqrt{l^2 - (l\theta)^2}]$$
$$= -Pl[(1 - \varepsilon^2)^{\frac{1}{2}} - (1 - \theta^2)^{\frac{1}{2}}] \tag{8.58}$$

Total potential energy of the system:

$$\Pi(\theta, P, \varepsilon) = \frac{1}{2}\alpha l^2(\theta - \varepsilon)^2 - Pl[(1 - \varepsilon^2)^{1/2} - (1 - \theta^2)^{1/2}] \tag{8.59}$$

The equilibrium condition can be expressed as:

$$\frac{\partial \Pi}{\partial \theta} = \alpha l^2(\theta - \varepsilon) - Pl\theta(1 - \theta^2)^{-1/2} = 0 \tag{8.60}$$

And there is:

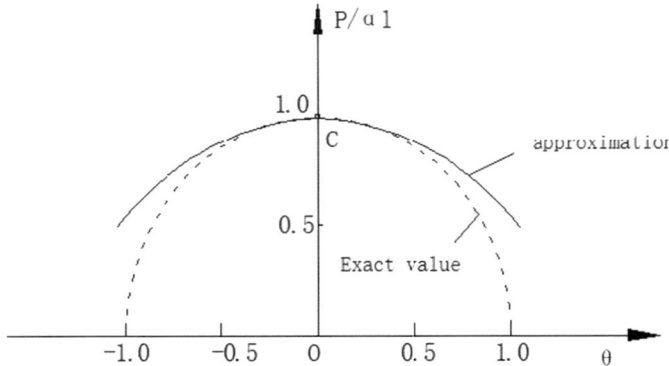

**Fig. 8.7** Symmetric unstable critical load bifurcation point and equilibrium principle

$$\frac{\partial^2 \Pi}{\partial \theta^2} = \alpha l^2 - Pl(1 - \theta^2)^{-1/2} - Pl\theta^2(1 - \theta^2)^{-3/2} \tag{8.61}$$

(1) Improving the system

This is obtained by substituting into Eq. (8.60): $\varepsilon = 0$.
Initial Balancing Path

$$\theta = 0 \tag{8.62}$$

and Posterior Flexion Balance Path

$$P(\theta) = \alpha l(1 - \theta^2)^{1/2} \tag{8.63}$$

These two paths intersect at the critical point C, as shown in Fig. 8.7, with the corresponding critical loads $P_{cr} = \alpha l$.
At this time,

$$\left(\frac{\partial^2 \Pi}{\partial \theta^2}\right)\bigg|_{\theta=0} = l(\alpha l - P) \tag{8.64}$$

Firstly, the stability of the initial equilibrium path is discussed. It is not difficult to see:

Below the branch point C $P\langle \alpha l, \frac{\partial^2 \Pi}{\partial \theta^2} \rangle 0$.

Stabilize the equilibrium; at the branch points C below $P\langle \alpha l, \frac{\partial^2 \Pi}{\partial \theta^2} \rangle 0$,

Above i.e. The branch point C $P > \alpha l$, $\frac{\partial^2 \Pi}{\partial \theta^2} < 0$, unstable equilibrium;

At the branch point C, there are, $P = \alpha l$, $\frac{\partial^2 \Pi}{\partial \theta^2} = 0$, critical equilibria.

The stability of the post-buckling equilibrium path is then analyzed by Eq. substituting (8.63) into Eq. (8.61) and using the series expansion to obtain:

$$\frac{1}{x+1} = 1 - x + x^2 - x^3 + x^4 + \cdots + (-1)^n x^{n-1} \cdots \quad (|x| < 1)$$

obtained by simplification:

$$\frac{\partial^2 \Pi}{\partial \theta^2} = \alpha l^2 - Pl[(1-\theta^2)^{-1/2} + \theta^2(1-\theta^2)^{-3/2}]$$

$$= \alpha l^2 - \alpha l^2[1 + \theta^2(1-\theta^2)^{-1}]$$

$$= -\alpha l^2(\theta^2 + \theta^4 + \cdots) < 0$$

This suggests that the posterior flexion equilibrium path is unstable everywhere.

The study continues with the critical equilibrium state, i.e., the branching point the stability of. When C $\theta = 0$ is, it remains to examine the extremes of third-order differentials or third-order partial derivatives: $\frac{\partial \Pi}{\partial \theta} = \frac{\partial^2 \Pi}{\partial \theta^2} = 0$

$$\frac{\partial^3 \Pi}{\partial \theta^3} = -3Pl\theta(1-\theta^2)^{-3/2} - 3Pl\theta^3(1-\theta^2)^{-5/2}$$

And so: $\left(\frac{\partial^3 \Pi}{\partial \theta^3}\right)\Big|_{\theta=0} = 0$.

Further write the following equation:

$$\frac{\partial^4 \Pi}{\partial \theta^4} = -3Pl(1-\theta^2)^{-3/2} - 18Pl\theta^2(1-\theta^2)^{-5/2} - 15Pl\theta^4(1-\theta^2)^{-7/2}$$

from this, we can see $\left(\frac{\partial^4 \Pi}{\partial \theta^4}\right)\Big|_{\theta=0} = -3\alpha l^2 < 0$.

Consider the Taylor series expansion of in the neighborhood of the branching point, where the first nonzero term is: $\Pi(\theta, P)$

$$\frac{1}{4!} \left(\frac{\partial^4 \Pi}{\partial \theta^4}\right)\Big|_{\theta=0} \theta^4 = -\frac{1}{8}\alpha l^2 \theta^4 < 0$$

Therefore, $\Pi$ is a local maximum at the bifurcation point C, which indicates that the critical equilibrium state is unstable.

The above conclusions can be shown in Fig. 8.7, where the stable equilibrium path is represented by a solid line and the unstable equilibrium path is represented by a dotted line. It should also be pointed out that one of the two curves shown in the figure is an exact solution, that is, the unit arc represented by Eq. (8.63), and the other is an approximate solution, that is, in the expansion of the bracket term in Eq. (8.63), only the $\theta^2$ term is taken and the high-order term is omitted.

(2) Defective systems
(3) Defect system

**Fig. 8.8** Symmetric unstable branching point principle

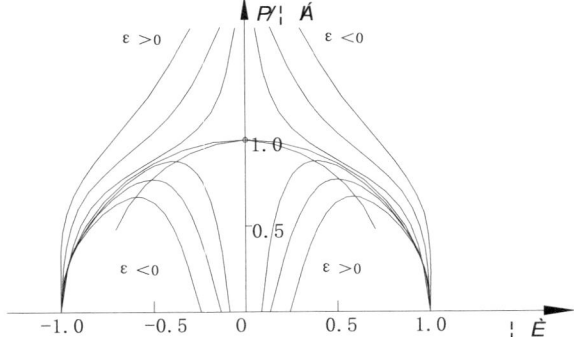

The equilibrium condition of the system is expressed as Eq. (8.60), which is expressed as:

$$P(\theta, \varepsilon) = \alpha l(1 - \theta^2)^{1/2}(1 - \frac{\varepsilon}{\theta}) \tag{8.65}$$

The equilibrium path of the perfect system ($\varepsilon = 0$) and the defect system ($\varepsilon = 0$) is plotted as the curve shown in Fig. 8.8 (symmetrical instability problem).

The extreme point load of the defective system $P_m$ can be by the condition determined, i.e. by Eq. (8.6): $\frac{\partial^2 \Pi}{\partial \theta^2} = 0$

$$\alpha l^2 - P_m l(1 - \theta^2)^{-1/2} - P_m l \theta^2 (1 - \theta^2)^{-3/2} \varepsilon = 0$$

Simplified, this gives:

$$P_m = \alpha l(1 - \theta^2)^{3/2} \tag{8.66}$$

According to Eq. (8.66), the curve of $P_m(\theta)$ can be drawn, that is, the ss curve shown in Fig. 8.8. The ss curve divides the whole region into a stable equilibrium region represented by a solid line and an unstable equilibrium region represented by a dotted line. Since the above curve family has the characteristics of symmetry and the critical equilibrium state is unstable, we call this type of stability problem a symmetric unstable bifurcation point problem (perfect system and defect system). By comparing Figs. 8.5 and 8.8, it can be seen that the important difference between symmetric stable bifurcation point problem and symmetric unstable bifurcation point problem is that the equilibrium path of the former is always rising and stable; the equilibrium path of the latter is rising at the beginning, but it turns to decline and becomes unstable after reaching the extreme point. It should also be pointed out that the problem of symmetrical unstable bifurcation points is often encountered in engineering structures. For example, the stability problem of thin cylindrical shells subjected to axial pressure belongs to this type of problem.

## 8.3 Nonlinear Stability Theory

The following is a brief introduction to the concept of defect sensitivity. In order to determine the relationship between the extreme point load Pm and the initial defect parameter e, the Eq. (8.66) is substituted into the Eq. (8.65) to obtain:

$$\alpha l(1-\theta^2)^{3/2} = \alpha l(1-\theta^2)^{1/2}(1-\varepsilon/\theta)$$

Simplified:

$$\theta = \varepsilon^{1/3} \tag{8.67}$$

Substituting Eq. (8.67) back to Eq. (8.66), the important relationship between $P_{cr}$ and $\varepsilon$ is obtained, that is:

$$P_m(\varepsilon) = \alpha l(1-\varepsilon^{2/3})^{3/2} \tag{8.68}$$

This formula shows the measurement of ' defect sensitivity ', and the curve is shown in Fig. 8.9.

It is easy to see that at $(P_m)_{\max} = P_{cr} = \alpha l$ and $\varepsilon = 0$, the slope of the $P_m(\varepsilon)$ curve is infinite, that is, the curve is tangent to the vertical axis at this point. This important feature means that the system is very sensitive to small defects. In other words, extremely small initial defects can significantly reduce the extreme point load $P_m$ borne by this mechanical structure, that is, the bearing capacity is weakened. This phenomenon is obvious in the thin shell stability problem, and the project is sometimes encountered, such as the abnormal pressure phenomenon and instability phenomenon of the working face roof.

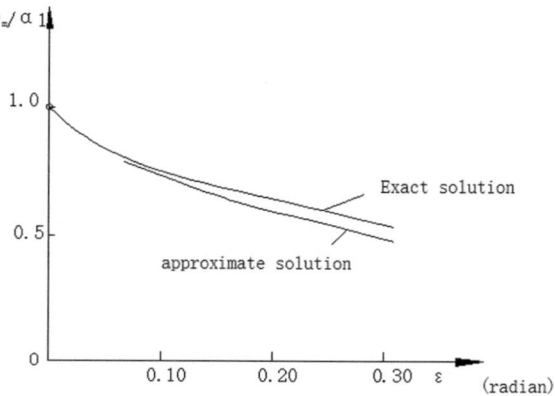

**Fig. 8.9** Schematic diagram of defect sensitivity principle

### 8.3.3 Jumping Phenomena in Flat Arches

Figure 8.10 shows a simplified model of a flat arch, which is a typical structural system model hinged by two linear springs. The spring can withstand the action of tension and pressure. The spring stiffness coefficient is $\alpha$, the initial inclination angle of the spring is $\theta_0$, the distance between the two points of AB is 2R, and a static load P in the vertical direction acts on the middle hinge C point. The stability of the model system is analyzed based on the energy principle criterion.

Here, only the case of symmetrical deformation of the above model is considered, so the model is a single-degree-of-freedom system, and the rotation angle $\theta$ is used as the system parameter.

Spring force potential energy:

$$\Phi(\theta) = 2 \cdot \frac{1}{2}\alpha \left( \frac{R}{\cos\theta_0} - \frac{R}{\cos\theta} \right)^2 \tag{8.69}$$

Assuming that the initial inclination angle $\theta_0$ and the post-deformation rotation angle $\theta$ of the flat arch structure are both small quantities, there are:

$$\cos\theta_0 = 1 - \frac{\theta_0^2}{2}$$

,

$$\cos\theta = 1 - \frac{\theta^2}{2}.$$

Substituting the second equation into Eq. (8.69) gives, after simplification:

$$\Phi(\theta) = \frac{1}{4}\alpha R^2 (\theta_0^4 - 2\theta_0^2\theta^2 + \theta^4) \tag{8.70}$$

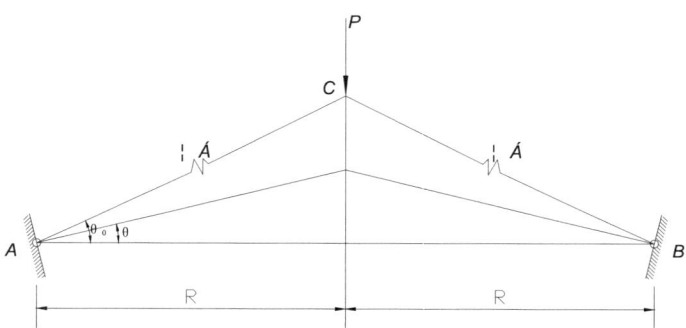

**Fig. 8.10** Jump phenomenon and system stability modeling

## 8.3 Nonlinear Stability Theory

Due to the deformation, the load drop displacement is:

$$\lambda(\theta) = Rtg\theta_0 - Rtg\theta = R(\theta_0 - \theta) \tag{8.71}$$

Total potential energy of the system:

$$\Pi(\theta, P) = \Phi + \phi_z = \frac{1}{4}\alpha R^2(\theta_0^4 - 2\theta_0^2\theta^2 + \theta^4) - PR(\theta_0 - \theta) \tag{8.72}$$

Thus, the following system (or structure) equilibrium conditions are obtained:

$$\frac{\partial \Pi}{\partial \theta} = \alpha R^2(-\theta_0^2\theta + \theta^3) + PR = 0 \tag{8.73}$$

This gives:

$$P(\theta) = \alpha R \theta (\theta_0^2 - \theta^2) \tag{8.74}$$

Let $P(\theta) = 0$, solve $\theta = 0$ and $\theta = \pm\theta_0$, that is, when the load is zero, the corresponding point has three equilibrium positions.

According to: $\frac{dP}{d\theta} = \alpha R(\theta_0^2 - 3\theta^2)$

$$\frac{d^2P}{d\theta^2} = -6\alpha R\theta$$

It can be known that when $\theta = \frac{\theta_0}{\sqrt{3}}$, $\frac{d^2P}{d\theta^2} < 0$, P takes the maximum value; when $\theta = -\frac{\theta_0}{\sqrt{3}}$, $\frac{d^2P}{d\theta^2} > 0$, P take the minimum value, the Eq. (8.74) is drawn into a curve, as shown in Fig. 8.11 (the extreme value function curve of the equilibrium system P) and Fig. 8.12. In order to analyze and discuss the stability of the equilibrium system, the second derivative of $\Pi$ is investigated:

$$\frac{\partial^2 \Pi}{\partial \theta^2} = \alpha R^2(3\theta^2 - \theta_0^2) \tag{8.75}$$

Then draw the above formula into a curve as shown in Fig. 8.13.

It can be seen that for two states $\theta = \pm\theta_0$ without force ($P = 0$), that is, B and F points shown in Figs. 8.11 or 8.12, $\frac{\partial \Pi}{\partial \theta} = 0$, $\frac{\partial^2 \Pi}{\partial \theta^2} > 0$, $\frac{\partial^2 \Pi}{\partial \theta^2} > 0$ equilibrium system is stable; however, for the non-load state $\theta = 0$, as shown in Fig. 8.12, point D. $\frac{\partial \Pi}{\partial \theta} = 0$, $\frac{\partial^2 \Pi}{\partial \theta^2} < 0$, $\Pi = \Pi_{\max}$, the equilibrium system is unstable; for the C and E points shown in Fig. 8.12, Because $\frac{\partial \Pi}{\partial \theta} = \frac{\partial^2 \Pi}{\partial \theta^2} = 0$ and note that the area B near the C and E points is changed, as shown in Fig. 8.13, the inflection points of C, E and $\frac{\partial^2 \Pi}{\partial \theta^2}$ correspond to each other. As shown in Fig. 8.13, therefore, the inflection points of points C, E and $\Pi$ correspond to each other, so that at C and E, $\Pi$ is not the minimum value, so the above two critical equilibrium states are also unstable. In

**Fig. 8.11** Extreme equilibrium function curves for a flat arch structure load P

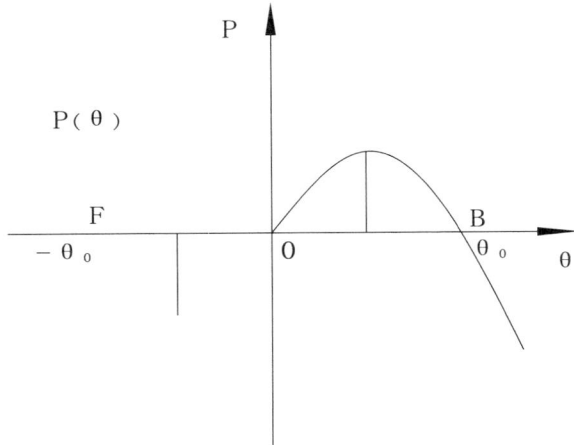

**Fig. 8.12** Load change states and stability of flat arches

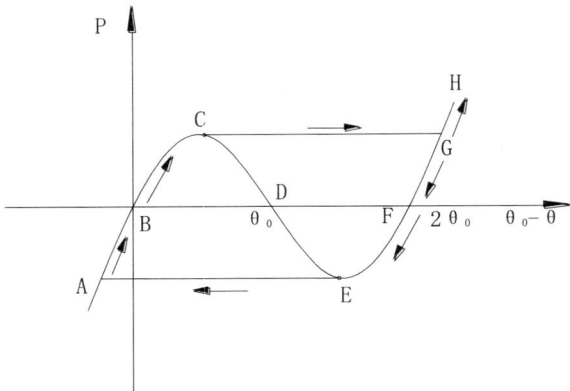

fact, Fig. 8.12 shows the general analysis method and results of equilibrium stability, where the solid line represents the stable equilibrium state of the structural system, and the dotted line represents the unstable equilibrium state of the structural system.

Combined with the curve shown in Fig. 8.12, the jump phenomenon (i.e., instability) of the flat arch shown in Fig. 8.10 is explained. If the load P gradually increases from zero, but is less than the load $P_{cr}$ corresponding to point C, the flat arch is in a stable equilibrium state represented by the curve BC. When $P = P_{cr}$, that is, the system is in an unstable critical equilibrium state point C, the flat arch will vibrate greatly under the constant load and jump to the stable equilibrium state G point. Due to the role of some damping dissipation energy generated during the CG displacement process, the system will stop at the point G of the equilibrium state. At this time, the flat arch first maintains the flipping form. If the loading continues, the system will be in a stable equilibrium state represented by the curve GH. A similar

## 8.4 Coal Rock Body Energy Principles

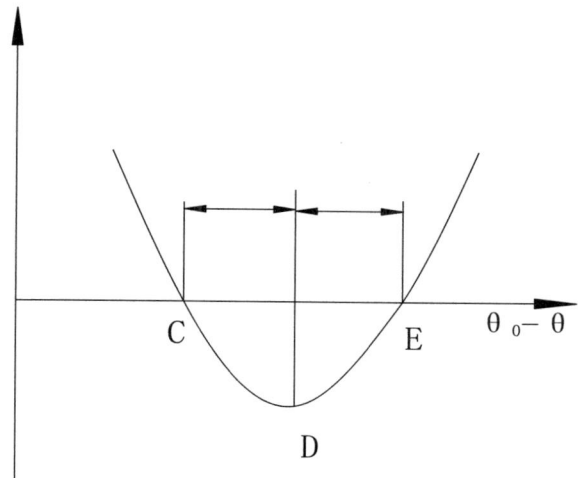

**Fig. 8.13** Extreme value and critical equilibrium principle

phenomenon occurs during unloading: the flat arch first maintains a flip state, and as the load decreases, it will be in a stable equilibrium state represented by the curve HGFE. Then E jumps to point A again at the point, and then the load is increased, the flat arch can return to the initial unstressed state B represented by point B.

It should be noted that the equilibrium path CE table shows the unstable equilibrium state. According to the conventional loading, this system equilibrium state will not be encountered. Only by using a certain control element can it be observed in the laboratory.

If the two springs in the model shown in Fig. 8.10 are changed into struts, the above analysis is also applicable, but the spring coefficient $\alpha = \frac{EA}{l}$ (where A is the area of the cross section of the strut and $l$ is the length of the strut). It should be pointed out that in the above analysis process, the problem of local instability, that is, the instability of each compression bar or spring due to compression, should not be considered. Otherwise, So far, the jump phenomenon is briefly analyzed and discussed. Although the selected example is a special case, it is necessary to study this similar engineering phenomenon. In fact, the jump phenomenon is not only one of the main contents of the shallow shell theory, but also an important research topic about the supporting structure and some displacement deformation phenomena.

## 8.4 Coal Rock Body Energy Principles

The basic criterion of the energy principle of coal and rock mass is that the elastic energy stored in the unit volume of coal and rock mass is a constant, which is independent of the stress state. If the elastic energy exceeds this constant, the coal and rock mass will be destroyed. The elastic strain energy is transformed from the work done by the external force in the case of elastic strain. In practice, sometimes coal rock mass has stored extremely high elastic strain energy in the process of geological formation, that is, elastic potential energy.

The work done by the external force and the principle of elastic strain energy conversion are explained by experiments. As shown in Fig. 8.14, object B is subjected to a gradually increasing load P, and a displacement $\Delta l$ is generated in B. If we assume that all the potential energy is converted into the elastic deformation energy of the object when the load sinks (the kinetic energy of the object, the energy consumption of the thermal energy and electromagnetic energy occurring with the deformation surface is neglected), the elastic deformation energy is numerically equal to the work done by the external force.

Within the elastic limit, when the external force gradually increases from zero to P, the displacement of the pressure P action point gradually increases from zero to $\Delta l$, as shown in Fig. 8.15, the work done by the P force on the displacement $\Delta l$ is:

$$W = \frac{1}{2} P \Delta l$$

Therefore, the elastic deformation energy of rock mass caused by P force should also be:

$$\phi = \frac{1}{2} P \Delta l \tag{8.76}$$

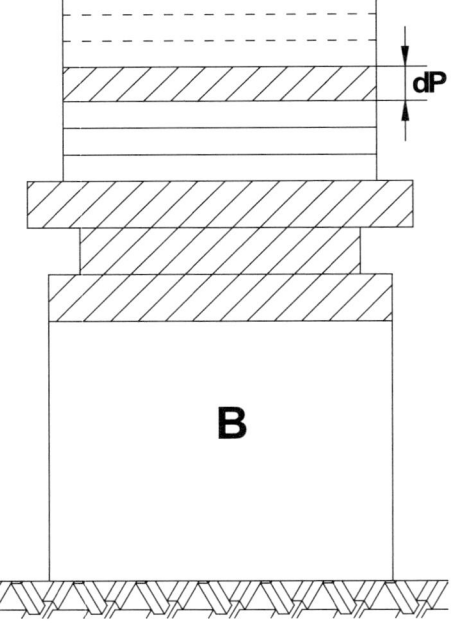

**Fig. 8.14** Object subjected to gradually increasing compression

## 8.4 Coal Rock Body Energy Principles

**Fig. 8.15** $P - \Delta l$ Work done on graphs

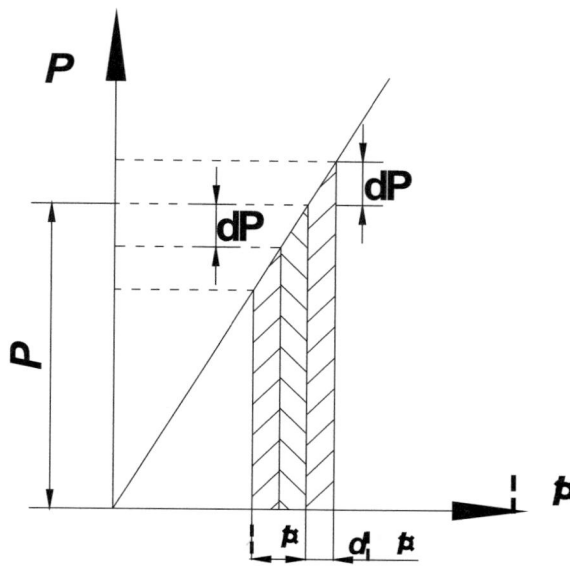

Now let the area of the object be A, the length be l, and the volume be V. Rewrite Eq. (8.76) as:

$$\phi = \frac{1}{2} \cdot \frac{P}{A} \cdot \frac{\Delta l}{l} \cdot V = \frac{1}{2} \sigma \cdot \varepsilon \cdot V$$

Then the polar elastic potential energy per unit volume can be found from the above equation:

$$\Phi = \frac{\phi}{v} = \frac{1}{2} \sigma \cdot \varepsilon \tag{8.77}$$

If a general elastic potential energy expression is used, according to Hooke's law the above equation becomes: $\sigma = E\varepsilon$

$$\Phi = \frac{\sigma^2}{2E} = \frac{1}{2} E \cdot \varepsilon^2 \tag{8.78}$$

In the three-way stress state, the cell is subjected to principal stresses of $\sigma_1$, $\sigma_2$, $\sigma_3$, and its principal strains are $\varepsilon_1$, $\varepsilon_2$, $\varepsilon_3$, using Eq. (8.78), the total potential energy is:

$$\Phi = \frac{1}{2} \sigma_1 \varepsilon_1 + \frac{1}{2} \sigma_2 \varepsilon_2 + \frac{1}{2} \sigma_3 \varepsilon_3 \tag{8.79}$$

Introducing the generalized Hook's Law (8.80) equation:

$$\left.\begin{array}{l}\varepsilon_1 = \frac{1}{E}[\sigma_1 - \nu(\sigma_2 + \sigma_3)] \\ \varepsilon_2 = \frac{1}{E}[\sigma_2 - \nu(\sigma_3 + \sigma_1)] \\ \varepsilon_3 = \frac{1}{E}[\sigma_3 - \nu(\sigma_1 + \sigma_2)]\end{array}\right\} \quad (8.80)$$

Substituting (8.80) into Eq. (8.79) collapses to:

$$\Phi = \frac{1}{2E}[\sigma_1^2 + \sigma_2^2 + \sigma_3^2 - 2\nu(\sigma_1\sigma_2 + \sigma_2\sigma_3 + \sigma_3\sigma_1)] \quad (8.81)$$

In the uniaxial compression or tensile test, when $\sigma_1 = \sigma_0, \sigma_2 = \sigma_3 = \sigma_0$ is broken, its elastic potential energy is:

$$\Phi_0 = \frac{\sigma_0^2}{2E} \quad (8.82)$$

The energy (intensity) criterion is written as:

$$\sigma_1^2 + \sigma_2^2 + \sigma_3^2 - 2\nu(\sigma_1\sigma_2 + \sigma_2\sigma_3 + \sigma_3\sigma_1) \leq \sigma_0^2 \quad (8.83)$$

Strictly speaking, the establishment of the energy criterion is conditional, only when $\sigma_0$ is always greater than P, otherwise the rock mass is subjected to equivalent compression $\sigma_1 = \sigma_2 = \sigma_3 = P$ (immediately), or when the P value is large, the deformation is also large, so the elastic deformation energy of the rock mass caused by the external force is even more than the extreme elastic potential energy of the rock mass shown in the Eq. (8.82), without any instability and fracture phenomenon. Therefore, the scholar Hu Bo suggested that the strength criterion should not be the extreme elastic potential energy shown in Eq. (8.81), but the elastic deformation energy (or work) consumed by the shape change after the volume change of the object should be subtracted from it, that is, the following formula is used:

$$\Phi_f = \Phi - \Phi_v \quad (8.84)$$

$$\text{(while) } \Phi_v = \frac{P_m \varepsilon_v}{2}$$

where $P_m$–the average stress, the value of which is $\frac{\sigma_1+\sigma_2+\sigma_3}{3}$
$\varepsilon_v$–volumetric strain, the value of which $\varepsilon_1 + \varepsilon_2 + \varepsilon_3$.
Substitute the values of $P_m$ and $\varepsilon_v$, and use the generalized Hooke's law (8.80) to simplify:

$$\Phi_v = \frac{1-2\nu}{6E}(\sigma_1 + \sigma_2 + \sigma_3)^2 \quad (8.85)$$

Substituting the value of Eq. (8.85) into Eq. (8.84), we can get:

## 8.4 Coal Rock Body Energy Principles

$$\Phi_f = \frac{2(1+v)}{6E}(\sigma_1^2 + \sigma_2^2 + \sigma_3^2 - \sigma_1\sigma_2 + \sigma_2\sigma_3 + \sigma_3\sigma_1) \quad (8.86a)$$

Or

$$\Phi_f = \frac{1+v}{6E}[(\sigma_1 - \sigma_2)^2 + (\sigma_2 - \sigma_3)^2 + (\sigma_3 - \sigma_1)^2] \quad (8.86b)$$

At this time, the elastic potential energy limit of rock mass ($\sigma_1 = \sigma_0, \sigma_2 = \sigma_3 = \sigma_0$) under uni-axial compression is:

$$\Phi_f = \frac{1+v}{6E}2\sigma_0^2$$

The energy (strength) criterion has another form of expression as follows:
When $P_m > 0$ (rock mass compression):

$$(\sigma_1 - \sigma_2)^2 + (\sigma_2 - \sigma_3)^2 + (\sigma_3 - \sigma_1)^2 \leq 2\sigma_0^2 \quad (8.87)$$

It can be further written in the form of shear stress:

$$\tau_{12}^2 + \tau_{23}^2 + \tau_{31}^2 \leq 2\tau_c^2 \quad (8.88)$$

where $\tau_c = \frac{\sigma_c}{2}$, $\tau_{12} = \frac{\sigma_1 - \sigma_2}{2}$, and so on.

When $P_m < 0$ (stretch), according to Hubble's opinion, the original energy (strength) Eq. (8.83) is still retained.

The energy (strength) criterion (8.87) is compared with (8.83). The characteristic of (8.87) is that it does not contain any elastic parameter hypothesis. When the rock mass is unstable and destroyed, it is not constrained by the premise of the elastic state (such as elastic deformation energy, Hooke's law, etc.). In this way, (8.87) can be used as a general criterion for the transition of rock mass from elastic state to plastic state. If the rock mass does not break and lose stability and can continue to undergo deformation, it can be used as a flow limit ($\sigma_c = \sigma_m$) analysis and discussion. Therefore, it can be used as a plastic condition of rock mass in the plastic deformation process:

$$(\sigma_1 - \sigma_2)^2 + (\sigma_2 - \sigma_3)^2 + (\sigma_3 - \sigma_1)^2 = 2\sigma_m^2$$

The results of the torque load stress test of thin-walled circular tubes made of copper, aluminum, mild steel and other materials show that the Eq. (8.87) also has a more common form of application, that is, the octahedral shear stress diagram is used. The octahedron is always found on the unit infinitesimal body an inclined plane, so that the normal of the inclined plane is equal to the three coordinate axes, that is, the direction cosine is equal to, $l = m = n = \frac{1}{\sqrt{3}}$. The unit micro-unit composed of this kind of inclined plane is octahedron. The normal stress and shear stress on the octahedron are:

$$\sigma_{oct} = \frac{1}{3}(\sigma_1 + \sigma_2 + \sigma_3) \tag{8.89a}$$

Or

$$\tau_{oct} = \frac{1}{3}[(\sigma_1 - \sigma_2)^2 + (\sigma_2 - \sigma_3)^2 + (\sigma_3 - \sigma_1)^2]^{\frac{1}{2}} \tag{8.89b}$$

It can be seen from the above formula that the normal stress on the octahedron is the average value of the three principal stresses. The shear stress of Eq. (8.89) is basically the same as that of Eq. (8.87), but the coefficients are different.

Based on the energy theory and viewpoint (considering stress and strain), the influence of three principal stresses is considered. Finally, it is expressed in the form of shear stress. This is the same as the Coulomb criterion, which is based on shear stress as the criterion criterion. In terms of value relationship, the following analysis and comparison can be made.

According to the Coulomb strength criterion, in the plane stress state, $\sigma_2 = 0$, then has:

$$\left.\begin{array}{l} -\sigma_0 \leq \sigma_1 \leq \sigma_0 \\ -\sigma_0 \leq -\sigma_3 \leq \sigma_0 \\ -\sigma \leq \sigma_3 - \sigma_1 \leq \sigma_0 \end{array}\right\} \tag{8.90}$$

The energy criterion is Eq. (8.87). Similarly, if it is a plane stress state, $\sigma_2 = 0$, then:

$$\sigma_1^2 - \sigma_1\sigma_3 + \sigma_3^2 \leq \sigma_0^2 \tag{8.91a}$$

If the $\sigma_1, \sigma_3$ coordinate systems are rotated by 45°, that is, $\sigma_1 = X\cos 45° - Y\sin 45°$ and $\sigma_3 = X\sin 45° + Y\sin 45°$ are substituted into Eq. (8.91a), and after sorting, we can get:

$$\frac{X^2}{\left(\sqrt{2}\sigma_0\right)^2} + \frac{Y^2}{\left(\sqrt{\frac{2}{3}}\sigma_0\right)^2} \leq 1 \tag{8.91b}$$

The Eq. (8.91b) is the ellipse of the spindle in the XY coordinate system, as shown in Fig. 8.16. It can be seen from the figure that they are not much different in terms of quantity.

The key problem of energy theory is as mentioned above. The premise is that the rock mass is an elastic body, that is, in an elastic state. However, in fact, when the coal rock mass is destroyed, it has gone beyond the scope of elastic identification and hypothesis. In addition, in the formula derivation, the use of tensile and compressive strength is equal, which is not in line with the actual rock material.

At present, the energy theory is less used in rock mechanics and mining rock mechanics due to the above reasons, and its application research is in the preliminary

## 8.4 Coal Rock Body Energy Principles

**Fig. 8.16** Comparison of Coulomb and energy criteria

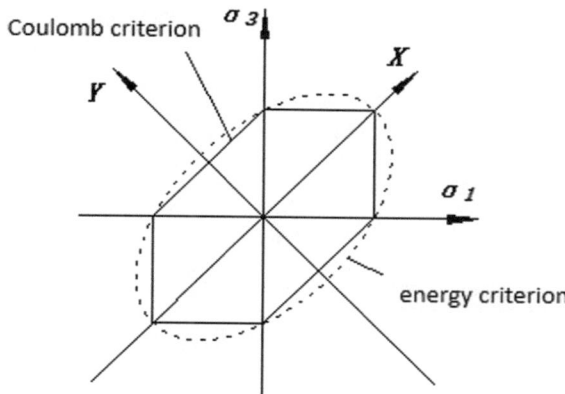

stage. However, in the rock mass with fluid and rock rheology as dynamic failure (such as rock burst, coal and rock outburst, high pressure water fine jet, etc.), the calculation energy is used more.

However, the energy criterion theory considers the problem more comprehensively. For example, the stress and strain are considered at the same time, and the influence of the three principal stresses is considered. In order to make the coal and rock mass unstable and fractured, it is necessary to overcome the inherent structural shape of the object and the inherent strength molecular force of the object. Therefore, energy must be consumed and work must be done. The application of energy theory is characterized by strong logic, profound, mysterious and clear interpretation of problems. In recent years, it has been favored by some experts, scholars and graduate students.

**Open Access** This chapter is licensed under the terms of the Creative Commons Attribution-NonCommercial-NoDerivatives 4.0 International License (http://creativecommons.org/licenses/by-nc-nd/4.0/), which permits any noncommercial use, sharing, distribution and reproduction in any medium or format, as long as you give appropriate credit to the original author(s) and the source, provide a link to the Creative Commons license and indicate if you modified the licensed material. You do not have permission under this license to share adapted material derived from this chapter or parts of it.

The images or other third party material in this chapter are included in the chapter's Creative Commons license, unless indicated otherwise in a credit line to the material. If material is not included in the chapter's Creative Commons license and your intended use is not permitted by statutory regulation or exceeds the permitted use, you will need to obtain permission directly from the copyright holder.

# Chapter 9
# Quarry Mine Pressure Monitoring Technology and Applications

## 9.1 Research Methodology for Mine Pressure Monitoring Summary

The mechanical behavior and action factors of the compressive stress of coal and rock mass and the corresponding supporting structure have the dynamic and timeliness of creep development, and its representation is complex. There are three kinds of research methods for mine pressure. One is the mathematical and mechanical simulation analysis method, such as computer simulation research using finite element or discrete element method (such as FLAC, 3DEC, UDEC, RFPA, ANSYS, ABQUS and other calculation software); the second is the physical mechanics analysis method, including the research method with the goal of experimental mechanics analysis, the physical similarity simulation experiment method, and the 3D printing technology. The third is the on-site target monitoring research method, which uses the corresponding instrument equipment to monitor the target parameters under the conditions of the production process on the spot. It is the simplest monitoring research method. This monitoring research method reflects the academic thought of simplifying the road. So far, the first two methods still need to simplify some complex and changeable factors, and the processing results are seriously affected by human subjective factors. The method is more suitable for academic discussion in theoretical research. As a qualitative analysis and result verification criterion, it is only limited to the test and parameter solution of simpler problems. At the same time, these two methods also require field observations to provide accurate parameters, otherwise the results will be seriously divorced from reality, and the practical value is extremely low.

The on-site monitoring research method is the measurement and recording of the whole process and phenomenon of the mine pressure appearance or mechanical behavior directly facing the production site. It uses various targeted instruments and tools to monitor the deformation and displacement of the surrounding rock of the stope, the damage of the roof and floor, the load of the support and the pressure of the roof. Then, from the dynamic analysis and research, the conditions, time and direction of rock movement and the distribution and change of abutment pressure around the stope are obtained, which have a significant impact on the mine pressure behavior of the stope. It provides a basis for solving the problem of on-site mine pressure control, such as the operation state and support force of the fully mechanized mining support, the ' prediction and prediction ' of the working face pressure, the prediction of the roof surrounding rock disaster, the support design and support selection, the selection of the roadway position, and the arrangement of the mining procedure.

Because the data obtained from the on-site production process monitoring reflect the interaction between dynamic pressure and support under the combined action of various engineering factors and the law of creep development under pressure, it is more reliable to use this data to demonstrate and analyze the problem, and the effect is remarkable. More and more for the production site technology management recognition and adoption. In addition, this model is conducive to the direct contact and participation of engineering and technical management personnel in mine pressure research.

The on-site mine pressure observation needs to have a clear purposeful target. Only by analyzing the measured data obtained under the guidance of professional theoretical knowledge in line with objective reality can effective application results be obtained. Therefore, on-site monitoring requires laboratory experimental research and numerical simulation analysis results of mathematical mechanics as theoretical guidance, so as to scientifically select target elements to improve monitoring methods and obtain correct information analysis data. The experimental research method and numerical simulation method need to be based on the data and data obtained from on-site monitoring, and their research conclusions also need to be monitored and tested on site. Therefore, the three types of research methods used in mine pressure complement each other and complement each other to achieve a breakthrough in theory and practice.

## 9.2 Purpose of Mine Pressure Monitoring in Quarries and Current Status of Technology Development

### *9.2.1 Purpose and Role of Mine Pressure Monitoring Studies*

Mine pressure monitoring and research must be closely related to the actual situation of the project, in order to solve the practical problems of production services. To sum up, its objectives and tasks are as follows:

(1) By analyzing and observing the monitoring information, understand the scope, conditions and time of the overlying strata movement under the specific mining methods and the corresponding analysis of the monitoring information, laws, characteristics and other information, strengthen the roof management of the

## 9.2 Purpose of Mine Pressure Monitoring in Quarries and Current Status ...

working face. It includes: ① By monitoring the variation law of working resistance of fully mechanized mining support, the roof pressure situation of working face is understood, the roof management of working face and the prediction of abnormal pressure are carried out to prevent the sudden pressure of large area roof; ② Understand the scope of the mine pressure control object, the actual support force of the working face support, the interaction between the support and the roof, the selection of the reasonable support and the accumulation of technical management experience; ③ According to the monitoring pressure of the support, the reflection law of the roof is judged, the operation state of the support in each area is analyzed and mastered, the process and labor organization are reasonably arranged, and the key areas of support are determined with the pressure change in the process of advancing the working face, so as to realize safe and efficient production; ④ The reliability of the operating conditions of the support, find and understand the existence of hidden dangers such as the sealing performance of the support, leakage of liquid, inclination, and empty roof; ⑤ Through monitoring to understand the relationship and law between the load force change of the bolt and anchor cable and the displacement deformation of the surrounding rock of the roadway roof, and to control and understand the response of the anchor net cable bearing structure of the roadway to the influence of mining and the influence of the roof collapse in the goaf.

(2) Monitoring and predicting the disasters induced by mining dynamic pressure disturbance, such as the deformation of roof (including floor deformation) and the deformation and displacement of the inner and outer sides of the roadway, which provides the basis and accumulated experience for reasonable roadway pressure control and support design, including:

① Firstly, the subsidence movement of the overlying strata of the working face roof and the goaf (including the adjacent goaf) induces the increase of the fluctuation of the compressive stress of the surrounding rock roof of the roadway (especially the roadway on the side adjacent to the goaf), that is, the formation of the dynamic pressure peak induces the fluctuation of the bolt and cable load of the surrounding rock roof of the roadway, and the creep instability and local catastrophic failure of the surrounding rock roof occur, resulting in the combined support bearing system composed of anchor net and cable. After the dynamic pressure, the violent dynamic pressure reaction occurs, resulting in local creep instability and anchorage force failure, so that the role of the support system or the combined arch beam creeps or loses stability and reduces the anchorage force. It even causes the bearing function of the supporting structure system to be transformed into a single suspension beam function form, and various appearances and laws (related parameters) in the process of change; ② The deformation effect of dynamic pressure on the floor of soft rock roadway and the propagation law of dynamic pressure and the monitoring of relevant parameters are studied, especially the analysis of the shear failure effect of the compressive stress transmitted from the roof and high rock strata to the roadway side; ③ With the help of monitoring means, on the basis of clarifying the distribution law and variation characteristics of the abutment pressure of the coal body in the working face, the advanced maintenance of the working face is reasonably carried out, and the location and time of the roadway excavation are determined. The control design of the size of the coal pillar and the mining procedure provides the basis; ④ The dynamic spread range of dynamic pressure in working face mining, such as the quantitative relationship (or function relationship) between the distance of working face advancing position and the roof subsidence and separation; ⑤ Monitor and understand the creep depth (range) and characteristics of surrounding rock loose circle caused by tunneling, and understand the law and phenomenon of decompression expansion displacement of original rock mass and catastrophic disasters.

(3) Mine pressure monitoring and research methods. The main purpose of stope mine pressure monitoring is to understand a series of mine pressure phenomena caused by the movement of overlying strata in the stope or mining area caused by mining. The monitoring means should include human experience and visual observation and test instrument observation.

> ① Empirical visual observation. Recording and describing the phenomenon of mine pressure in stope by visual observation based on experience depends on the accumulation of long-term engineering experience of human beings, including the comparison by engineering analogy method. The main contents of visual observation include: deformation and failure of advanced roadway, rib spalling of coal wall, roof subsidence and crushing, local caving, fully mechanized mining support damage (location and time) and gangue caving status. In addition, with the advance of the working face, it is necessary to record and describe the geological changes, and predict some possible problems, such as the change of roof lithology and performance strength and coal seam thickness, joint fissures and faults. ② Instrument monitoring. Instrument observation can quantitatively describe the phenomenon of mine pressure, so as to reveal and record the law of mine pressure more deeply and directly. Mine pressure instruments mainly include monitoring displacement, deformation and monitoring pressure. The monitoring process is the mine pressure behavior parameters obtained under the advancing state of the stope. According to these parameters, it is used as the basis for analyzing and studying the problems: the relative displacement and velocity of the roof and floor at different positions of the advancing roadway with the advancing of the stope, and the distance from the working face L and the mine pressure behavior parameters. The positional relationship or functional relationship between; roadway roof subsidence, subsidence velocity, deformation and velocity of surrounding rock, and its relationship with the working face distance L; the rationality between the calculated value of the support bearing pressure and the periodic weighting of the roof, as well as the relationship between the support bearing pressure and the roof caving in the goaf.

## 9.2.2 Current Status of Technology Development

The monitoring technology of mine pressure in stope includes the monitoring of basic parameters such as working pressure of fully mechanized mining support, pressure and displacement deformation of surrounding rock and roof of two roadways.

① Fully mechanized mining support is the main equipment of fully mechanized mining face, which provides safety guarantee for coal mining operation, and can effectively prevent roof strata caving from endangering the safety of people and equipment. The support quality of fully mechanized mining support is usually evaluated by indexes such as initial support force and working resistance. The former can reflect whether the initial support force (support force) of the support is strong, and the latter can reflect the dynamic change law of the pressure of the roof to the support during the coal mining process.

② The monitoring of roof pressure and displacement deformation of roadway surrounding rock is to measure the displacement deformation of surrounding rock roof by monitoring the load force or anchoring force of anchor cable and the displacement deformation of surrounding rock roof by drilling multi-point displacement meter abscission layer instrument or cross method measuring rod, and evaluate the stability of surrounding rock roof of stope according to these two indexes. The cross method can only measure the surface deformation, and it is inconvenient to understand the deep displacement of surrounding rock, which is basically eliminated.

In recent years, China's stope mine pressure monitoring methods and technical management level have developed rapidly, especially for the support quality monitoring technology of coal mining face and roadway. Whether it is mechanical instruments or electronic instruments, the quality and quantity have been greatly improved compared with the past five years. A direct-reading pressure (deformation) measuring instrument with stable performance and convenient operation and a monitoring system composed of data storage records and semi-automatic data acquisition and transmission and computer software based on single-chip microcomputer as the core have emerged. Based on years of experience, the technical indicators, advantages and disadvantages of the main widely used fully mechanized mining support and roadway support quality inspection (monitoring) equipment are listed in Table 9.1.

In general, the two types of monitoring equipment listed in Table 9.1, both in terms of application quantity and structure, are gradually developing towards practicality and high technical content. Some equipment with troublesome technical operation and poor stability is gradually replaced and eliminated. The monitoring equipment with high practicability and integration is gradually accepted and recognized. However, as far as the existing situation is concerned, the overall performance price ratio of monitoring equipment is relatively low compared with other fields, and its function is not strong. It has not yet involved the technical application of automatic control. The monitoring index function is single, and the information

**Table 9.1** Main technical specifications of several types of quarry support monitoring equipment

| Types of instruments | | Display mode | Data acquisition modes | Data application | Stabilization | Working life/Month | Handling labor | Real-time | Performance price |
|---|---|---|---|---|---|---|---|---|---|
| Integrated mechanical coal mining support monitoring | Vibration-resistant mechanical gauge | Analog pointer | Manual periodic readings | General | Poor performance | 2–5 | General | Poor performance | General |
| | Digital pressure gauge | Liquid crystal | Manual periodic readings | General | Good stability | 6–12 | Low maintenance effort | Poor stability | High cost-effectiveness |
| | Real-time micro-controller-based data logger with storage | Liquid crystal | Micro-controller-based data logger with storage and wireless transmission data acquisition via collectors | Good | Good | 6–16 | General | Good | High cost-effectiveness |
| | Computer on-line monitoring system | Liquid crystal | Ground computer-based online monitoring and data processing system | Good | General | 6–12 | Comparatively large | Good | General |
| Roadway surrounding rock stability monitoring | Mechanical monitoring instrument | Pointer, scale formula | Manual periodic readings | General | General | 2–6 | Few | Poor stability | High cost-effectiveness |
| | Digital monitoring instruments | Liquid crystal | Manual periodic readings | General | Good stability-good | 6–12 | Few | Poor stability | General |
| | Real-time monitoring of monolithic memory loggers | Liquid crystal | Micro-controller record storage, collector infrared wireless data collection micro-controller-based data logger with storage and wireless transmission data acquisition via collectors | Good | Good | 6–16 | General | Good | High cost-effectiveness |
| | Computerized on-line monitoring system | Liquid crystal | Ground computer-based online monitoring and data processing system | Good | General | 6–12 | Comparatively large | Good | General |

utilization rate is very low. Especially for the computer online monitoring system of fully mechanized mining support with advanced technology, its monitoring index and engineering application need to be further strengthened and improved in the future to meet the needs of engineering technology management.

The significance and application of mine pressure monitoring should be reflected in three aspects: ① Online real-time monitoring and display of the current data information, in order to facilitate the production staff, technical personnel, safety inspectors semi-random, intuitive view of the current support quality, support system conditions, roadway deformation and other parameter information; ② The engineering and technical management personnel use the basic parameter information of these monitoring to predict and analyze the data information related to safety management, such as the law of roof pressure and the deformation law of surrounding rock roof, and to predict the abnormal information such as the early warning of surrounding rock roof disaster; ③ Establish monitoring data information files, accumulate learning experience for engineering technology management and improve the ability to solve practical technical problems. However, as far as the current situation is concerned, there are many unsatisfactory phenomena. Due to the lack of specialized technical personnel, it is caused by the lack of formal monitoring, lack of scientific objectivity, and failure to play the role of safety management and abnormal information prediction and early warning. As far as the national situation is concerned, there is a large gap in the level of technological development and utilization in this area, and there is a lack of unified guidance text. The fundamental reason for this situation is that the research on the related mine pressure is not deep enough and not systematic enough. The engineering utilization and operability of the research results are not strong, and there is a lack of practical innovation. In particular, there is a lack of practical and guiding engineering monitoring technical standards for the industry departments to carry out targeted monitoring data information. This paper puts forward some ideas on the application of mine pressure monitoring data information based on a large number of monitoring data information accumulated for many years.

## 9.3 Information on Mine Pressure Observation Data for the Two Roadways in the Quarry

### 9.3.1 Arrangement of Mine Pressure Monitoring Zones in the Two Runways of the Quarry

The dynamic parameters of roadway surrounding rock mainly include the arrangement and application of data information of surrounding rock deformation and surrounding rock compressive stress. At present, the deformation of surrounding rock roof and the compressive stress of surrounding rock roof are generally (for a long time) used to indicate the stability of surrounding rock roof by using the coordinate curves (displacement deformation, time or distance) and (load or pressure,

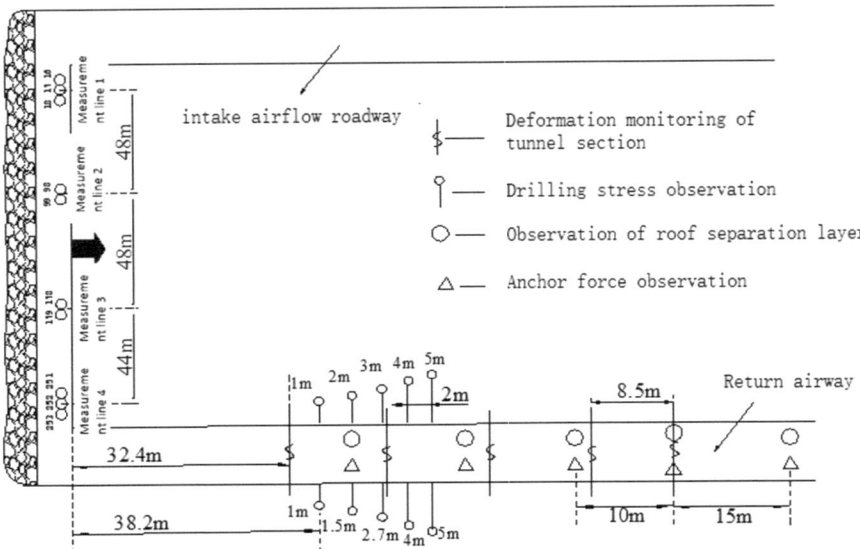

**Fig. 9.1** Schematic layout of the measuring area for dynamic monitoring of the roof of the roadway

time or distance) in the form shown in Fig. 9.1. This representation method can only one-sidedly explain a quantification at this time or location and a trend of change at a stage. In the case of inconvenience to give a safety threshold, it is considered from the perspective of engineering technology management needs. It is impossible to see that the monitored surrounding rock roof is stable and safe or abnormally dangerous at that time, and it is inconvenient or impossible to give an effective and substantive intuitive warning or indication to the overall creep state or abnormal state of the surrounding rock support body, which can provide insufficient information for comprehensive evaluation and analysis. It is easy to appear that the disaster risk of the support body cannot be found in time, resulting in one-sided engineering understanding of the deformation of the surrounding rock roof, and the engineering significance of the monitoring project should not be obtained.

## 9.3.2 Dynamic Monitoring Curve Application for Cave Surrounding Rock

The dynamic parameters of roadway surrounding rock mainly include the arrangement and application of data information of surrounding rock deformation and surrounding rock compressive stress. At present, the deformation of surrounding rock roof and the compressive stress of surrounding rock roof are generally (for a

9.3 Information on Mine Pressure Observation Data for the Two Roadways … 245

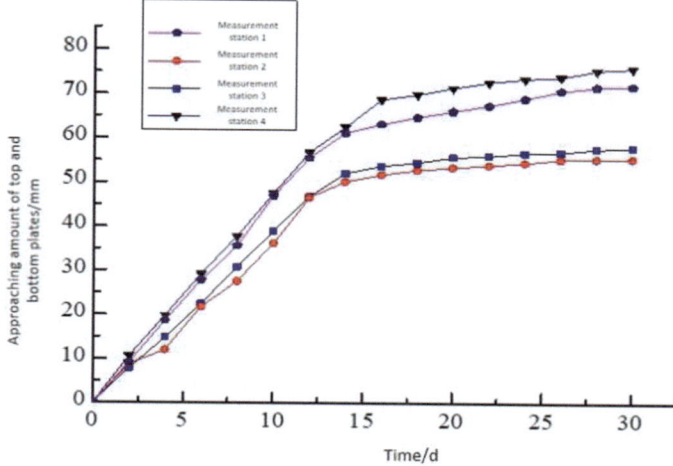

**Fig. 9.2** Commonly used dynamic curve models

long time) used to indicate the stability of surrounding rock roof by using the coordinate curves (displacement deformation, time or distance) and (load or pressure, time or distance) in the form shown in Fig. 9.2. This representation method can only one-sidedly explain a quantification at this time or location and a trend of change at a stage. In the case of inconvenience to give a safety threshold, it is considered from the perspective of engineering technology management needs. It is impossible to see that the monitored surrounding rock roof is stable and safe or abnormally dangerous at that time, and it is inconvenient or impossible to give an effective and substantive intuitive warning or indication to the overall creep state or abnormal state of the surrounding rock support body, which can provide insufficient information for comprehensive evaluation and analysis. It is easy to appear that the disaster risk of the support body cannot be found in time, resulting in one-sided engineering understanding of the deformation of the surrounding rock roof, and the engineering significance of the monitoring project should not be obtained.

The author suggests and proposes a new type of roof load force–deformation curve or load–displacement curve as shown in Fig. 9.3. This representation method can comprehensively reflect the stability or abnormality of the surrounding rock roof. The load force (or anchoring force) of the roof support body of the roadway surrounding rock and the displacement deformation have significant creep or jump interaction. Within the control range of the anchor cable or in the same roadway section: ① If the development trend of the coordinate monitoring curve is stable or the fluctuation is small, it indicates that the working condition of the support system is effectively given within the control range; ② If the fluctuation trend of the development trend of the monitoring curve is lower than the starting point of the

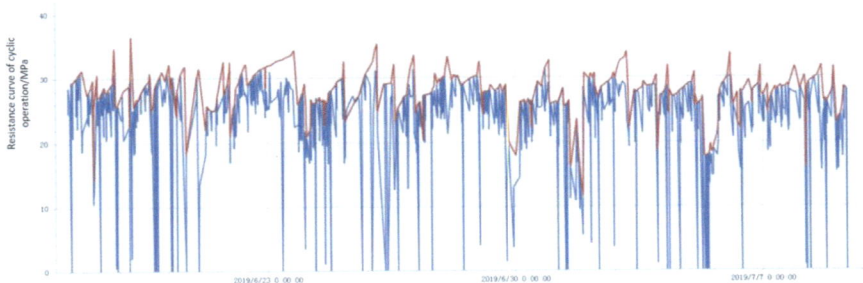

**Fig. 9.3** New dynamic curve model

ordinate, it indicates that the real-time anchoring force of the anchor cable is less than the initial anchoring force (or pre-tightening force), and the effectiveness of the supporting system in the control range is reduced, which is in an abnormal dangerous state. Therefore, it is of great engineering significance to develop and use a sensing device with simultaneous monitoring and recording of compressive stress (load) and displacement deformation to monitor the working conditions of the supporting system of the surrounding rock roof of the roadway. This method combines the force–deformation of the surrounding rock mass in the same control range or the same roadway section in the same space–time scale to study the application and guide the engineering management. This paper uses the classical rock mechanics method to study the dynamic change of the surrounding rock mass (or support system) of the roadway in the stope, which has more engineering significance for the prediction and prediction of the disaster of the surrounding rock of the roadway in the stope.

## 9.4 Integrated Mining Support Pressure Monitoring Methods

### 9.4.1 Bracket Pressure Monitoring Measurement Area Arrangement

The computer on-line monitoring communication system of fully mechanized mining support pressure is used to monitor the working pressure of hydraulic support in real time, so as to understand the roof weighting step, periodic weighting and grasp whether the roof of goaf collapses on time (if it is not collapsed or difficult to collapse, artificial control method can be carried out, such as anchor cable retreating, roof picking and other measures to help the roof collapse in time according to the set step), as well as the working condition of hydraulic system of support (such as

## 9.4 Integrated Mining Support Pressure Monitoring Methods

leakage, liquid string, empty roof, inclination, etc.). The specific monitoring method is shown in Fig. 9.4. Figure 9.4 shows that three measuring areas are set up with a surface length of 150 m. If the length of the working face exceeds 200 m, five measuring areas can be set up, and three to five monitoring extensions can be set up in each measuring area. The arrangement and analysis of monitoring data can use the supporting computer data processing software system provided, and refer to the supplied data and other mine pressure technical data.

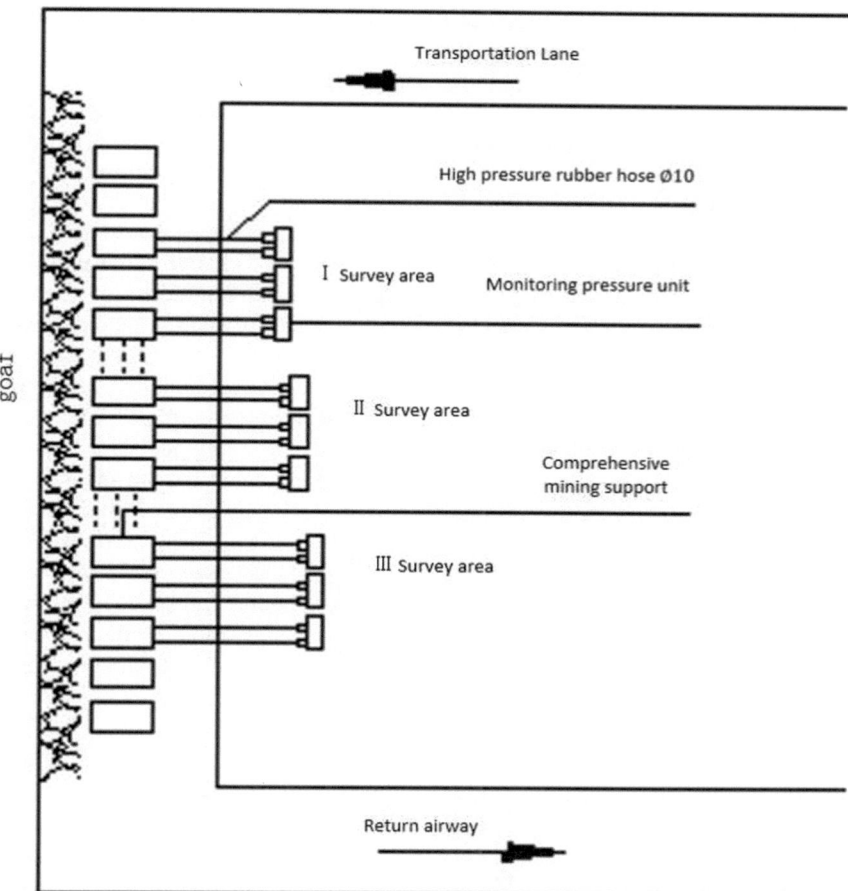

**Fig. 9.4** Schematic layout of the roof dynamic monitoring measurement area

## 9.4.2 Methods of Analyzing Pressure Curves for Integrated Mining Supports

The roof pressure of the working face of the stope is mainly reflected by monitoring the working resistance of the hydraulic support. The pressure intensity or size of the roof to the support is affected by the pressure of the overlying strata subsidence deformation of the goaf to the roof (direct roof). For example, the thick and hard roof shows significant mechanical and physical properties, such as elastic properties, easy or not easy to fall. The current roof pressure monitoring curve model is shown in Fig. 7.8(a–c). In the project, the roof periodic pressure (including the initial pressure step) law, working resistance, initial support force and other parameters can be obtained by monitoring the support pressure curve to meet the needs of engineering and technical management (which is also a general routine requirement). The author systematically analyzed and summarized the accumulated tens of thousands of on-site monitoring data, and found that the amplitude and frequency of the change of the support pressure curve can be roughly divided into three types shown in Fig. 7.8: three types of curves: dramatic change, moderate dramatic change and insignificant change. Based on the analysis and summary of academician Song Zhenqi's theory of 'transfer rock beam' and the characterization of mechanical and physical properties (properties, hardness, thickness, etc.) of coal and rock mass, it is proved that the roof pressure curve is the embodiment of the inherent mechanical and physical properties of the roof. It is a comprehensive reflection result under complex stress state; therefore, the characterization of the support pressure curve fully reflects the characteristics and laws of the creep reflected by the roof pressure during the advancement of the support. If the goaf, support and roof are regarded as a space–time structure relationship, the support pressure monitoring curve and the roof type (mechanical and physical performance characterization) are classified accordingly. Therefore, the transfer rock beam roof or overlying rock roof is divided into three categories: high-strength transfer rock beam roof (referred to as high-strength roof), medium-strength transfer rock beam roof (referred to as medium-strength roof), weak-strength transfer rock beam roof (referred to as weak-strong roof), as shown in Fig. 7.8. The high-strength roof and the support are rigid roof contact coupling, and the elastic roof contact characteristics are very obvious. The support is often placed in an unstable motion situation, and the pressure period (step distance) is long and not easy to fall. In the project, the pressure curve of the thick and hard roof is corresponding; the medium-strong roof and the support are medium-contact coupling, and the support effect is ideal; the weak strong roof and the support are coupled to the top, and the pressure period is not obvious or there is no periodic pressure phenomenon. The support effect is ideal, but the phenomenon of roof falling and gangue falling is easy to occur.

## 9.5 Monitoring Equipment Performance Classification and Applications

### 9.5.1 Lane Displacement and Deformation Monitoring Instrumentation and Applications

(1) Monitoring principle and application

The development process of any kind of creep instability and catastrophe representation form of coal and rock mass in stope follows the common change rule criterion: the creep of compressive stress or elastic potential energy inside coal and rock mass is the beginning of displacement creep caused by strength weakening, cohesion and friction angle change of coal and rock mass. The essence of displacement deformation is the displacement and surface deformation that occur first in the coal and rock mass. The roof of roadway, surrounding rock (two sides of roadway), coal seam, coal pillar, floor and other places prone to displacement and deformation need to be continuously monitored. The deformation displacement is caused by the decompression expansion of deep rock mass, and the action direction of force or elastic energy is the generation of displacement deformation. For example, the dominant stress of roadway floor heave deformation is the effect of high-level compressive stress of roof. In all kinds of mine roadways, the roadway of working face is most prone to deformation and failure due to the influence of mining and overlying strata movement in goaf (including adjacent goaf). Based on safety considerations, the absolute displacement deformation of the roadway should be controlled within 80 mm.

Displacement deformation monitoring is the most basic, most important and most complex content of mine pressure monitoring research. Taking the two-point displacement separation instrument as an example, the displacement deformation of surrounding rock and the deformation monitoring principle of floor heave are explained. The displacement and deformation monitoring principle of surrounding rock of roadway is shown in Fig. 9.5, usually using two scales of No.1 and No.2, No.1 monitors the displacement and deformation scale reading of coal and rock mass in the range of L1 length (or depth) of deep base point, No.2 monitors the displacement and deformation scale reading of coal and rock mass in the range of L2 length (or depth) of shallow base point, (L1-L2) represents the displacement and deformation of rock strata between deep and shallow base points (or crack development). If the value of L2 appears, it indicates that the rock mass in the control range of bolt or anchor cable has displacement loosening, and the anchoring force of bolt or anchor cable decreases significantly. The length selection of No.1 and No.2 scale, that is, the length of deep base point L1 and shallow base point L2, should consider the purpose of monitoring, such as the stability of roof surrounding rock, the design of bolt and anchor cable support or the study of pressure change, stress field and rock burst, and determine the number and length (depth) of selected measuring points. Based on the analysis and summary of the experience of anchor cable support in the field, considering that the bonding force between the borehole wall and the anchoring agent is the most likely to be destroyed in the project, the author suggests that the selection method of the borehole depth L1 and L2 monitored by the abscission layer instrument is: L2 = (1.1 ~ 1.3) × anchor (anchor cable) length, L1 = (1.6 ~ 2.0) × L2, borehole diameter φ = 28, take the lower limit, borehole diameter φ = 40, take the upper limit.

(2) Monitoring instruments

In recent years, the instruments suitable for monitoring displacement and deformation of stope have developed rapidly, forming a scale mechanical type, electronic (optical control) digital display type, real-time online record storage type, real-time online long-distance computer monitoring system. Details are shown in Table 9.2.

(3) Common installation monitoring methods
 ① Displacement and deformation of roadway roofs and roadway gangs

As shown in Fig. 9.6, the mechanical scale roof displacement separation monitoring instrument (ordinary type). Figure. (a) The ZKBY-IIB roof separation monitor is used. This form of separation monitor is specially developed for the deformation of roadway roof separation. It is especially suitable for large-scale roadway roof deformation monitoring, and the indication is easy to observe and read. Figure (b) ZKBY-IIA surrounding rock displacement abscission layer monitor, this form of abscission layer instrument roadway roof, roadway deformation monitoring, indication reading. Figure. (c) ZKBY-IIC roadway displacement monitor. Figure (d) ZHWY-IIB type surrounding rock displacement multi-point monitor. The common roof surrounding rock separation monitoring instrument is practical and reliable, and

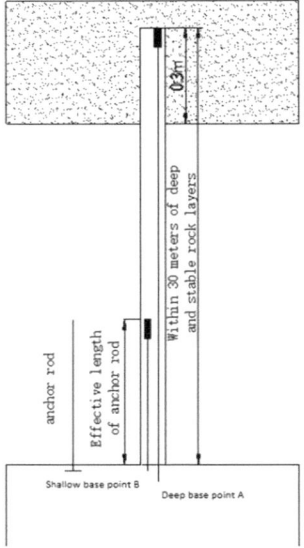

**Fig. 9.5** Schematic diagram of the principle of monitoring the displacement and deformation of the top plate surrounding rock

## 9.5 Monitoring Equipment Performance Classification and Applications

**Table 9.2** Commonly used roof displacement monitoring instruments

| Serial number | Model name | Functional principle uses | Display recording method | Manufacturer (of a product) |
|---|---|---|---|---|
| 1 | YHW150 roof displacement dynamic instrument (digital display) | Real-time monitoring of roof and floor displacement, record type, digital display type | New type of support monitoring stabilization | Shandong Keda Mine Monitoring Equipment Co., Ltd. and other manufacturers |
| 2 | KY82 roof dynamic instrument | Easy to operate, detection indicator pointer type | Real-time monitoring | |
| 3 | ZHWY-IIB type roadway displacement multi-point monitor | Multi-point displacement monitoring of surrounding rock and floor heave deep, scale measurement, 1 borehole 4 and 6 point type, depth self-determination | Real-time monitoring | |
| 4 | ZHWY-II A multi-point displacement monitor for roof rock layer | Roof multi-point displacement separation, scale measurement, 1 borehole 6-point type, depth self-determination | Scale indication formula | |
| 5 | ZKBY-IIB roof abscission layer monitoring indicator | Roof deformation, direct reading and color indication warning. 1 borehole L1, L2 measuring points, depth self-determined | Reflective three-color display | |
| 6 | ZKBY-IIA parallel surrounding rock separation indicator | Roof, roadway displacement deformation, direct reading. 1 borehole L1, L2 measuring points, depth self-determined | Reflective color display | |
| 7 | ZKBY-IIC type direct reading roadway wall separation indicator | The displacement deformation of the roadway side is read directly. 1 borehole L1 measuring point, depth self-determined | Reflective color display | |

(continued)

**Table 9.2** (continued)

Commonly used roof displacement monitoring instruments

| Serial number | Model name | Functional principle uses | Display recording method | Manufacturer (of a product) |
|---|---|---|---|---|
| 8 | YHW150 intrinsically safe roof displacement monitor | Roof, roadway displacement deformation, direct reading. 1 borehole L1 and L2 measuring points. Digital display or software analysis | Light control start, 1 borehole digital display warning or continuous storage records | |
| 9 | YHW150 intrinsically safe roof displacement monitor and recorder | Top plate and road gang displacement and deformation, showing the value of direct reading. 2 drill holes L1, L2 measurement points. Software analysis | Optically controlled start, set of 2 drilled holes with four measuring points Digital storage | |
| 9 | YHW150C intrinsically safe roof displacement monitoring recorder | Top plate and road gang displacement and deformation, showing the value of direct reading. 3 drill holes L1, L2 measurement points. Software analysis | Optical activation, sets/6 measurement points | |
| 11 | ZHC-II types mine movable measuring tape | Portable, direct-reading, retractable length | Scale | |
| 12 | CBL-II portable ultrasonic rangefinder | Portable digital display, measuring distance 0–30 m | | |

has the characteristics of high practical reliability and price ratio. It is suitable for a large number of large-scale comprehensive monitoring and monitoring installation. It can not only meet the requirements of production safety management, but also meet the requirements of monitoring and research, provide the basis for the design of support parameters, and analyze and summarize the failure law of surrounding rock mass (layer) under pressure.

Real-time on-line monitoring of roadway gangs (monolithic continuous recording and storage type)

As shown in Fig. 9.7, YWH150 dynamic monitoring and recording device for top plate rock displacement and deformation is used, and Fig. 9.7a represents two ways of real-time monitoring, namely, real-time monitoring optical numerical display type and real-time on-line monitoring storage and recording type using borehole, and Fig. 9.7b represents utilizing one real-time on-line monitoring and storage and

## 9.5 Monitoring Equipment Performance Classification and Applications

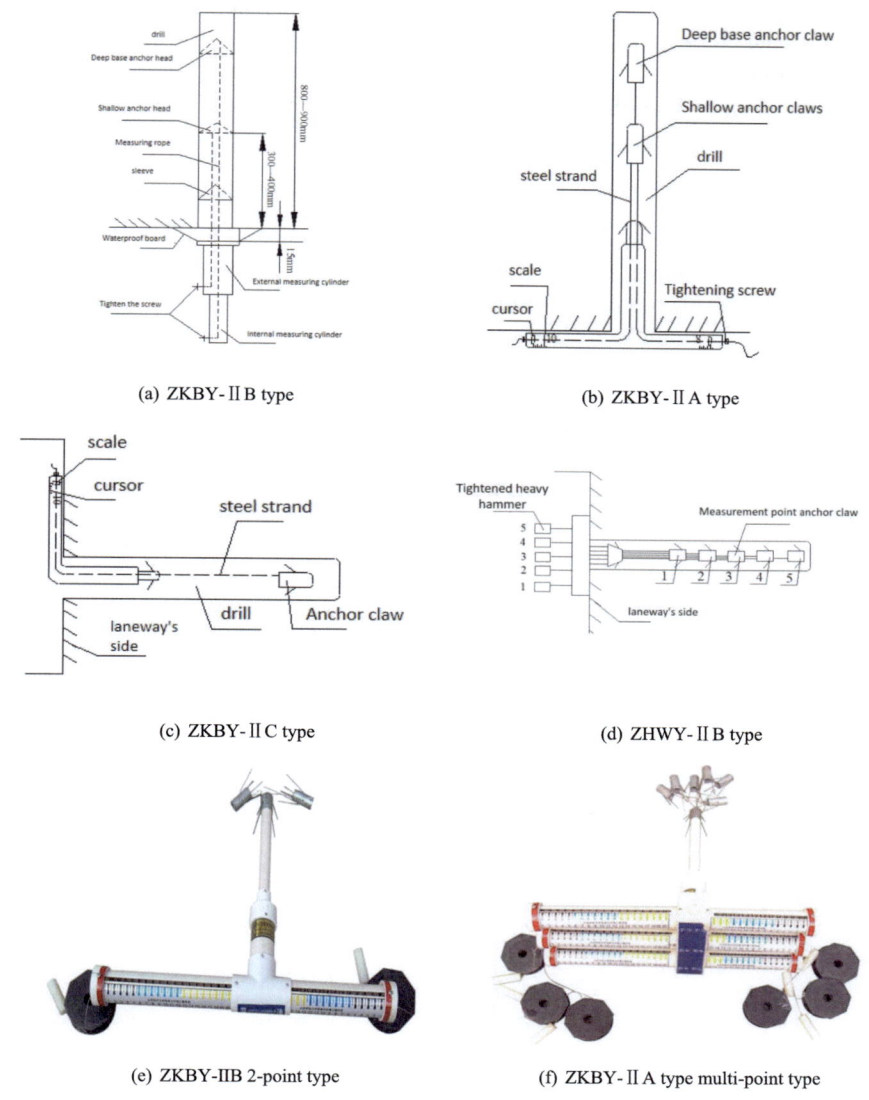

**Fig. 9.6** Roadway surrounding rock displacement separation indicator structure and installation diagram

recording type boreholes. Figure 9.7a shows the top plate separation monitor in the microcomputer programming considered two functions, namely, over-limit alarm and record storage function. Figure (b for the perimeter rock separation monitor shown in, the circuit is designed with real-time continuous (online) monitoring storage and recording function and systematic monitoring mode. The data stored in the microcomputer can be timed regularly using a portable wireless (non-contact) collector once several monitoring recorder stored data through wireless communication one by one collection, brought back to the ground input computer for data analysis and summary of the system workflow as shown in Fig. 9.8).

Principles and methods of monitoring the bottom dropsy in alleyways

Roadway dropsy monitoring is soft along an important part of roadway monitoring. According to the monitoring results, analyze and study the development of dropsy in the roadway. According to the results of the study to develop appropriate countermeasures, these countermeasures need to be based on the characteristics of the roadway bottom plate, the surrounding rock deformation and pressure characteristics (especially the roof), roadway use, the use of the service cycle and other factors to determine the synthesis.

The measurement method of dropsy is the same as the method of off-layer and displacement. According to the width of the roadway to determine the number of roadway cross-section measurement points, the number of measurement points on

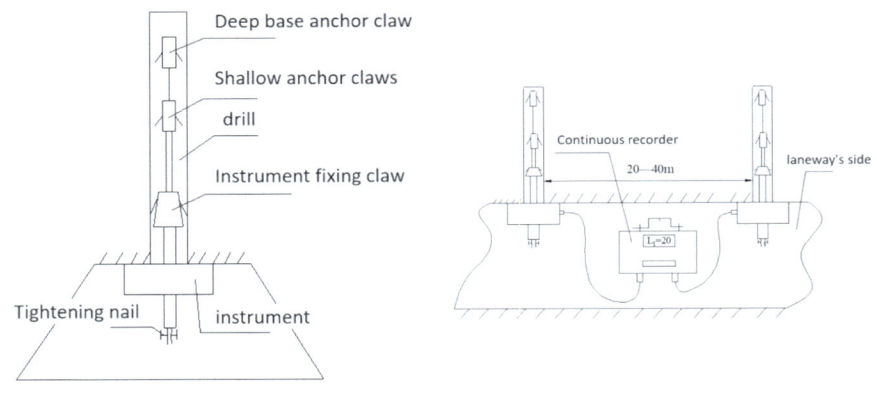

(a) YHW150 Intrinsically Safe (1 drilled hole)    (b) YHW150 Intrinsically Safe (2 drilled holes)

**Fig. 9.7** Real-time storage record alert displacement offset monitor

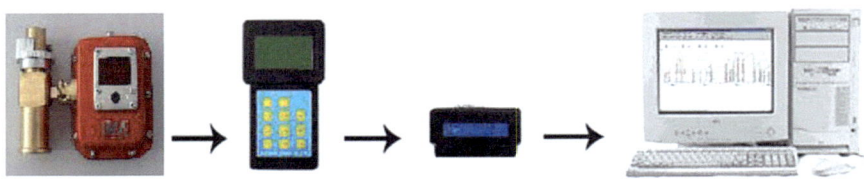

**Fig. 9.8** Real-time record storage and information acquisition monitoring system flowchart

## 9.5 Monitoring Equipment Performance Classification and Applications

the cross-section: 1-point or 3-point type, 1-point type to set the drill hole in the center of the bottom plate; 3-point type drill holes were set in the bottom corner of the inner gang, the outer gang and the center of the center of the 3 points.

Measuring instrument according to the principle of multi-point displacement meter development, more ZHWY-IIB-type floor multi-point displacement bottom dropsy deformation monitor (2, 3, 4, 6 points). Borehole four measuring point type, for example, its working principle is: in the bottom plate drilling 8 m deep borehole, buried four steel strands. One end of the strand is connected to an inverted trident, one end of the strand is fixed at a certain depth of the base plate, and the other end is exposed in the roadway. When the bottom plate dropsy, the length of the strand exposed in the roadway will be reduced, and the length of the strand reduced is measured as the displacement between a point under the bottom plate and the bottom plate of the roadway.

The amount of change in the exposed strand in the channel is measured and recorded periodically so that the amount of displacement relative to the surface of the floor at a point under the floor can be obtained.

In order to prevent the drilled holes from being blocked by the cinders on the roadway floor or artificially damaged, a drilled is put on the upper part of the drilled holes steel plate with small holes, so that 4 strands are threaded out from the holes, the exposed strands are protected by a hose, and a small groove is cut in the bottom plate of the roadway and the hose is fixed and led to the gang of the roadway with a special U-type card so that the strands are avoided from being thought to be damaged, and also it is more helpful to measure and keep the data accuracy. The principle of the method and the implementation plan are shown in Fig. 9.9.

The real-time online monitoring storage and recording type displacement and deformation monitoring mode has the function of regular storage of monitoring data and manual collection and output at any time, and the density of monitoring data is high, so it is very suitable for summarizing and analyzing the law of displacement and deformation, which is a monitoring and research instrument. As shown in Fig. 9.7. YHW150 intrinsically safe roof rock displacement monitoring recorder (referred to as monitoring extension) systematic information monitoring and information analysis process mode as shown in Fig. 9.8, with real-time on-line monitoring and computer data analysis system functions, to achieve the full guarantee of the integrity of the monitoring data, is not easy to produce the monitoring of intermittent and data loss, with a very high degree of engineering practicability and monitoring research. In the current situation, the monitoring process model shown in Fig. 9.8 in terms of cost-effectiveness and operational reliability is the most ideal way to monitor.

(4) Data application

The monitored displacement deformation is usually analyzed and applied in combination with variables such as time, distance, and force. It is used to examine the anchoring force of anchor rods and anchor cables, the design of anchor rods and anchor cables support parameters, and the creep destabilization depth (thickness) range of the coal rock body. Summarize the displacement and deformation of the coal rock body monitoring data application of the data are mainly three aspects: ①

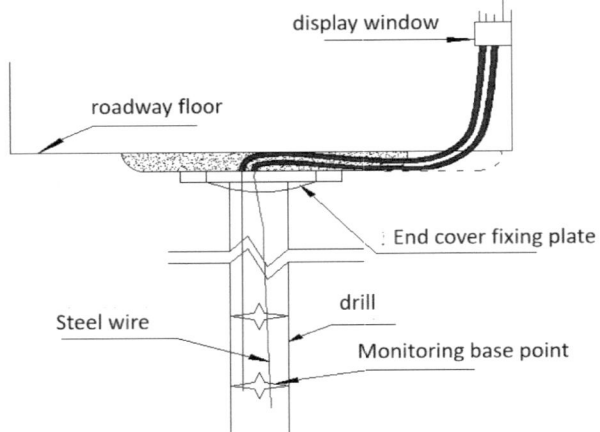

(a) Principle of monitoring and installation of bottom dropsy

(b) Representation of the bottom dropsy of the bottom plate in relation to time

**Fig. 9.9** Principles and methods of monitoring deep displacement bottom dropsy in roadway floor slabs

roadway peripheral rock roof stability and destructive; ② assessment and analysis of anchor rods and anchor cable support design parameters reasonableness; ③ investigation and research of deep coal rock body displacement and compressive stress and compressive stress shear damage and development and change law, especially the roof and even the overlying rock layer (high-level rock layer) to the roadway gangs transfer to the compressive stress of the shear damage law on the efficacy of the anchor reduces the role.

## 9.5.2 Lane Roof Surrounding Rock Pressure Monitoring Instrumentation and Application

(1) Monitoring Principles and Applications

Quarry coal rock body and the corresponding support structure formation of the mechanical behavior and the role of elements with the development of creep dynamics and timeliness, and its characterization of the complexity of the appearance. Pressure monitoring is required in the places where deformation and damage of coal and rock bodies in the mining area is easy to occur, such as the top plate of the roadway, the surrounding rock (two gangs of the roadway), the coal seam, the coal pillar, the bottom plate and other places where the pressure increases easily and triggers the displacement and deformation damage of the coal and rock bodies, as well as the large area of the thick rock layer over the coal seam of the working face to pressure triggered the storage of very high elastic potential energy of the coal body of the mining to form a tendency to impact or impact pressure, and also such as the decompression and expansion of the coal and rock bodies at the depth to trigger an increase in compressive stress, as well as the deep coal and rock bodies to decompress. Expansion of the deep coal rock body decompression triggers the increase of compressive stress, and then the compressive stress monitoring of materials such as filling paste in the mining hollow area. The direction of force or elastic energy is exactly the direction of deformation and damage, so it is often necessary to monitor and study the compressive stress and displacement and deformation one-to-one correspondence. In all kinds of uses of mine roadway, the quarry roadway by the working face mining, mining airspace and adjacent to the airspace of the overlying rock movement, the most prone to pressure changes, a wide range of compressive stress and stress concentration. According to the purpose of monitoring is divided into: ① application of safety warning random monitoring instrumentation, such as ordinary mechanical anchor cable force gauge, pressure box (including electronic digital display), monitoring to play a role in production safety warning checks; ② real-time recording and storage of continuous monitoring instrumentation, monitoring data density, with real-time on-line monitoring of the function of the role of the data with the help of the computer to facilitate the analysis of the research and summing up of the law, belongs to monitoring Research instrumentation.

(2) Monitoring Instruments

In recent years, applicable to the quarry support load pressure, or compressive stress and triggered by rock displacement deformation of compressive stress monitoring instrumentation development is very fast, the formation of mechanical vibration meter type, electronic (optical control) digital display type, real-time on-line recording and storage type, real-time on-line long-distance computer monitoring system. Specific details are shown in Table 9.3

**Table 9.3** Commonly used quarry perimeter rock roof pressure monitoring instruments

| Serial number | Model name | Functional principle | Display recording method |
|---|---|---|---|
| 1 | KJ22 roof dynamic parameter monitoring system | Real-time online computerized monitoring and communication of roof dynamic parameters | Computerized data analysis and data management |
| 2 | ZKY-I hydraulic drilling pressure gauge | Pressure pillows to monitor drilling pressure | Mechanical watch |
| 3 | ZKGY-II A drilling manometer sensor | Drilling pressure pillow sensors | Digital display collector |
| 4 | YHY60-IIA coal rock impact continuous monitoring recorder | Borehole compressive stress sensor with real-time monitoring and storage of three measurement points | Computerized processing of data collected by the collector |
| 5 | GP15 compressive stress sensor | Compressive stress monitoring | |
| 7 | YHY60A anchor strength monitoring and recording instrument | Real-time monitoring of stored triple measurement points | Computerized processing of data collected by the collector |
| 8 | KY-II type coal mine special oil pressure pillow | Adaptation of bearing pressure monitoring of support bodies | Digital display, shock-resistant pressure gauge |
| 9 | ZMC-II A hydraulic anchor stress gauge | Anchor rods and cables load monitoring | Hydraulic digital display or shock-resistant pressure gauge |
| 10 | DZD-II A portable top and bottom plate pressure gauge | Compressive strength test and calculation of base plate | Shock resistant pressure gauge |
| 11 | YHY60B pressure pillow sensor monitoring recorder | Real-time monitoring of filling body compressive stress continuous recording, three-channel type | Manual randomized acquisition and computerized data processing |

(3) Commonly used monitoring methods
① Roadway anchor rods and anchors

As shown in Fig. 9.10, Fig. 9.10a adopts. ZMC-IIA type anchor rod and anchor cable force gauge, and Fig. 9.10b GP15 type pressure sensor monitoring. The Fig. 9.10a shown applies to manual real-time observation of the pressure gauge's indication value, comparing the previous value changes, and analyzing the real-time support strength of the anchors. Figure 9.10b adopts higher resolution strain sensors, fixed between the tray and the nut labor can always carry a special collection meter connected to the sensor plug connector to read the pressure value at that time.

## 9.5 Monitoring Equipment Performance Classification and Applications

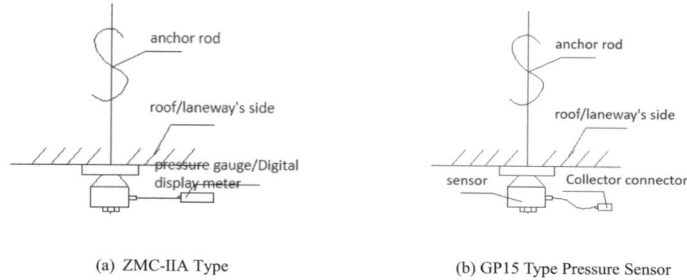

(a) ZMC-IIA Type        (b) GP15 Type Pressure Sensor

**Fig. 9.10** Hydraulic force gauge and strain gauge force transducer monitoring methods

Real-time online monitoring records (continuous recording storage type)

As shown in Fig. 9.11, Fig. 9.11a YHY60A pressure monitoring recorder is used to monitor the bearing capacity of anchor rods and cables on the cross-section of the roadway, and Fig. 9.11b YHY60B pressure monitoring recorder is used to monitor the compressive stress on the coal rock body at the nodal surface of the coal seam and the bottom plate through the pressure pillows in the drilling holes, and the shapes of the pillows are divided into circles, squares and rectangles. Figure 9.11c is used to monitor the pressure of the filling body and the compressive stress given to the coal seam by the roof plate at the joint surface of the coal rock seam.

(4) Ultimate tensile strength anchor rods and cables for anchoring force testing of

Anchor rod and anchor cable ultimate pull force testing is different from real-time online monitoring of anchor rod and anchor cable anchorage force. Anchor limit pulling force detection is the use of anchor pulling force meter on the installed anchor to implement the maximum pulling test, in order to examine the bonding force between the anchoring agent and the anchor body, the bonding force between the anchoring agent and the wall of the drilled hole, and the tensile breaking force of the anchor body, which is a destructive test. With the improvement and perfection of anchor installation technology and anchoring agent production technology, the role of this testing method should be gradually diluted or abandoned. In addition, through long-term monitoring research found that the anchor rods, anchor cable function failure is mainly reflected in the roof plate and even the overlying rock layer transferred to the road gang of compressive stress on the anchorage area (anchorage rock layer) coal rock body of the shear destructive effect, including the neighboring hollowing area of the overlying rock movement brought about by the destructive effect, resulting in cracks and cracks, deformation of the gang of the road gang and triggered by the road gang of the original rock decompression and expansion of the displacement of the formation of the strength of the creep, causing The bonding force between the anchoring agent and the drill hole wall is reduced. therefore, the development of technology no longer requires us to pay attention to and consider the ultimate tensile strength of the anchor, strengthen the real-time online monitoring of

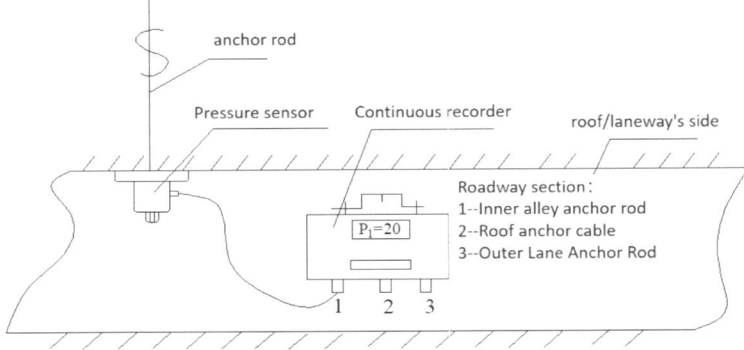

(a) YHY60A type anchor(cable)load force recorder

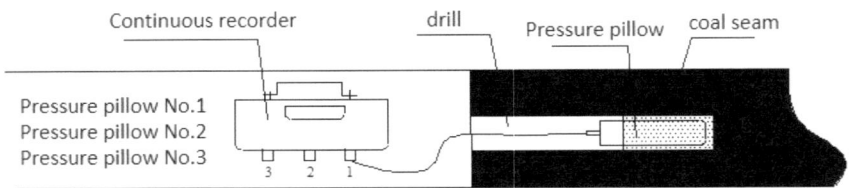

(b) YHY60B Pressure Pillow Sensor Monitoring Recorder

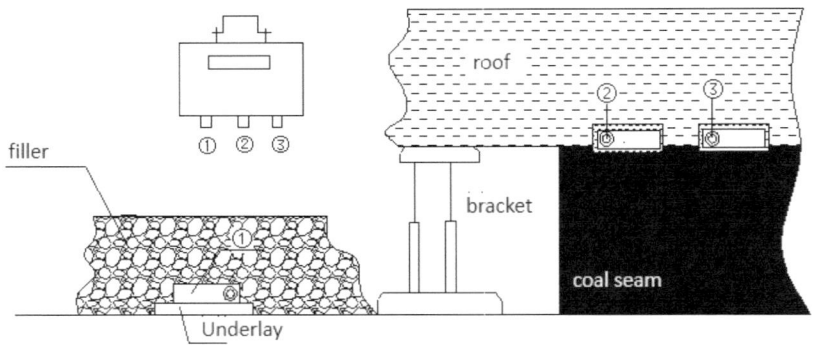

(c) GP25 Compressive Stress Induction Pillow Monitoring and Recording Extension

**Fig. 9.11** Working principle of real-time record storage monitoring method for pressure monitoring extension

the anchor, anchor cable, and control the deformation of the surrounding rock is the key to the problem.

(5) Data application

① monitor and study the change rule of roof pressure and influencing factors; ② understand the development law of peripheral rock compressive stress and displacement deformation and influencing factors; ③ master the creep track characteristics of the bearing capacity of the roadway anchor rods and anchor cables and the prevailing law; ④ monitor and analyze the change rule of the supporting pressure of the coal body and the correlation between the role of the roof compressive stress; ⑤ roadway roof peripheral rock wall surface morphology of the creep and the mining pressure, the mining area of the movement of the overburden of rock layers and its correlation characteristics; ⑥ apply the change rule characteristics of coal body impact tendency and protrusion tendency of coal and rock bodies. Characteristics of the change rule of the surface morphology creep of the roadway roof rock wall with the mining pressure and the movement of the overlying rock layer in the mining hollow area and its correlation; Characterization of the creep trajectory analysis of the compressive stress and displacement deformation in relation to the propensity to impact and protrusion of coal and rock bodies.

### 9.5.3 Bracket Pressure (Operating Resistance) Monitoring Instruments and Applications

(1) Monitoring Principles and Applications

Bracket pressure monitoring includes working face synthesized mining bracket and working face over-above-supported monolithic hydraulic pillar, which is an important content of mine pressure monitoring research. Through monitoring the working pressure of the stent to understand and analyze the principle of pressure change of the roof plate and cause and effect relationship, as well as the prediction and forecasting analysis and research of abnormal pressure coming from the roof plate, to maintain the safety of the roof plate, and to accumulate experience in the selection of stent and the management of mine pressure.

(2) Monitoring Instruments

In recent years, applicable to the stent pressure (load force, work resistance) monitoring instrumentation equipment development is rapid, the formation of ordinary mechanical vibration meter type, electronic digital display type and real-time online record storage type, real-time online remote computer monitoring system. At present, the real-time online record storage type and computer communication system more than 5 min record or communication once. Specific details are shown in Table 9.4.

**Table 9.4** Commonly used stent work resistance monitoring instruments

| Serial number | Model name | Functional principle | Display recording method |
|---|---|---|---|
| 1 | YHY50 type synthesized mining stent pressure digital display pressure gauge | Optical digital display (1, 2, 3 measurement points), suitable for a wide range of monitoring | Electronic light control, digital display |
| 2 | YHY60 coal mine pressure continuous monitoring recorder | Real-time monitoring of single-chip microcomputer storage bracket bearing pressure<br>The measuring points (1–3 channels) are optional | Data acquisition computer data processing, 5 min storage once |
| 3 | KJ22 stent pressure monitoring system | Real-time online monitoring and computer communication once in 0–5 min. Suitable for the pressure monitoring of synthesized mining stent | Computer interface operations and data processing queries |
| 4 | ZBYJ-II types shock-resistant synthesized mining bracket working resistance pressure gauge | Seismic and shock resistant online monitoring of stent pressure | Direct-reading 1 or 2 measurement points optional |
| 5 | SY-40W type pressure gauge type monolithic column working resistance detector | Hand wheel hydraulic principle single column pressure monitoring | Digital display with direct reading of pressure gauge |
| 6 | Hydraulic pressure box | Put in the top of the monolithic column to touch the top to monitor the roof pressure | Pressure gauge, digital meter type direct reading |
| 7 | Strain relief pressure sensors | Put in the top of the monolithic column to touch the top to monitor the roof pressure | Acquisition of data by the collector |
| 8 | SY-40 digital display pressure measuring instrument | Portable hydraulic principle single column pressure monitoring | Digital display and pressure gauge type direct reading |

(continued)

## 9.5 Monitoring Equipment Performance Classification and Applications

**Table 9.4** (continued)

| Serial number | Model name | Functional principle | Display recording method |
|---|---|---|---|
| 9 | YHW150 intrinsically safe compression displacement recorder | Real-time monitoring of single-chip storage | Acquisition of data by the collector |
| 10 | KJ22 coal rock pressure online monitoring system | Real-time online monitoring and computer communication and data processing analysis, applicable to coal body pressure monitoring | Computer interface data query and summary abnormal pressure alarm |

(3) Common installation monitoring methods

① The basic monitoring method of the pressure of the synthesized mining stent is Figure shown in, which gives the use of 9 sets of 2 measuring points of real-time online continuous storage and recording type pressure monitoring extensions, and adopts the monitoring process mode as shown in Fig. 9.8 to store and record the data with every 5min inspection. According to the 3-measurement zone arrangement method, that is, 3 zones × 3 racks of measurement × 9 units; 10 monitoring extensions can also be used 5-measurement zone method, that is, 5 zones × 2 racks of measurement × 10 units. Pressure monitoring machine adopts 2 measuring points (channels) to meet the requirements. Some mines under Shenhua Ning Coal Group Company adopt the monitoring method of 2 racks × 2(or 3) air × 2 racks based on the management experience, with a wide range of monitoring and control.2-pillar bracket connects the left and right 1 and 2 measurement points of the pillars, and 4-pillar bracket connects the front and rear pillars corresponding to the left and right 1 and 2 measurement points. Table 9.4

② Ordinary type synthesized mining stent pressure gauge, real-time online monitoring and displaying synthesized mining stent working pressure. The pressure gauge is divided into: mechanical vibration and shock resistant pressure gauge (single and double measuring point); electronic digital pressure gauge (single and double measuring point). This type of pressure gauge is suitable for comprehensive shelf monitoring or large-scale installation monitoring. Electronic display window is large, can facilitate intuitive warning stent in the process of fast-moving frame pressure changes and running process pressure value, for the operator to prompt the stent ascending frame is sufficiently strong or the existence of leakage, string of liquid phenomenon, such as the picture shown in Fig. 9.12.

③ Commonly used monitoring instruments for monolithic hydraulic pillar pressure are shown as items 5, 6, 7 and 8 in Table 9.4. At present, it is mostly

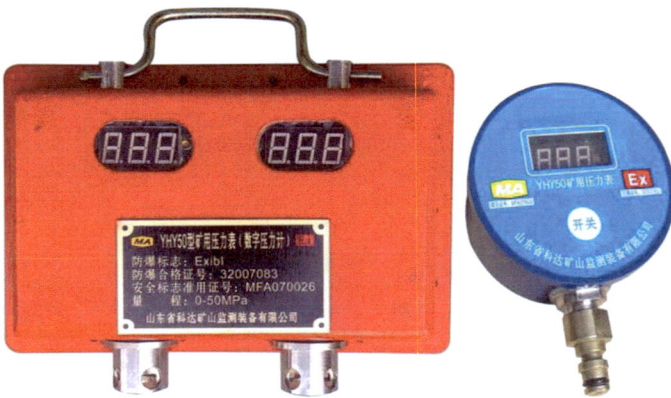

**Fig. 9.12** YHY50 electronic digital pressure gauge picture

used for the monitoring of the overrunning monolithic column support to understand the range of influence of mining and roof cycle pressure.

④ Monitoring the top and bottom plate movement of the roadway A new type of dynamic meter installation monitoring method, as is used for shown in Fig. 9.13. The improvement of this method better solves the shortcomings of unstable space occupation and inconvenient installation that existed in the previous deployment method. Another advantage of using dynamic instrument monitoring is that the installation is simple and convenient, without the need to install the off-layer instrument as troublesome as drilling holes and the installation operation laborious. It is especially suitable for the dynamic monitoring of the roof of large roadway, such as the dynamic monitoring of the roof of the roadway without railroad car transportation. This kind of dynamic meter monitoring mode has pointer type, electronic digital display type, continuous record storage type monitoring mode, monitoring data management process as shown in Fig. 9.8.

(4) Data application

Through monitoring the stent bearing pressure or working resistance analysis summarizes and analyzes the top plate pressure law, and provides reference basis for maintaining the stability of the stent and the safety of the top plate. Bracket bearing law is equivalent to the law of top plate pressure, and the law of change of bracket pressure is the characterization of the top plate state and form. therefore, the brace pressure monitoring curve is the basis for further understanding the movement characteristics and manifestation law of the roof plate to the overlying rock layer in the working face and the mining area, the intensity of the incoming pressure, as well as the physical phenomena produced by the roof plate and the brace triggered by the incoming pressure. The normal information provided by the bracket pressure curve mainly includes: initial support force, working resistance, lowering and raising pressure in

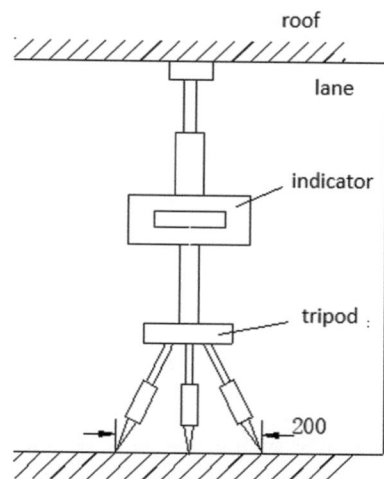

**Fig. 9.13** New roof dynamic instrument installation diagram

the process of moving the bracket, incoming pressure cycle, leakage of string liquid (no pressure), and the quality of opening and unloading the safety valve.

The basic characteristics of the load bearing pressure monitoring curve of the stent are shown in Fig. 9.14. The curve given in the figure is the pressure curve of the whole stent and the related information, and the initial bracing force: $P_c$ = working resistance × (65–85%) or estimated by the pressure of the pumping station minus the loss of pipeline. Usually, the cycle to pressure step used to adopt the distance L, and can also be used to time interval $\Delta_t$ said to pressure cycle. The pressure decreases and increase in the figure indicates the process of moving the frame. In some mines, the sound of roof breaking (vibration phenomenon) can sometimes be heard when the roof comes under pressure. At present, from some literature can be found, some manufacturers of such monitoring instrumentation, data processing and analysis of the curve given by the deformed, very inconvenient to view the basic information that should be reflected (Fig. 9.14).

### 9.5.4  Real-Time Online Monitoring Systems and Applications

(1) Monitoring Principles and Applications

At present, the real-time online monitoring system mainly refers to: ① roadway dynamic (compressive stress, deformation) parameter monitoring, to realize the displacement and deformation of the roadway roof and surrounding rock (two sides of the roadway) to continuously monitor the changes in the amount of the phase. ② Mining face stent work resistance for staged, continuous real-time monitoring, to realize the monitoring of the working face roof to the development and changes in the law of pressure. At present, it is customary to call the way of monitoring through

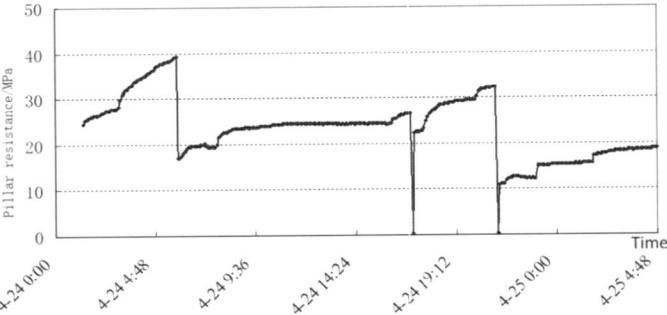

**Fig. 9.14** Principle of representing information about stent pressure monitoring curves

ground computer long-distance cable communication as real-time on-line computer monitoring system, which is based on the on-line monitoring of computer. Based on the monitoring data continuity integrity considerations, the authors advocate that the monitoring instrument with real-time monitoring record storage function, relying on the use of manpower to carry a special collection meter will be stored in the real-time monitoring instrument data collected one by one and return to the ground to transmit to the computer for data processing of a class of monitoring instrument, this way to obtain the data continuity integrity is good and the time scale and the amount of data, as shown in Fig. 9.8 shows the monitoring process way. Figure 9.8 shows the way of monitoring process, which can ensure online monitoring for a sufficient period of time, are categorized as real-time online monitoring system type of instrumentation.

(2) Monitoring instruments

In recent years, applicable to the quarry monitoring displacement deformation of the instrumentation technology is developing rapidly, real-time record storage type, real-time online remote computer monitoring system. Specific details are shown in Table 9.5.

(3) Common installation monitoring methods

At present, the computer data analysis system applicable to the systematic mode of mine pressure monitoring is mainly embodied in the following: on-line real-time monitoring of stored data and computer data processing and analysis system, such as the monitoring and collection process shown in Fig. 9.8, as well as the computer monitoring and communication shown in Fig. 9.15. Its key technical elements are reflected in the stage continuity and integrity of the data to meet the requirements of the monitoring study and monitoring data processing engineering professional practicality and high utilization of information analysis, the results of the express simple and intuitive. Based on this principle, the following monitoring methods are proposed to meet the practical requirements of quarry engineering. The mine pressure monitoring system should not be as complicated as the safety and gas monitoring system, and the engineering problems caused by the complexity of the monitoring

## 9.5 Monitoring Equipment Performance Classification and Applications

**Table 9.5** Real-time online monitoring systems for commonly used quarry roadway workings

| Serial number | Model name | Functional principle | Display recording method |
|---|---|---|---|
| 1 | KJ22 on-line computerized monitoring system for pressure of stent of comprehensive mining | Real-time computerized monitoring of communication bracket pressure, monitoring of research-based | Computerized data processing abnormal stress alarm |
| 2 | KJ12 roof surrounding rock dynamic monitoring system | Real-time online computerized roof dynamic monitoring, monitoring research type | Computerized data analysis and data management to analyze anomaly alarms |
| 3 | YHY60 coal mine pressure continuous monitoring recorder | Real-time monitoring of single-chip microcomputer storage bracket pressure (1–3 channels) Optional | Data acquisition computer data processing analysis abnormal alarm |
| 4 | YHY60-IIA coal rock impact continuous monitoring recorder | Borehole compressive stress sensing pressure pillows, real-time monitoring storage triple measurement points, monitoring research type | Collector collects data computer processing and analyzes abnormal alarms |
| 5 | YHY60A anchor strength monitoring and recording instrument | Real-time monitoring and storage of three measurement points (three anchor rods and anchor cable loads), monitoring and research type | Collector collects data computer processing and analyzes abnormal alarms |
| 6 | YHW150 intrinsically safe roof displacement monitor and recorder | The displacement and deformation of the top plate roadway gang is continuously recorded and stored.2 drill holes L1, L2 four measurement points | Digital display storage, collector collects data computer processing and analyzes abnormal alarms |

system should be simplified. Therefore, it is recommended to adopt the systematic monitoring extension series monitoring method as shown in Figs. 9.8 and 9.15a, b, followed by the combined monitoring method shown in Fig. 9.15d, c. That is, the series combination of the system monitoring method shown in Fig. 9.7 or Fig. 9.11 and the system monitoring method shown in Fig. 9.15a, b. Monitoring of two trough roadways or track roadways and belt roadways. If there is no special consideration, the monitoring extension should first consider installing the trough on the side of the adjacent goaf that is prone to damage. It is not necessary to use the same monitoring method and the same parameters for both trough (ordinary monitoring instruments can be used).

(4) Performance of computerized monitoring systems and communication technologies

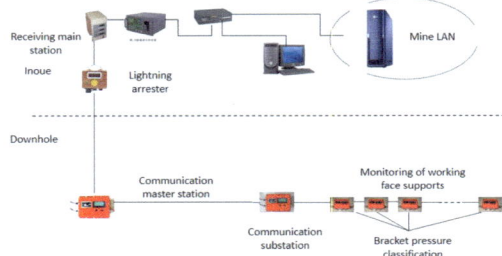

(a) Real-time on-line computerized monitoring system for KJ22 type integrated mining supports

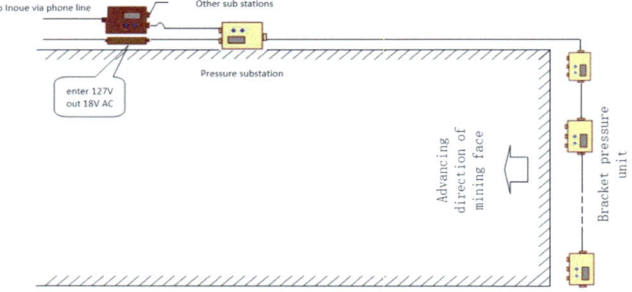

(b) KJ12 computerized monitoring system for dynamic parameters of roadway perimeter rock

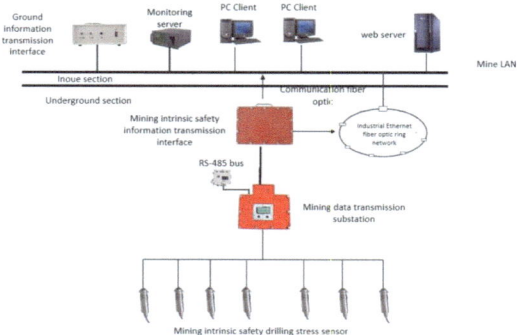

(c) Tandem combination of wireless acquisition of monitoring storage and KJ22 computerized monitoring

**Fig. 9.15** Real-time online monitoring and computerized communication processing system

## 9.5 Monitoring Equipment Performance Classification and Applications

### 1. KJ22 online monitoring system

KJ22 type mine hydraulic bracket pressure online monitoring wireless communication system is a computerized intelligent system used for monitoring pressure parameters of coal rock body pressure, bracket pressure and other monitoring objects in underground coal mine. It adopts the most practical product design technology point. Already such systems to rely on cables between the monitoring branch to achieve communication power supply, the system is composed of wireless communication pressure monitoring branch (positive pressure monitoring sensors),communication substations, mining communication converter interface, system monitoring computer, printer, etc., with a small size, simple structure, easy to operate, easy to master the implementation of technology and subsequent data processing and utilization of strong features. Products in line with the requirements of the production technology management process and technical management specifications, in the choice of function does not blindly add impractical functions, resulting in confusion and fuzzy. Also highlights the technical advancement and rationality of the characteristics of the ideal pressure monitoring products. The monitoring extension (positive pressure monitoring sensor) adopts wireless radio frequency transmission to design the communication circuit, and the communication sub-station is matched to carry out the data monitoring and communication. KJ12 system gives the analysis of the cycle pressure law of the roof plate, mathematical statistics and the analysis and summary of the initial bracing force through the monitoring of the computer on the ground and data processing, especially in the computer data processing aspect to strengthen the analysis and statistics of the abnormal pressure and the query processing, and it has the function of intuitive result display and prediction alarm. It has the function of displaying and predicting alarm.

Main technical characteristics of the KJ22 system

(1) Measuring range: 0–60 MPa
(2) Measurement error: ± 1.0 MPa
(3) Resolution: 0.5 MPa
(4) Display mode: 3 1/2-digitLED
(5) Means of communication:

① Communication mode: underground wireless communication, the maximum distance of wireless transmission between monitoring extensions is 50 m.
② The communication baud rate of the monitoring extension (monitoring sensor) is 9600 bps.
③ The baud rate of communication between the communication substation and the ground computer is 600, 1200 and 2400 bps.

(6) System backbone functions:

① Analog acquisition and display: GPD50 monitoring extension sensor acquires analog and output values, and converts the data for storage and display.

② Automatic control: GPD50 monitoring extension sensors are automatically powered up and working every 5min interval.
③ Storage and query: During the working process of the system, the system software will automatically store the collected data into the database.
④ Data backup: the system software stores the data information into the database, and the operating system software interface backs up the data to the storage device.
⑤ Data protection: Restore historical data for querying through the data restore function of the system software interface.
⑥ Drawing curve: The system software automatically generates the (pressure–time) curve.
⑦ Display and print tabulation: When the system is working, the screen displays the collected data and curves in real time, and the data and curves are printed through the system software.
⑧ Man–machine dialogue function: call various menus with man–machine dialogue, and carry out various operations according to the menu. The monitoring extension regularly collects pressure signals and wireless transmission data, and processes and analyzes the monitoring data through the computer; the communication substation receives the monitoring extension signals in real time and transmits the signals to the ground computer.

(7) Power supply:

① Communication substation power supply:18 V DC power supplied by explosion-proof power supply.
② Monitoring sensor power supply: two 3.6 V lithium batteries in parallel package.

(8) Cable between the communication substation and the ground: two cores, cross-sectional area $S \geq 1$ mm$^2$/core (e.g. MHYVR $1 \times 2 \times 7/0.43$).

2. KJ12 online monitoring system

KJ12 system development based on the principle and design ideas is mainly the quarry roadway coal roof enclosure (stress–strain) off the design of the integration of monitoring sensors, can synchronize intuitively view the (stress–strain) between the relationship between, and then for specific engineering examples of scientific research and daily production of roof management to provide accurate data and rule of change.

The system consists of two parts of equipment: ground and underground. It is an intelligent system monitoring). It is composed of mainly used for deformation and displacement of roof and surrounding rock and anchor rods in underground coal mines dynamic parameters (pressure, deformation of monitoring objects such as bearing capacity of and cables dynamic parameter sensor, communication sub-station, mining communication converter, system computer, printer, etc. It has the characteristics of small volume, simple structure, convenient operation, easy to realize the mastery of technology and strong function of subsequent data

## 9.5 Monitoring Equipment Performance Classification and Applications

processing and utilization. Dynamic parameter monitoring sensor is using wireless radio frequency transmission design of the communication circuit, and communication substation supporting data monitoring communication. The ground computer gives the command request, receives the data of each monitored sub-computer cyclically through the communication sub-station, and carries on the data processing, and gives the processing result conveniently so as to consult and browse. For example, there are display methods: digital two types of type and curve type. According to the speed of data change, it can further judge the change of roof enclosure database trend of, which plays a good guiding role for production safety. The computer can also set different time intervals for data storage, so as to be used in the future for historical and anchor bearing capacity data analysis and summarize technical experience. The product meets the requirements of production technology management process and technology management standard, removes the impractical cell phone text messaging function, strengthens the abnormal data discernment query and information sharing network technology, and highlights technological advancement the characteristics of, which is the ideal roof plate monitoring product. And reasonableness.

KJ22 system main technical standards

(1) Measuring range: displacement:0–300 mm, load force:0–25 MPa
(2) Measurement error: displacement ± 1.0 mm, load force ± 0.1 MPa
(3) Resolution:1
(4) Display mode:3 1/2-digit LEDs
(5) Means of communication:

　① Communication mode: wireless communication of, the between monitoring extensions monitoring extensions maximum distance of is wireless transmission 50 m.
　② The monitoring the wireless communication baud rate of is 9600 bps.sensor
　③ Communication substations and ground computers use cable communication baud rate of 600,1200, 2400 bps(optional).

(6) System backbone functions:

　① Analog acquisition and display: DJ-W monitoring extension (monitoring sensor) acquires analog and output values, and converts the data for storage and display.
　② Automatic control: DJ-W monitoring extensions every interval 5min automatically power up the work.
　③ Storage and query: During the working process of the system, the system software will automatically store the collected data into the database.
　④ Data backup: the system software stores the data information into the database, and the operating system software interface backs up the data to the storage unit.
　⑤ Data protection: Historical data can be through the data restoration function of the system software interface restored and queried.
　⑥ Plotting curves: The system software analyzes and generates (displacement, time) and (load force, time) curves.

⑦ Display and print tabulation: When the system is working, the collected data and curves are displayed in real time, and the data and curves are printed through the system software.

⑧ Man–machine dialog function: call a variety of menus with man–machine dialog, and in accordance with the menu for a variety of operations.

Such as monitoring sensor has timing acquisition and wireless transmission of data. Through the computer processing and analysis of monitoring data of the work.

(7) Power supply:

① Communication substation power supply:18V DC power supplied by explosion-proof power supply.

② Displacement parameter sensor power supply: two 3.6V lithium batteries in parallel package.

(8) Cable: use two cores with a cross-sectional area Between the substation and the ground communication $S \geq 1$ mm$^2$/core.

3. KJ system main communication function

❶ System generation function

① Able to define the name and installation location of the equipment monitored by the system.

② Complete the definition of parameters such as the name of the sensor, the installation location, the unit, and the nature of the measurement point of the substation connected to the system.

③ Complete the definition of the graphic screens in the system.

④ Complete the generation of monitoring data reports in the system.

⑤ Information sharing(networking)of computer networks in mining areas.

❷ System communication functions

① The ground host shall process the analog data sent from each substation in real time.

② The substation should carry out real-time inspection of the connected sensors and transmit the collected data to the central station in a timely manner.

❸ Computer on-screen display function

① Instantaneous value and history display for each measured value.

② Sub-station inspection map and transmission system fault status display.

③ Query and alarm display of abnormal signals at each measurement point.

④ Curve display of the monitored signal parameters.

❹ Storage Function

The system is connected to the sensor monitoring data under normal conditions every two minutes to store the disk (stored at any time in the alarm state), and to

## 9.5 Monitoring Equipment Performance Classification and Applications

the maximum value, the average value and the minimum value of the formation of historical curves, curves can be zoomed in or out; at the same time recording the alarm and the time to lift the alarm. All data can be stored for more than one year (Excel conversion).

❺ Print function

① Output data analysis graphs and tables in real time.
② Real-time browsing and querying of data analysis graphs and tables.

❻ Self-diagnostic function

When the sub-station, sensors and transmission cables are faulty, the system should be able to diagnose the fault with corresponding display and alarm.

❼ System software self-monitoring function

The system software can correctly monitor the task operation status of each target monitoring sensor.

❽ System transmission function

The system protocol is open to access the underground Ethernet, and can transmit relevant data in the ground LAN.

4. Main technical indicators

❶ Transmission metrics

System architecture: Main control computer ← Sorter ← 1–4 substation ← 10–40 monitoring extensions per station (monitoring sensor);
　　System capacity: Multi-serial protocol communication meets single transmission of dynamic parameters and integrated (4-way) transmission mode;
　　Sub-station transmission: ① CAN wired upload mode, transmission rate 1200–4800bps, ground CAN reception; ② CAN wired downlink mode, transmission rate 4800–9600bps;
　　Monitoring sensor transmission: wireless upload method, transmission rate 4800–9600bps, sub-station reception;
　　Ground master control interface: CAN-RS232C signaling mode, master control computer serial port receiving and receiving.

❷ Ground master information station

Monitoring host: industrial control computer; processor 1.0G or more/memory 4G/ hard disk 160G.
　　Printer: Color Laser.
　　Ups:Ac220v/2kva;

Monitor: LCD 17″.

Communication mode: RS232C-serial port (with serial port expansion function requirements);

Underground CAN upload: CAN-RS232C interface receiving and receiving, wired transmission rate 1200–4800bps;

CAN-RS232C interface serial port to master computer.

Integrated dynamic parameters pressure sensor or displacement sensor ≤ 60/group.

❸ Underground communication substations

Display mode: liquid crystal display;
Communication protocol: the ground computer sets the working parameters;
Explosion-proof power supply: AC 127V/DC 18 V;
Communication distance to ground: ≥ 25 km, 2-core cable;
Signal transmission: ① CAN wired upload method, transmission rate 1200–4800 bps;

② CAN wireless reception downlink method, transmission rate 4800–9600 bps;
③ Wireless reception sensitivity: − 107dbm.

❹ Monitoring sensors (monitoring extensions)

Explosion-proof type: Intrinsically safe for mining, marked as "Exib I";
Monitoring extension capacity: ① pressure monitoring extension: with 1–4 channels; ② roof monitoring extension 1–4 channels;
Measuring range: ① 0–60 MPa; ② 0–25MP; ③ 0–300 mm;
Response time: ≤ 3s;
Alarm point: can be set arbitrarily;
Alarm mode: red light flashing;
Transmission method: wireless infrared uploading method, 4800–9600bps;
Wireless communication distance: 5–50m;
Power supply: DC 3.6V lithium battery.
Hole diameter: Φ28, Φ40, hole distance ≤ 50m.
Communication substation transmission cable: MHYVR 1 × 2 × 7/0.43/ intrinsically safe.

In general, the real-time online monitoring computer communication and data processing system, as shown in Fig. 9.15a, b, and real-time online monitoring continuous storage and recording and acquisition computer data processing system, as shown in Fig. 9.8 systematic monitoring mode. These two monitoring modes can realize the completion of a phase (or a period) of real-time online continuous monitoring and data storage and computer data analysis applications, to achieve a systematic monitoring and research-based technology, real-time online monitoring and continuous storage and recording of monitoring data information systematic monitoring mode to obtain a large amount of information, continuity and good, easy to find the laws and changes in the characteristics of the data, especially in the data processing and analysis applications can be substantial and convenient. Analysis and application of data processing can be substantial and convenient to achieve prediction and forecasting, providing information processing results intuitive, simple and clear, To achieve the basic mode of engineering explicit reading, both to achieve the monitoring of the research and close to the engineering practice. In view of the current development, the research and development of monitoring instrumentation should focus on the analysis and utilization of monitoring information and guidance engineering, rather than developing ways to expand the scale and capacity of the monitoring system, resulting in the complication of the high failure rate and low utilization rate.

## 9.6 Coal Rock Mass Catastrophe Prediction Methods

The basic idea of coal and rock body catastrophe prediction is that before the occurrence of sudden catastrophe in the coal and rock body of the quarry, there exists a course of development that can be monitored and controlled, and the physical signal variables released at any point on the trajectory of the development of this course of course of development can be obtained through the method of monitoring or observation. The development of catastrophic characterization variables of coal and rock bodies from creep to drastic change to catastrophe (e.g., the time interval [T]:$\{T_1, T_2, T_3, \ldots, T_m\}$) is characterized by the presence or emergence of a group of variables or a concomitant array of variables, e.g., the off-gradient variables of the coal and rock bodies [L]:$\{L_1, L_2, L_3, \ldots, L_m\}$, Acoustic Emission Signal [AE]:$\{AE_1, AE_2, AE_3, \ldots, AE_m\}$, Compressive Stress Variable [P]:$\{P_1, P_2, P_3, \ldots, P_m\}$, Anchor Rod, Anchor Cable Load F]:$\{F_1, F_2, F_3, \ldots, F_m\}$, etc.

The most difficult and critical method of disaster prediction of coal rock body is how to combine with engineering practice, how to choose or determine a variable value (also called threshold value) or a variable interval, as the judgment value of disaster or anomaly signals, as the variable signals for disaster (hazard warning) prediction, here the author puts forward a few variable signal's application and prediction methods for reference and reference. Here the author proposes several variable signal application and prediction methods for reference. The author advocates that each mine combine their own situation (such as the size of the roadway section, the use of the roadway, the size of the coal pillar, the strength performance of the coal and rock body, the technical factors, the thickness of the coal seamed.) to determine the threshold value of the variable of the anomaly signal for their mines.

(1) Anomalous signal prediction method for the amount of deformation of the top plate enclosing rock away from the layer

At present, there is no abnormal evaluation index $L_y$. Abnormal signal variable value, about the unified deformation amount of the roof and surrounding rock of the roadway. However, there are many mines in the mining area that combine their own conditions to formulate the corresponding maximum permissible deformation value $L_y$, which has a positive and effective effect on the management of engineering technology. Combined with the strength properties of coal and rock bodies and experience from the field, it is suggested that: soft rock and very soft rock roadways, such as railroad roadways and belt roadways adopt $L_y = 70$mm or $L_y = 80$mm deformation as the anomalous signal value or early warning value; and the roof enclosing rock reaches medium hardness or above adopts $L_y = 60$mm or $L_y = 70$mm deformation as the anomalous signal variable value to deal with. And if there is a development trend of continuous increase of deformation amount in [L] and the change speed becomes faster, it is considered that there is an abnormal hazardous situation, as shown in Fig. 9.16. When choosing to determine the abnormal value of the roadway deformation or the danger warning threshold, the strength of the coal rock body of the roadway surrounding rock layer and the size of the roadway cross-sectional area scale are fully considered, referred to as the strength effect and the scale effect. The signal acquisition time interval designed by the current application of off-story monitoring instrument is long, so the monitored data analysis curve changes are generally smoother, and the drastic displacement and deformation can hardly be monitored. If applied to forecasting, the signal sampling density should be increased.

(2) Anomalous signal prediction method for anchoring force decline of anchor rods and cables

The change of anchorage force (bearing capacity) of anchor rods and cables is used to express the force that triggers the deformation and damage of surrounding rock, and the force effect of anchor rods and cables is applied to directly express the compressive stress of the roadway top plate and roadway gang. The early warning threshold value of the real-time anchorage force change of anchor rods and cables adopts one half of the pre-tensioning force (also known as initial anchorage force) 1/2 of $P_0$ the anchor rods and cables when they are installed as the anomalous value, which is expressed by $F_y$, and the anomalous value of $F_y = F0/2$, as shown in Fig. 9.17. When the real-time online anchorage load force of continuous anchor rods and cables falls to the value of $F_y$, attention should be paid to it, and it should be managed as a warning threshold, strengthen the visual observation density of the perimeter rock top plate, and analyze the possible phenomenon of the top plate and

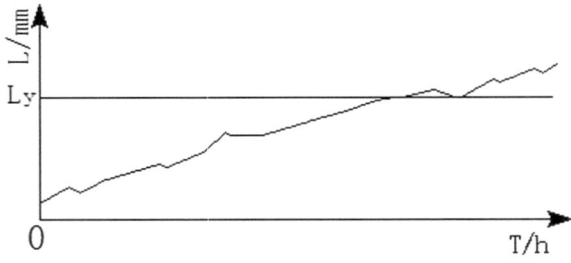

**Fig. 9.16** Principle of signal prediction of anomalous deformation signals in the lane perimeter rock off the layer

## 9.6 Coal Rock Mass Catastrophe Prediction Methods

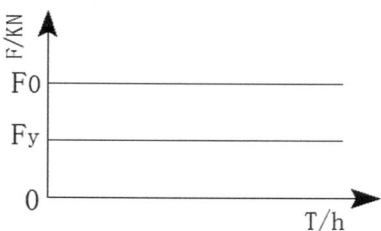

**Fig. 9.17** Principle of predicting anomalous signals of load force in anchor rods and cables

slice gangs in combination with the observation of the deformation situation. Due to the role of the surrounding rock integrity of the anchor network cable support structure, in the absence of the formation of a range of single anchor rods, anchor cables appear anchoring force decline to abnormal values, usually does not appear obvious deformation damage. In general, when the deformation reaches a certain amount, the anchorage force will drop extremely fast.

(3) Application and prediction methods for stent pressure anomalies

The real-time working pressure or load bearing capacity or working resistance of the mining support is obtained by means of a pressure monitoring and recording instrument (memory acquisition or computer communication monitoring mode), which records the pressure $P_i$ corresponding to time $T_i$ in a time interval $[T]:\{T_1, T_2, T_3, \ldots T_i, T_m\}$, there is Pressure change interval $[P]:\{P_1, P_2, P_3, \ldots, P_i, P_m\}$. Stent pressure., roof pressure, so the prediction of abnormal value of stent pressure includes the meaning of both stent and roof, abnormal pressure value triggered by stent's own system performance, and ② The abnormal roof pressure is the abnormal pressure value. The monitoring data of stent pressure is summarized and analyzed in the form of diagrams and tables, and the diagrams are divided into three forms as shown in Fig. 9.18a–c.

Figure 9.18a shows the frequency of occurrence of each pressure interval by using a single stent as the statistical unit, giving all the [P] values monitored for a given stent in the [T] interval, and applying the percentile statistics method. Similarly, all the [P] values of all the monitored stents of a working face, or the percentage statistics of all the [P] values of the monitored stents in the unit of measurement area, are also available.

Figure 9.18b, c for the storage acquisition type instrumentation to monitor the stent pressure change curve, Fig. 9.18b for a stent two measurement points after the integrated curve, and give (from the data analysis system software interface selection) set the upper limit of the stent pressure allowable upper limit and lower limit [38, 18], in the upper and lower limits outside of the unqualified, should be analyzed for reasons. This is conducive to observation (reading) to analyze the stent running in the pressure tendency and found that the stent initial support force is sufficient to give force, whether the pressure continues to rise, the emergence of anomalies, such as Fig. 9.18c in the time coordinate 17–21-time interval of the pressure continues to rise, and then identify the front column system joints leakage repair normal.

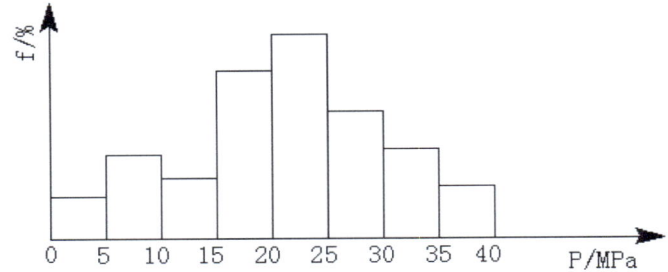
(a) Percentage statistics indicating the frequency of occurrence of pressure intervals

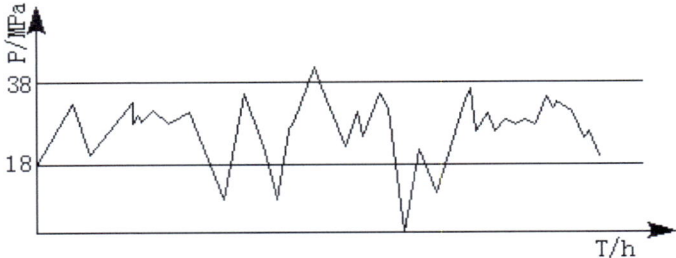
(b) Positive frame synthesized curve with method of setting upper and lower limits of the frame

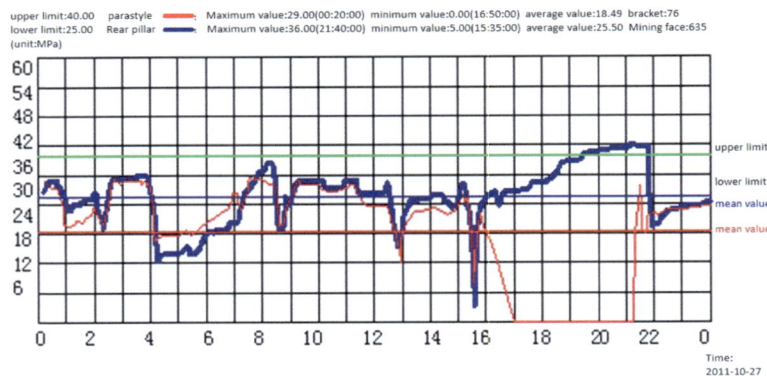
(c) The measured pressure change rule curve

**Fig. 9.18** Methods of analyzing pressure monitoring data from the IMM bracket

Figure 9.18c shows the curve of pressure change rule of the front and rear pillars (four-pillar stent) of the No.76 synthesizing stent in a mine (the horizontal coordinate curve in units of 1h is obtained after software processing of the data signal stored and recorded once in 5 min for the monitoring machine). By observing the curve in the figure, it shows that the roof plate of the mine is a weak and strong type of roof plate, and the pressure continues to rise in the interval from 16 to 22 h, and the pressure

## 9.6 Coal Rock Mass Catastrophe Prediction Methods

from the roof plate is once borne completely by the rear column, and it starts to be normal in the 21h after the shift change and timely repair, and it will be extremely high chance for the occurrence of a large area of roof plate pressure if it cannot be dealt with in time or if it encounters a thick and hard top plate or if the top plate of the working face appears in a fissure structure from the top plate to a high level (top plate crushing, faulting and other fissure structures). If you can't deal with it in time, or if you encounter thick and hard roof, or if there is a fracture structure from the roof to the high rock layer (roof breakage faults, etc.), then the chance of a large area of roof pressure will be high.

(4) Methods for determining and predicting the threshold of acoustic emission signals

Acoustic emission signal threshold value A $E_y$ determination has a double engineering significance, acoustic emission monitoring instrumentation often requires large storage capacity to store the collected signal [AE], continuous monitoring of the collection and storage of data recording density, In order to eliminate the interference of unwanted vibration signals at the monitoring site, increase the credibility of the monitoring signals, reduce the amount of data storage and analysis, take the monitoring method to obtain the threshold value of abnormal signals, the method is to monitor the recording instrumentation after the installation of the sensor probe in the selection of the monitoring site to monitor the collection of vibration signals that cover the entire normal production process, take the normal acoustic emission signal collected in the 12h or 24h time-weighted average (or the weighted average of acoustic emission signals collected during the time). Take the weighted average of the normal acoustic emission signals collected during a 12h or 24h period (or take a slightly higher value) as the acoustic emission signal threshold value A $E_y$, as shown in Fig. 9.19, the threshold value is different from the abnormal signal value. Acoustic emission signal monitoring instrument is mainly used to monitor the impact, protruding pre-coal rock body creep process of sound or vibration omen signal, in the application of signal analysis to study the impact, protruding tendency, the same should pay special attention to the continuous change of the signal development tendency, focusing on the analysis of the frequency of its release signal and the amplitude of the signal change, when there is a frequency of low amplitude of the high should be alerted to the emergence of low frequency and high amplitude, such as Fig. 9.19 in the $\Delta t_1$ and $\Delta t_2$ two For example, compared with the two intervals of $\Delta t_1$ and $\Delta t_2$ in Fig. 9.19, the low frequency and high amplitude of the $\Delta t_2$ interval is more destructive. Also pay attention to its intensity and signal release intermittent characteristics, such as the appearance of pre-signal frequency is high, late (follow-up) signal amplitude high frequency low phenomenon, and should be combined with the impact of ground pressure, prominent occurrence of the principle of analysis and evaluation.

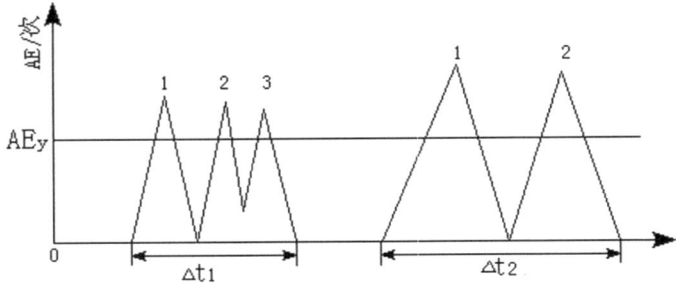

**Fig. 9.19** Schematic of frequency vs. Amplitude and threshold application methods

## 9.7 Creep Monitoring Equipment Development Directions for Coal Rock Bodies

In the process of creep development to damage of coal rock body, parameters (compressive stress and displacement deformation) are two key variables. Can be divided into two cases to consider, general common coal rock body creep destabilization damage, compressive stress of the development of the creep characteristics of continuity, the role of the intensity of the frequency is very low, the acoustic emission signal generated by the frequency and amplitude is not high(5–500Hz within the low-frequency signal proportion is large), that is, the role of the force and the energy elements do not produce drastic changes in a period of time, with a continuous and stable changes in the development of the characteristics of the destabilization damage. The process is easy to predict and master control. Comparatively speaking, the resulting creep displacement damage is also a slow manifestation of the characteristics of a class, such as common mine pressure displacement deformation damage phenomenon, which produces physical changes and signals with the continuity of the development of the rule of change of creep. Coal rock body itself has a very high stress concentration area $\Omega$ or storage of high energy (potential energy and kinetic energy) area $\Omega$ or coal rock body has to withstand very high pressure stress performance (such as thick and hard top plate-inherent induced impact pressure or impact tendency), the development process of creep release of acoustic emission signal intensity of randomness, intermittent characteristics of the obvious, the signal frequency is very high, the amplitude of the change is large, the coal rock body produces vibratory The creep and drastic changes in the strong (including compressive stress, displacement deformation and other signals of creep development), the generation and release of physical signals with the regularity of the pulse signal characteristics of the manifestation of the acoustic emission signals released at a high frequency(25–1000Hz). Once the destruction of this phenomenon, should be used as a precursor to the occurrence of impact pressure and gas protrusion, or impact, protruding tendency to deal with the problem, when the application of the principle of energy to analyze and discuss the development process of catastrophic change is more objective, convenient and practical. Therefore, the monitoring equipment technology and data processing analysis should fully understand and pay attention to the two-creep development situation of the community, which has an important engineering significance for the development of creep of coal and rock body and instability signal recognition and prediction forecasting.

In the future, the development of monitoring instrumentation equipment ① should be closely surrounded by the various physical phenomena released by the development process trajectory of the characterization of the state of appearance of the creep to destabilization of the coal and rock body of the quarry. Depending on the creep to the drastic changes in the development of the drastic degree of consideration of the dominant choice of monitoring and research mode (including the concept). But the dominant model should be: real-time, continuity, systematic, information signal regularity, monitoring and research conceptualization, the scientific objectivity of the physical information, engineering practicality of the basic ideology. Should recognize the value of the systematic monitoring model shown in Fig. 9.8 of the engineering practical. Strengthen the development of computer software function enhancement, maximize the use of monitoring information for the prediction of coal and rock body disaster prediction and disaster expression ability of research and development work. Monitoring instrumentation equipment ② development should be developed in the direction of integrated application, mine pressure of some parameters of the monitoring research application and awareness of the concept of ideas, such as pressure, displacement and deformation monitoring methods and information analysis methods, using the impact of ground pressure, prominent prediction and forecasting of the monitoring research. Reach only pay attention to the analysis of coal rock body creep to drastic changes in the development process of physical information released during the manifestation of the law and characteristics of mine pressure and impact protrudes the manifestation of the law and characteristics, and the application of this manifestation of the characteristics of the trajectory of the movement and change law.

**Open Access** This chapter is licensed under the terms of the Creative Commons Attribution-NonCommercial-NoDerivatives 4.0 International License (http://creativecommons.org/licenses/by-nc-nd/4.0/), which permits any noncommercial use, sharing, distribution and reproduction in any medium or format, as long as you give appropriate credit to the original author(s) and the source, provide a link to the Creative Commons license and indicate if you modified the licensed material. You do not have permission under this license to share adapted material derived from this chapter or parts of it.

The images or other third party material in this chapter are included in the chapter's Creative Commons license, unless indicated otherwise in a credit line to the material. If material is not included in the chapter's Creative Commons license and your intended use is not permitted by statutory regulation or exceeds the permitted use, you will need to obtain permission directly from the copyright holder.

# Chapter 10
# Principles and Applications of Acoustic Monitoring of Coal Rock Bodies

## 10.1 Introduction to Coal Rock Acoustic Monitoring Technology

The applied research of acoustic wave technology for coal rock body in China began in the 1960s. Its start drew on metal ultrasonic detection and underwater acoustic detection technology, and it took more than 40 years from instrument development, transducer imitation to research and development, on-site in-situ detection and indoor specimen test methods, which was accomplished by the efforts of a generation of scientific and technological workers in a multidisciplinary group.

To today, the testing instrument from the first generation of electron tube type, the second generation of transistor type, the third generation of small-scale integrated circuit type, the development of today's fourth generation, that is, by the acoustic wave emission circuit, large-scale integrated circuit of the data acquisition system, the computer embedded motherboard, operating system software, signal analysis and processing software, etc., to become the acoustic wave detection analyzer with a certain degree of intelligent analysis function. At the same time by the longitudinal wave application test development to the transverse wave test; by the application of acoustic parameter time, to the application of wave amplitude and frequency.

At present, acoustic monitoring technology has been incorporated into a number of protocols and specifications of different industries, and the development of this technology is becoming more and more mature. However, the development of acoustic monitoring applied to mining engineering is relatively backward, still in the stage of experimental research, with fewer research results and instrumentation equipment. Strictly speaking, acoustic monitoring technology is divided into two aspects of the background technology: ① coal and rock body due to force or other conditions change the role of the rock body itself produces or releases acoustic emission signals and acoustic monitoring and research activities, such as roof to pressure, coal and rock body deformation of the perimeter rock, the coal seam power impact effect phenomena, usually referred to as the rock acoustic emission monitoring, this type of monitoring with the nature of the rock body disaster prediction and forecasting purpose; ② in the detection of research for the purpose of acoustic monitoring, coal rock body due to geological formation and crustal changes, human factors and other reasons for the formation of various types of tectonic zones, cavities, groundwater, high-temperature points (zones), abnormal rock stratum lithology and defects, etc. Detection of activities, which usually need to follow the characteristics of the acoustic emission of sound waves to the area to be measured to the one side of a specific frequency of the sound waves (most of them are ultrasonic) or multi-frequency sound waves, and receive sound waves in the direction of the other side of the sound waves, and the sound waves in the direction of the other side, and the sound waves received. The acoustic signal is received on the other side. For example, monitoring activities such as geo-radar, elastic wave CT, ultrasonic monitoring, etc. All belong to the category of acoustic wave detection (or monitoring) technology activities.

## 10.2 Laws of Sound Wave Propagation in Coal Rock Bodies

### 10.2.1 Acoustic Waves in Coal Rock Bodies

Rock body acoustic monitoring (inspection) technology can be widely used with a perfect physical foundation. First, we discuss the relationship between the sound velocity of coal rock bodies and the physical properties of coal rock bodies. In view of the structural characteristics of the coal rock body and the monitoring objects are both large rock bodies and rock specimens of small size. From the solution of the fluctuation equation in solids, it is known that the relative relationship between the geometry of rock bodies with different properties and the wavelength of sound waves is different, the boundary conditions are different, and the expressions for the speed of sound are not the same, so it is necessary to discuss them separately.

(1) Speed of sound in an infinite (unbounded) solid medium

Infinite body (medium) refers to the medium size is much larger than the wavelength $\lambda$, theory and experimental evidence that when the medium and the acoustic propagation direction perpendicular to the size of D, the existence of $D > (25)\lambda$, visible so-called infinite is also a special case of the finite, at this time the medium can be considered as an infinite body, then the infinite body longitudinal acoustic wave propagation speed:

$$c_P = \sqrt{\frac{E}{\rho} \frac{(1-\mu)}{(1+\mu)(1-2\mu)}} \quad (10.1)$$

## 10.2 Laws of Sound Wave Propagation in Coal Rock Bodies

Velocity of acoustic propagation of transverse waves in infinite bodies:

$$V_S = \sqrt{\frac{G}{\rho}} \sqrt{\frac{E}{\rho}\frac{1}{2(1+\mu)}} \tag{10.2}$$

where

E-Modulus of elasticity (Pa);

G-Shear modulus (Pa);

μ-Poisson's ratio (dimensionless);

ρ-Density (kg/m³).

(2) Velocity of sound in a finite solid medium
(I) Velocity of sound for a one-dimensional rod. (①) Boundary conditions for one-dimensional rods. A solid medium is called a one-dimensional rod when its dimensions and wavelength satisfy the following relationship:

$$\lambda > 2D$$

$$D < \frac{1}{5}L$$

where

λ-wavelength;

D-Diameter of a one-dimensional rod;

L-length of a one-dimensional rod;

②Longitudinal sound velocity in the direction of the axis of a one-dimensional rod:

$$V_B = \sqrt{\frac{E}{\rho}} \tag{10.3}$$

Obviously, $V_B$ differs from the longitudinal sound velocity of the infinite body $\sqrt{\frac{(1-\mu)}{(1+\mu)(1-2\mu)}}$, see Eq. (10.1), when $\mu = 0.2$–$0.25$.

$$V_P = (1.05 - 1.1)V_B$$

The purpose of analyzing and discussing the speed of sound of one-dimensional rod is that, when measuring the speed of sound of coal rock specimen, the rock specimen may be a cylinder or a rectangular body, so it is not possible to process the dimensions of the rock specimen as a one-dimensional rod, because the speed of sound measured at this point in time is the speed of sound of a one-dimensional rod of Eq. (10.3), not the infinite body of the speed of sound, and the value of which is not representative of the infinite body of the speed of sound measured at the site, and it can't be used as a value of the calculation of the rock integrity index. The rock integrity indexes the value of is not representative of the sound velocity of infinite bodies measured in the field, nor can it be used as a $V_{PR}$ value for calculating. However, this does not mean that the one-dimensional rod does not have physical significance with engineering value.

However, if the rock specimen is intentionally machined into a one-dimensional rod and its axial sound velocity is measured, the modulus of elasticity of the rock can be measured in accordance with Eq. (10.3).

The speed of sound of a two-dimensional plate. When the dimensions of the rock body satisfy the boundary of the two-dimensional plate, i.e., the dimensions in the x and y directions are much larger than the dimensions in the z direction, and the dimension L in the z direction $L_z < \lambda$, the speed of sound of the two-dimensional plate in the x and y directions is as follows:

$$V_P = \sqrt{\frac{E}{\rho}} \sqrt{\frac{E}{\rho} \frac{1}{2(1+\mu^2)}} \tag{10.4}$$

For acoustic wave detection of laminated rock mass, for the longitudinal wave speed of sound perpendicular to the thickness direction, it should be considered according to Eq. (10.4), and the same speed of sound can be used to determine its integrity and dynamic elastic mechanical properties.

## 10.2.2 Reflection, Refraction, and Waveform Transformation of Acoustic Waves in Coal Rock Bodies

The reflection, refraction and wave-type conversion of sound waves in coal rock bodies is an important theoretical basis for acoustic monitoring of coal rock bodies.

(1) Reflection and transmission of perpendicular incidence. When the solid medium is discontinuous, such as the existence of wave impedance boundaries (The definition of wave impedance is the product of medium density $\rho$ and sound velocity c, namely $Z = \rho c$), as shown in Fig. 10.1, when the sound wave propagation of the $n = x$ sound line to with perpendicular the interface the, known as perpendicular incidence. At this interface, a vertical reflection will occur with the following reflection coefficient:

$$\left. \begin{array}{l} R_P = \frac{P_1}{P} = \frac{\rho_2 c_2 - \rho_1 c_1}{\rho_2 c_2 + \rho_1 c_1} = \frac{Z_2 - Z_1}{Z_2 + Z_1} \\ R_V = \frac{V_1}{V} = \frac{\rho_1 c_1 - \rho_2 c_2}{\rho_1 c_1 + \rho_2 c_2} = \frac{Z_1 - Z_2}{Z_1 + Z_2} \end{array} \right\} \tag{10.5}$$

## 10.2 Laws of Sound Wave Propagation in Coal Rock Bodies

**Fig. 10.1** Incidence of an acoustic plane wave

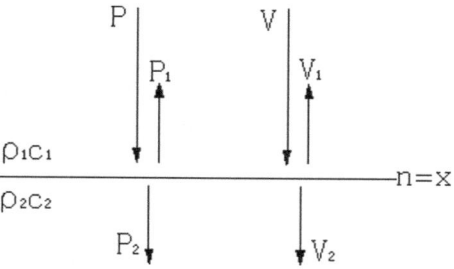

where $R_P$ is called the sound pressure reflection coefficient, which describes the stress relationship of the vibration of the mass during the reflection; $R_V$ is called the vibration velocity reflection coefficient. Similarly, the sound pressure transmission coefficient can be deduced:

$$R_T = \frac{P_2}{P} = \frac{2Z_2}{Z_2 + Z_1} \tag{10.6}$$

Vertical reflections are simpler and do not produce waveform transitions.

(2) Reflection, refraction and wave transition at oblique incidence. If the incident acoustic wave at the wave impedance interface is obliquely incident, it will produce reflection, refraction and wave transition, and its law is shown in Figs. 10.2 and 10.3.

The laws of reflection and refraction follow Snell's law as follows:

$$\frac{\sin \alpha_L}{V_{P_1}} = \frac{\sin \alpha'}{V_{P_1}} = \frac{\sin \beta_L}{V_{P_2}} = \frac{\sin \beta_t}{V_{S_2}} = \frac{\sin \alpha'_t}{V_{S_1}} \tag{10.7}$$

The meaning of is shown in Fig. 10.3, $\alpha_L, \alpha', \beta_L, \beta_t, \alpha'_t$ $V_{P_1}$ is in Fig. 10.3c1, $V_{P_2}$ is in Fig. 10.3c2, $V_{S_1}$ is the transverse acoustic wave of the upper medium in Fig. 10.3, and $V_{S_2}$ is the transverse acoustic wave of the lower medium in Fig. 10.3.

There is a special kind of incidence, as shown in Fig. 10.3b, and a special angle of incidence, called the first critical angle can be obtained from Eq. (10.7): $\alpha_I$,

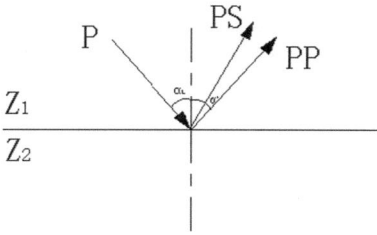

**Fig. 10.2** Reflection at oblique incidence

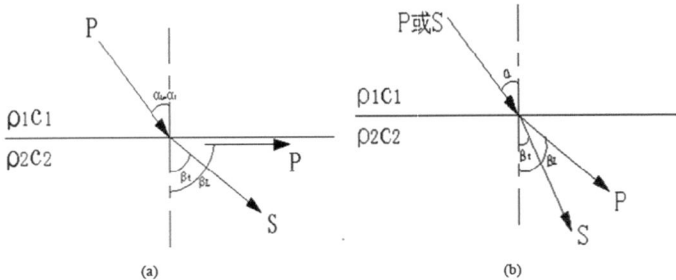

**Fig. 10.3** Refraction at oblique incidence

$$\alpha_I = \arcsin \frac{V_{P_1}}{V_{P_2}} \tag{10.8}$$

It illustrates that when the speed of sound in the first medium is higher than that in the second medium and the angle of incidence of the longitudinal wave is equal to the first critical angle, the angle of refraction is equal to 90°, i.e., the refracted wave glides over the surface of the second medium. The obliquely incident sound wave produces a conversion of the wave pattern in the rock mass, i.e., a refracted transverse wave is produced, as shown in Fig. 10.3a.

Reflection and refraction coefficients for oblique incidence. The reflection coefficient for oblique incidence of longitudinal waves as shown in Fig. 10.2 $R_P$ is as in Eq. 10.9, and the transmission coefficient for longitudinal waves as shown in Fig. 10.3a $R_T$ is as in Eq. (10.10).

$$R_P = \frac{Z_2 \cos \alpha_L - Z_1 \cos \beta_L}{Z_2 \cos \alpha_L + Z_1 \cos \beta_L} \tag{10.9}$$

$$R_T = \frac{2 Z_2 \cos \alpha_L}{Z_2 \cos \alpha_L + Z_1 \cos \beta_L} \tag{10.10}$$

where $Z_1 = \rho_1 c_1$, $Z_2 = \rho_2 c_2$, is the wave impedance of the upper and lower media, respectively.

## 10.2.3 Propagation of Sound Waves in Relation to Coal Rock Body Structure

The structure of the rock body is complex, according to the structure of the rock body, it can be categorized into block structure, laminated structure, fractured structure, and bulk structure.

## 10.2 Laws of Sound Wave Propagation in Coal Rock Bodies

Acoustic wave in massive coal rock mass. Obviously, the propagation of sound wave in coal rock mass with block structure can be approximately regarded as uniform. The propagation in infinite medium is relatively simple, such as granite, thick limestone, sandstone, marble, thin coal seam and so on.

Sound waves in layered structures Layered rock bodies can be modeled as shown in Figs. 10.4, 10.5, and 10.6. It can be seen that the propagation of acoustic waves in this type of rock is a bit more complicated. The wave impedance of the rock layers in the above figures are $Z_1, Z_2, Z_3$, and $Z_1, Z_2, Z_3$ are not equal. It can be seen that the vertical wave impedance incident longitudinal wave $P$, the propagation of sound waves is relatively simple, as shown in Fig. 10.4; oblique incident longitudinal wave as shown in Fig. 10.5, to be a little more complex. There is waveform conversion at the interface, refraction of $PP$ wave to $Z_2$ and $Z_3$ interface again refraction and waveform conversion, so the receiving point will receive $PPP$ wave and wave $PPS$ (converted wave at $Z_2, Z_3$, the interface of the converted wave will not be discussed for the time being), not only the receiving wave group will become more complex, the wave propagation time will be lengthened; if you consider the more complex rock, as shown in Fig. 10.6, the three-layer model of the refraction of the waves between and wave impedance interface and the conversion of waveform. Considering all of them, the waves arriving at the receiving will be: The first wave arriving at the $Z_1 \ Z_2 \ Z_2 \ Z_3$ ng Poe receiving point will be. The first to arrive is $PPP$, $PSP$, $PPS$ and $PSS$ $PPP$ wave, the last to arrive is $PSS$ wave, the other two groups of waves in the middle of the successive arrival. This not only makes the received wave group complex and elongated, but also lengthens the sound time. This kind of rock body such as gneiss, shale, as well as by the laminae, lamellae. Fissures separated by the block rock body.

Acoustic waves in fractured rock body The fractured rock body can be simulated by the model used in Fig. 10.7. Due to the development of lamination, joints, lamellae, and fissures in the rock body, the rock body is cut into fragments, and the wave impedance of each block is respectively $Z_1$, $Z_2$, and $Z_3$ ... The longitudinal wave incident at the transmitting point $P$, if only consider the refraction at the interface of the interface of wave impedance (without considering the reflection) in the conversion of the wave type, it is considered to be as complicated as that shown in Fig. 10.7, and it can be seen that the wave received at the receiving point has the following wave groups: wave, $PPP$ $PSP$ wave, $PPS$ wave, and $PSS$ wave. If the reflection at the wave impedance interface is taken into account, it

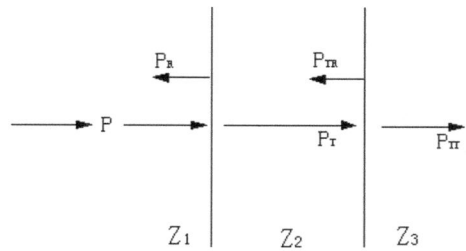

**Fig. 10.4** Incidence at the vertical structural plane of the layer structure

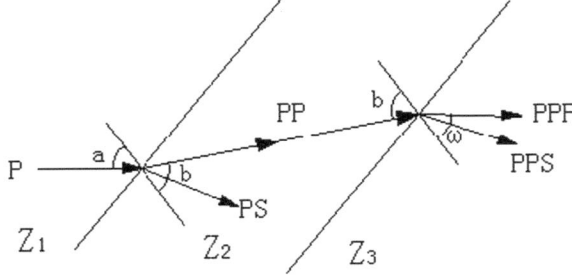

**Fig. 10.5** Layer structure oblique structure surface incidence

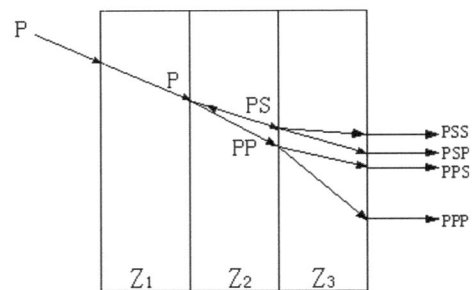

**Fig. 10.6** Oblique incidence of multi-layer structural plane

can be imagined that the problem will be very complex, and it is also very difficult to express it graphically. As shown in Fig. 10.7 is still a simple simulation, the actual rock body is more complex, it can be inferred that its receiving wave group is more complex, but we can focus on the characteristics of a certain wave group. For example, the first arriving at wave will have a longer acoustic time (due to the reduced speed of sound), and the wave group will be more complex and the wave train more elongated. However, it is good to know that we are interested in the first arrivals, and we can ignore the next waves.*PP* ... *P*.

Acoustic waves in bulk structures Bulk structure rock bodies, such as fault fracture zones, tectonic and weathering fissures dense structural surfaces and the combination of intricate rock reinforcement of wind zones. This kind of rock body if there is a certain amount of water content, low-frequency sound waves (such as a

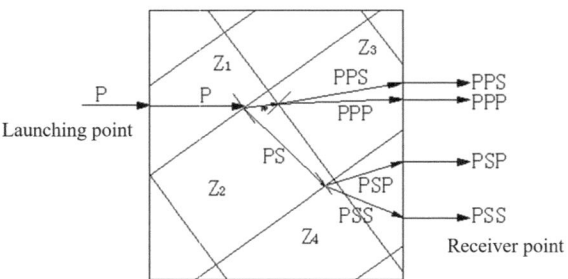

**Fig. 10.7** Propagation of acoustic waves in a massive rock mass

## 10.2 Laws of Sound Wave Propagation in Coal Rock Bodies

few hundred to a few kilohertz) can be penetrated, but requires a larger emission energy, while the ultrasonic frequency of the elastic wave is not able to help, there are used to emit high-frequency electromagnetic waves.

### 10.2.4 Acoustic Wave Bypassing and Scattering in Coal Rock Bodies

The theory of bypassing and scattering is detailed in the previous chapters of this book. The winding of sound waves in a rock mass is concerned, as shown in Figs. 10.8 and 10.9. For a schematic of scattering, see the diagram shown in Fig. 10.10.

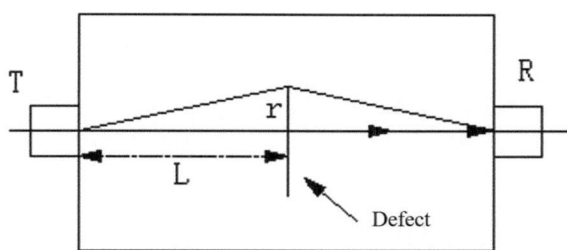

**Fig. 10.8** Bypassing of a sound wave

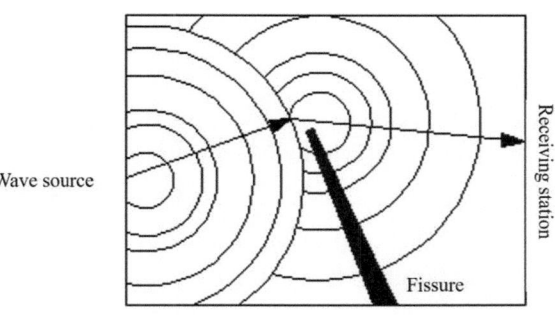

**Fig. 10.9** Acoustic wave bypassing in cracks

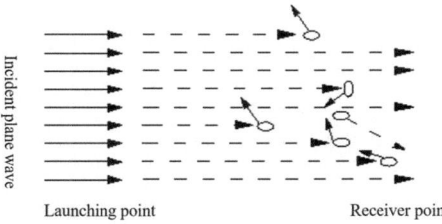

**Fig. 10.10** Schematic of acoustic scattering

## 10.2.5 Correlation of Coal Rock Body Traits with Acoustic Velocity and Applications

(1) Sound velocity and elastic mechanics parameters

The degree of pore and fissure development and structural properties of a coal rock body have a great influence on the propagation speed (wave velocity) of sound waves. In short, a coal rock body with high compressive strength has a relatively high propagation velocity of sound waves, which is usually expressed and illustrated by elastic mechanical parameters

Poisson's ratio:

$$\mu = \frac{(V_P/V_S)^2 - 2}{2[(V_P/V_S)^2 - 1]} \quad (10.11)$$

Shear modulus:

$$G = V_S^2 \rho \times 10^{-6} (GPa) \quad (10.12)$$

Modulus of elasticity:

$$E = V_P^2 \rho \frac{(1+\mu)(1-2\mu)}{(1-\mu)} \times 10^{-6} (GPa) \quad (10.13)$$

where

$V_P$ is the longitudinal sound velocity (m/s), $V_S$ is the transverse sound velocity (m/s), and $\rho$ indicates density of geomaterials (kg/m$^3$).

(2) Evaluating coal rock mass quality using $V_P/V_S$

Poisson's ratio $\mu$ reflects the elastic property of the coal rock body, i.e. the "soft" and "hard" degree of the rock body. Poisson's ratio has a close relationship with the ratio of longitudinal and transverse wave speeds, so the acoustic monitoring of the rock body is commonly used longitudinal and transverse wave speed ratio to reflect the physical properties of the rock body. The relationship between the ratio of longitudinal and transverse wave velocity $V_P/V_S$ and Poisson's ratio $\mu$ is shown in Table 10.1.

**Table 10.1** Relationship between the ratio of longitudinal and transverse waves $V_P/V_S$ and Poisson's ratio $\mu$

| $\mu$ | 0.20 | 0.22 | 0.25 | 0.27 | 0.30 | 0.33 | 0.35 | 0.40 | 0.49 |
|---|---|---|---|---|---|---|---|---|---|
| $V_P/V_S$ | 1.63 | 1.67 | 1.73 | 1.78 | 1.87 | 1.98 | 2.08 | 2.45 | 7.55 |

## 10.2 Laws of Sound Wave Propagation in Coal Rock Bodies

Obviously, the larger the value of $V_P/V_S$, the "softer" the body. For an intact and dense body $V_P/V_S = 1.73$. A large number of statistics show that the value of $V_P/V_S$ correlates with the degree of integrity of the body as shown in Table 10.2.

(3) Velocity of sound and coal-rock body integrity indices

The mass of a rock of the same volume is much greater than the mass of the rock mass, so the rock can be used as the evaluation criterion. Evaluation of the mass of the rock body can only use the longitudinal wave speed. For example, according to the "Engineering Rock Classification Standard" (GB 50218-1994), the rock body integrity index $K_V$ can be calculated by using the longitudinal velocity and the longitudinal sound velocity of the rock, as shown below:

$$K_V = \left(\frac{V_{pm}}{V_{pr}}\right) \tag{10.14}$$

Obviously, the rock mass contains more fissures and joints than rocks of the same volume, so $K_V < 1$. It can be seen that Eq. (10.14) reflects the degree of integrity of the rock mass. The engineering mechanical properties of the rock mass can be categorized by the integrity index, as shown in Table 10.3 for another evaluation classification method.

(4) Acoustic monitoring of rock grading indicators

Because the size of wave velocity can reflect the quality (strength) of coal rock body, so the sonic monitoring can be applied to the strength of the surrounding rock grading index, due to the different acoustic test equipment and working conditions, the test method of rock body elastic longitudinal wave velocity is not the same

**Table 10.2** Measurements of and the degree of rock integrity $V_P/V_S$

| $V_P/V_S$ | $\mu$ | Rock mass condition |
|---|---|---|
| 1.73 | 0.25 | Better quality of intact rock |
| 2.35–2.45 | 0.35–0.4 | Quality deterioration and gradual crack development |
| 2.45–7.55 | 0.4–0.48 | The rock mass ranges from fractured to very fractured |

**Table 10.3** Classification of rock bodies in a department of the corps of engineers

| $K_V$ | < 0.45 | 0.45–0.75 | > 0.75 |
|---|---|---|---|
| > 5.5 | Class III: hard rock of poor integrity | Class II: hard and relatively intact | Class I: hard and intact rock |
| 5.5–3.5 | Category IV: harder, poorer integrity | Class III: harder good integrity | Class II: firmer and intact |
| 3.5–1.5 | Category V: soft rock, poor integrity | Class iSOFT rock, good integrity | Class III: soft rock, good integrity |

between the domestic industries, there are mainly cross-hole test method, single-hole logging method, hammering method and other monitoring methods. The results of different testing methods are slightly different, and the calculated $K_V$ differ from each other by plus or minus 10%, but they can still quantitatively evaluate the degree of integrity of the rock body.

Table 10.4 shows the evaluation criteria for the railroad industry established by the Southwest Branch of the Iron Academy of Sciences in 1978 on the basis of experimentally measured elastic wave parameters.

At present, many industries have evaluation standards based on the elastic wave velocity to classify the strength classification of the surrounding rock in roadways and tunnel chambers. Taking highway as an example, Table 10.5 shows the classification of surrounding rock by applying longitudinal elastic wave velocity in highway tunnels.

(5) Acoustic detection of the effect of surrounding rock grouting reinforcement

The effect of grouting is mainly the coincidence of the actual distribution state of the slurry in the formation with the designed intended injection range and the composite soil parameters after grouting (shear strength, bearing capacity, elastic modulus. Permeability parameters, etc.), and the improvement of the overall physical and

**Table 10.4** Evaluation of elastic wave parameters and rock mass classification

| Rock type | | I | II | III | IV | V | |
|---|---|---|---|---|---|---|---|
| Elastic wave parameter | Longitudinal wave velocity $K_m$ | 4.0–6.0 | 3.0–4.0 | 2.0–3.5 | 1.0–2.5 | 1.00 | |
| | Integrity factor $K_v$ | 0.75 | 0.50–0.75 | 0.35–0.50 | 0.20–0.35 | 0.20 | |
| | Split coefficient $K_j$ | 0.25 | 0.25–0.50 | 0.50–0.65 | 0.65–0.80 | 0.80 | |
| | Weathering factor $K_w$ | 0.10 | 0.10–0.20 | 0.20–0.40 | 0.40–0.60 | 0.60–1.00 | |
| | Longitudinal and transverse sound velocity ratio | 1.7 | 2.0 to 2.4 | 2.5–3.0 | 3.0 | | |
| Characteristics of the rock mass | Whole, hard fresh | Layered and massive, slightly fissured, slightly weathered | Fractured, fissure development, weathering | Loose, very developed fissures, strongly weathered | Dispersed plastic, extremely fissured, severely weathered | | |
| Stability evaluation | Stable | Basically stable | Less stable | Precarious | Highly unstable | | |

## 10.2 Laws of Sound Wave Propagation in Coal Rock Bodies

**Table 10.5** Classification of the surrounding rock in tunnels

| Rock type | VI | V | IV | III | II | I |
|---|---|---|---|---|---|---|
| Longitudinal wave velocity of elastic waves $V_p$ (km/s) | > 4.5 | 3.5–4.5 | 2.5–4.0 | 1.5–3.0 | 1.0–2.0 | < 1.0 |

mechanical indicators (common stress, uniformity and thickness of the composite soil layer). The effect of water plugging (strength, thickness, homogeneity, permeability coefficient) is often included in tunneling projects.

The relationship between acoustic wave and elasticity coefficient of rock body, using the acoustic wave method to obtain the longitudinal and transverse wave propagation velocity $V_p$ and $V_s$, a series of dynamic elasticity parameters can be derived from the following equations:

Dynamic Poisson's ratio: $\mu_d = [1/2(V_P/V_S)^2 - 1]/[(V_P/V_S)^2 - 1]$
Dynamic modulus of elasticity: $E_d = \rho V_S^2(3V_P^2 - 4V_S^2)/(V_P^2 - V_S^2)$

$$E_d = \rho V_P^2(1 + \mu_d)(1 - 2\mu_d)/(1 - \mu_d)$$

Dynamic shear modulus: $G = \rho V_S^2$
Dynamic Laméconstant: $\lambda = \rho(V_p^2 - 2V_S^2)$
Dynamic bulk modulus: $K = \rho(V_p^2 - 4V_S^2/3)$.

These parameters can be used as indicators of the mechanical properties of the rock mass. The rock mass and quality can also be evaluated on the basis of the aspect ratio and the dynamic Poisson's ratio.

The modulus of elasticity of rock used for general engineering design is based on the static method, which establishes a normalized standard, i.e., the static modulus of elasticity; instead, the dynamic modulus of elasticity measured by the acoustic wave method is converted into a standard value.

According to the relevant data introduced by the test and some measured data at home and abroad show that: in terms of rock specimens, the dynamic and static modulus of elasticity of the value is basically close to, with the improvement of the sampling technology of the rock body, the value of the two may be significantly different; and in terms of the rock body according to the domestic and foreign 175 comparative data on the statistical structure, more than 85% $E_d$ is 1–10 times for $E_s$.

Accordingly, the data measured by the acoustic wave method can be converted to roughly estimate the static modulus of elasticity to be used for design, as shown in the following equation:

$$E_{ms} = j \times E_{md}$$

where

**Table 10.6** Rock integrity discount factors

| Rock integrity $(V_m/V_r)^2$ | 1.0–0.9 | 0.9–0.8 | 0.8–0.7 | 0.7–0.65 | <0.65 |
|---|---|---|---|---|---|
| Discount factor $j$ | 1.0–0.75 | 0.75–0.45 | 0.45–0.25 | 0.25–0.20 | 0.20–0.10 |

$j$-discount factor;

$E_{ms}$-Static modulus of elasticity of rock;

$E_{md}$-Modulus of dynamic elasticity of a rock mass.

The discount factor may be selected according to the method in Table 10.6.

## 10.2.6 Acoustic Frequency and Received Signal Frequency in Coal Rock Bodies

(1) Acoustic emission and reception

Acoustic monitoring (detection) of man-made emission of pulse waves, by spectral analysis can be seen by the pulse is composed of a number of different frequencies of sinusoidal waves, of which more high-frequency pulse waves, that is, the emission of the universal frequency wave. In the rock body with the propagation distance increases, or due to the development of rock fissures, weathering degree of different, to be received by the high-frequency components of the pulse wave attenuation faster, so that can be received for the main frequency of the signal (the most energy-rich frequency) to reduce. However, it is usually the low-frequency and high-amplitude signals that can be propagated and received, so the frequency characteristics of the received acoustic signal can reflect the physical properties of the rock body. The frequency of the received signal can be fully reflected by the absorption attenuation spectrum described above. For example, acoustic wave (ultrasonic) monitoring technology, electromagnetic wave monitoring technology, this type of monitoring mode with the detection of the nature of the search for the monitoring, the purpose of which is to monitor or detect the existence of some kind of defects in the coal rock body, anisotropic rock layers, caves and so on.

(2) Acoustic monitoring (reception)

Here the main acoustic wave refers to the coal rock body itself due to the force, high temperature influence, changes in the environmental conditions, coal rock body deformation movement and other circumstances induced by the coal rock body, rock layer, the crust to release the elastic potential energy and produce acoustic signals (i.e., acoustic emission), the resulting sound wave has a certain release emission frequency. The frequency of such sound waves is relatively low, and monitoring studies are often carried out for one frequency band. The purpose of monitoring is often in the prediction of coal and rock damage forecast, Table 10.7 for different engineering and research fields from the engineering practice summarized in the

## 10.2 Laws of Sound Wave Propagation in Coal Rock Bodies

**Table 10.7** Common band values (frequency windows)

| Frequency Hz | $10^{-2}$ | $10^{-1}$ | $10^0\ 10^1\ 10^2$ | $10^3\ 10^4\ 10^5\ 10^6$ |
|---|---|---|---|---|
| Research | Tele seismic research | Earthquake exploration | Recent laboratory and field experiments on geologic materials Early studies of geological materials | Limited acoustic emission studies of geologic materials, metal Acoustic Emission Studies of Materials |

common frequency band, is in different research purposes and conditions through the monitoring research and the monitoring-type acoustic wave frequency band window.

### 10.2.7 Acoustic Emission Phenomena and the Kaiser Effect

When the rock mass is subjected to external forces, such as residual stresses in the ground, stress concentrations caused by man-made or natural disturbances to the rock mass, and other phenomena. Once the external force exceeds the strength of the rock mass, the rock mass will be damaged internally. This damage tends to go through a process, which starts with localized micro-ruptures and the appearance of some new cracks, and when the external stress increases, the number of such ruptures continues to increase, and the newborn fissures increase and extend. If the external force increases further, the above phenomena will develop cumulatively and eventually cause the overall rock body to break down and destabilize. In the process of the above rock body damage, every time a rupture occurs, energy is released and converted into a pulse fluctuation, forming a group of acoustic impulses, which is called "acoustic emission". Each acoustic emission is called an acoustic emission "event".

Acoustic emission phenomenon produced by the frequency of the pulse acoustic wave is very rich (such as hammering mode of emission of acoustic waves), according to foreign literature and domestic research units, the upper limit of its frequency to the megahertz, the lower limit of the frequency to the kilohertz. therefore, the distance from the acoustic emission point (source) dozens of meters away from the reception of such acoustic emission signals, so the general reception of the instrument to receive the main frequency of several kilohertz below the acoustic emission pulse wave group. The number of acoustic emission events, the number of events per unit of time, and the amplitude and intensity of the acoustic emission signals are used to forecast the instability of coal and rock bodies.

Rock acoustic emission phenomenon, there is a special effect discovered by Kaiser, so it is named "Kaiser effect". That is, when the rock body in the history of the maximum stress value, the emission phenomenon can be memorized by the maximum stress value. Can be taken from the rock body of a complete rock specimen, placed on the material testing machine to slowly apply pressure, in the added pressure

**Fig. 10.11** Schematic diagram of the Kaiser effect

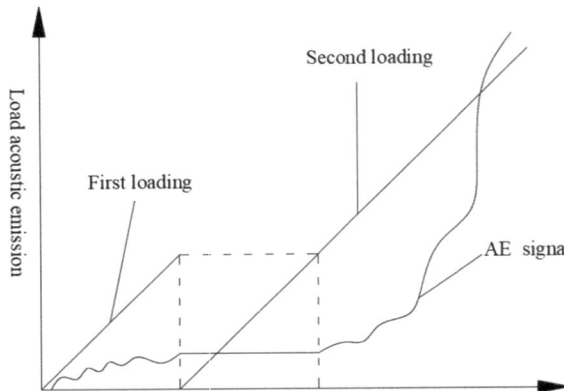

is not more than the history of his memory of the maximum pressure received before the acoustic emission phenomenon does not occur. Thus, the first level of pressure from the time the pressure is applied until the acoustic emission phenomenon occurs is the maximum stress applied to the rock mass, as shown in Fig. 10.11.

For example, some such as leaks in press systems, yielding processes of materials, noise from hydraulic and rotating machinery are signals of continuous acoustic emission. Continuous acoustic emission is characterized by the absence of large undulations in the wave amplitude and the high frequency and low energy of the emission (as shown in Fig. 10.13).

Burst-type acoustic emission signals are generated by the creation and expansion of cracks in metals, composites, geological materials, etc., as well as by impacts on the material. Burst-type signals exhibit a pulsed waveform, and the peak value of the pulse may be large, but it decays quickly, as shown in Fig. 10.13a.

It should be noted that the classification of acoustic emission signals into continuous and burst is not absolute; when the frequency (amplitude) of the burst signal is large, its form is similar to that of the continuous type. In addition, the actual measured acoustic emission signal is very complex and may be the result of the composite of two types of basic signals. Experiments have shown that acoustic emission signals are usually characterized by the following:

(1) The acoustic emission signal is an oscillating pulse signal with a short rise time of $10^{-8}$ to $10_{-4}$ s. The signal has a high repetition rate.
(2) Acoustic emission signals have a wide frequency range, typically from infrasound frequencies to 30 MHz ultrasound frequencies.
(3) The acoustic emission signal is generally irreversible, that is, the acoustic emission signal has the phenomenon of non-repetition. The same specimen under the same conditions to produce only one acoustic emission, which is known as the Kaiser effect, as shown in Fig. 10.11.

## 10.2 Laws of Sound Wave Propagation in Coal Rock Bodies

As can be seen from Fig. 10.11, acoustic emission is generated when loading is started on the specimen. When the load is removed and loaded for the second time, there is no acoustic emission signal when the load does not exceed the maximum load of the first loading, and the acoustic emission phenomenon begins to occur only when the load of the second loading reaches and exceeds the value of the maximum load of the first time, a phenomenon known as the irreversible effect of acoustic emission. This phenomenon is determined by the irreversibility of material deformation and crack extension. It should be noted, however, that the Kessel effect does not exist if the two loads are applied in different ways and directions. In addition, some materials do not have a permanent effect of the Kessel effect and may be left for a period of time with internal structural changes that will restore the acoustic emission.

(4) The generation of acoustic emission signals is not only related to external factors, but also to the internal structure of the material. Due to the complexity of the factors affecting the acoustic emission signal has a random nature, that is, the same type of specimens in the same conditions for observation, the distribution of the data obtained may vary greatly.

(5) The mechanisms of acoustic emission are varied and have a wide range of frequencies, making the acoustic emission signal somewhat ambiguous.

The above properties of the acoustic emission signal are mainly determined by the strength, strain rate, crystal structure, and temperature of the material.

### *10.2.8 Acoustic Emission Signal Detection and Processing*

The basic process of acoustic emission signal monitoring (detection) is shown in Fig. 10.12. The sensor is used to collect the received AE signal; preamplifier on the sensor output of weak electrical signals (sometimes only a dozen microvolts) to amplify to achieve impedance matching; filter is used to select the appropriate frequency window in order to eliminate the impact of noise; the main amplifier on the filtered acoustic emission signal for further amplification, while recording, analysis and processing.

Basic characteristics of acoustic emission signals. As we know, acoustic emission is a transient elastic wave released by an object when it is subjected to external conditions (e.g., force, temperature change, etc.) that cause its state to change. Its waveform can be divided into two categories: continuous and burst, as shown in Fig. 10.13.

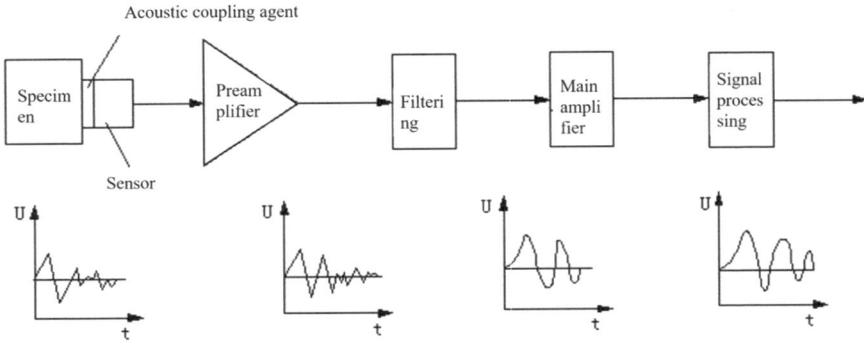

**Fig. 10.12** Principles of the acoustic emission signal monitoring process

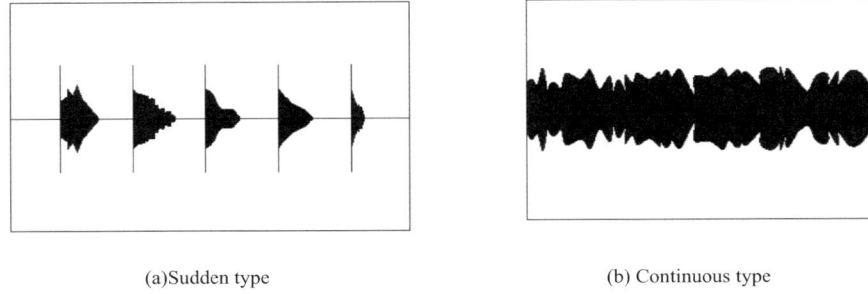

(a) Sudden type  (b) Continuous type

**Fig. 10.13** Typical waveforms of acoustic emission signals

## *10.2.9 Relationships Between Coal Rock Acoustic Emission and Rock Stress*

(1) Relationship between acoustic wave velocity and rock stress

The rock mass is compressed under compressive stress, which closes the fissures, decreases the porosity and increases the sound velocity. As shown in Fig. 10.14, the relationship curves between sound velocity and stress for different rock specimens under uniaxial pressure are shown. From Fig. 10.14, it can be seen that the acoustic velocity increases as the stress increases.

In the low stress conditions sound wave velocity change fast, when the stress is larger, the stress change caused by the sound wave velocity change is smaller. The acoustic wave velocity is also different in different pressurized directions, and the acoustic wave velocity in the horizontal direction changes more than that in its vertical direction. A similar correspondence is obtained between the measured acoustic wave velocity and stress in the rock mass.

## 10.2 Laws of Sound Wave Propagation in Coal Rock Bodies

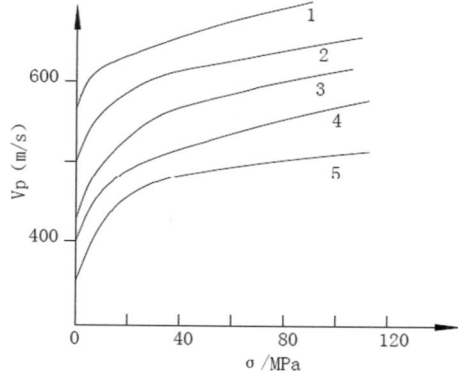

**Fig. 10.14** Transverse waves of acoustic velocity versus rock stresses

### (2) Coal rock body acoustic emission law

The formation of acoustic emission is a mechanical phenomenon, and for the lowest energy signals, the source of acoustic emission is fracture, as it results in the study of metals. According to the fracture theory of plastic deformation, the emission of elastic waves occurs only when the fracture velocity changes, i.e., during acceleration or retardation of the fracture. For higher energy acoustic emission signals, brittle breakage of the rock mass, intergranular slip, and edge regions of plastic slip and plastic deformation occur. Whereas, even larger energy signals are generated by macroscopic breakage of the rock mass or displacement of different parts of the rock mass. Figure 10.15 shows the relationship with the acoustic emission count rate during constant rate loading of a raw coal with a uniaxial loading rate of 0.1 mm/mim. The acoustic emission rate exhibits a significant increase before the main rupture, but the acoustic emission rate is very small in the initial strain stage of loading, which is probably related to the Kaiser effect.

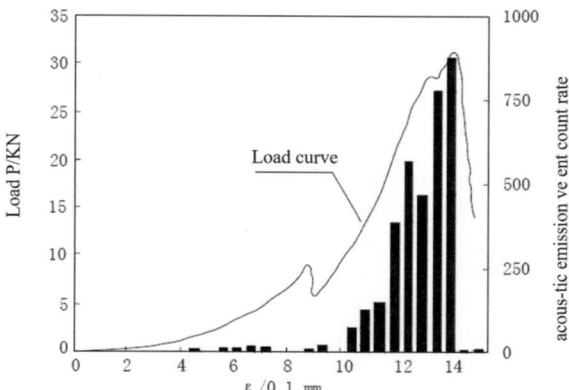

**Fig. 10.15** Acoustic emission versus loading

As can be seen from Fig. 10.15, the macroscopic destruction of the coal rock mass lags behind the peak of the acoustic emission activity, and the decline and equilibrium after the peak of the acoustic emission activity does not indicate that the rock mass is stabilizing, but rather that structural destruction of the rock mass has occurred. Thus, it is possible to monitor the creep development process and the law of the propensity of the coal rock mass to power impacts by determining the strength of the acoustic emission phenomenon and using the analysis to realize it.

## 10.3 Creep Destabilization Signals of Coal Rock Mass and Monitoring Principles

However, the coal-rock body faced by mining engineering is a discontinuous, on-homogeneous and anisotropic body formed under complex and specific conditions (geostress, groundwater, geothermal temperature, etc.), and its mechanical form or ontological relationship shows a very complex nonlinearity. At the same time, the mechanical structure formed by the mining project itself, construction factors and so on have special characteristics, which will make it difficult to get the quantitative results of the stability in line with the engineering reality through the calculation method for a long period of time. therefore, the mining project is faced with the coal rock body morphology of creep to drastic changes in the destabilization damage must be predicted through the field monitoring information signals, forecasting.

The actual design and management of the project requires direct mechanical parameters and mechanical models, and it is inconvenient to use the fuzzy method to deduce the mechanical results, and monitoring is the most direct and reliable method to obtain the information of creep and drastic change of coal and rock bodies. From the development direction of the monitoring method, on the one hand, we should strengthen the monitoring method and the application research of monitoring results processing, on the other hand, we should pay attention to the monitoring results and the theoretical calculation and analysis of the results of the comparative study, to find out and find out the difference between the two, and then according to the monitoring results of the use of inverse analysis method, to seek for the corresponding value of the original parameters, and then for the correction of the original parameters, to improve the original design of the errors and imperfections provide a basis, and so on. This will provide a basis for correcting the original parameters and improving the errors and imperfections in the original design, and so on, making the system design gradually conform to the engineering reality.

In terms of the current stage of development of monitoring technology, combined with the specificity of the environmental conditions of the mine, the changes that occur and the physical information signals released when the coal rock body is subjected to the wave energy of the force or elastic stress wave can be categorized into two orders of magnitude of physical quantities:

First order of magnitude physical quantities: information signals generated and released with significant linear correspondence to the magnitude of the force or the intensity of the wave energy (energy) and with good event correspondence in real time. Such physical quantity signals are most suitable for monitoring, especially for real-time monitoring methods, e.g., compressive or load forces, displacement deformation quantities (especially surface deformation), acoustic emission and microseismic signals, and borehole dust quantities.

The second order of physical quantities: the physical signals of this order of magnitude are more suitable for the detection of static objects for the purpose of monitoring the nature of the performance of the shape of the study, is not suitable for disaster prediction forecasting of key information, the amount of change in the emergence of information signals and the role of the force or energy has a corresponding relationship, but corresponding to the event of the real-time is relatively weak and poor, which is to say that even if no event has occurred, there is also information, or the emergence of information signals are weak. Information signals are weak, for example, holes, fissure changes, resistivity (electrical conductivity), the amount of change in temperature, fluctuations in magnetic field strength. In comparison, the selection of a second order of magnitude of physical quantities as monitoring signals is extremely demanding on monitoring circuits, sensor elements, and information processing, and is highly susceptible to signal recognition difficulties or erroneous information analysis results.

### 10.3.1 Acoustic Emission Principles and Monitoring of Coal Rock Bodies

Among the physical quantities of the first order of magnitude, the monitoring technologies and methods of compressive or loading forces, deformation displacement quantities, and drilling dust quantities (for impact ground pressure and protrusion), as well as the monitoring instrumentation and equipment are at a higher level of development, and are more objective and scientific in their ideological and cognitive concepts, as well as unified in their direction of development. therefore, the focus is on the principles, methods and applications of acoustic emission and microseismic signal monitoring.

#### 10.3.1.1 Concepts of Acoustic Emission

Coal rock body materials or structures subject to external or internal force perturbation effect produce deformation or fracture, or material internal defects and potential defects in the external conditions (such as electromagnetic radiation, vibration, etc.) under the action of perturbation to change the state of the moment, the material to give the response: to the elastic wave form of the phenomenon of the release of energy is known as acoustic emission (commonly known as micro-earthquake), acoustic emission is commonly used in the AE symbols to indicate the acoustic emission can be used in the counting method of acoustic emission to illustrate the intense The acoustic emission is commonly represented by the symbol AE, and the intensity of the acoustic emission can be described by counting methods, such as amplitude, frequency, probability, and unit time.

A series of acoustic emission phenomena accompany the dislocation movement in the internal crystal of coal and rock materials, the formation and development of cracks, the internal friction of the material, the slip and phase change of the particles in the gel, as well as the internal rupture and fracture of the material caused by the above reasons. Acoustic emission signal intensity, frequency and other characteristic indicators with the type of acoustic emission source, state and material properties of different, but also with the form of the role of the external force and the strength of the change and change. therefore, the acoustic emission characteristics is a characterization of the nature and state of the material, acoustic emission monitoring technology is the use of acoustic emission characteristics of the material force, the internal damage to the state and the history of the force to determine the method, which is not only able to check the internal defects of the material (monitoring), but also reflect the formation of defects in the material, the development of the destabilization of the entire dynamic process of destruction. Figure 10.16 shows the results of a typical rock acoustic emission experiment and the corresponding stress–strain curve.

Acoustic emission is a common physical phenomenon, a variety of materials acoustic emission signal frequency range is very wide, from a few hertz of infra-acoustic frequency, 20 Hz–20 kHz sound frequency to several megahertz of ultrasonic frequency; acoustic emission signal amplitude of the range of changes is also very large, from 10 to 13 μm microscopic dislocation movement to the 1 m magnitude of the seismic wave. If the strain energy released by acoustic emission is large enough,

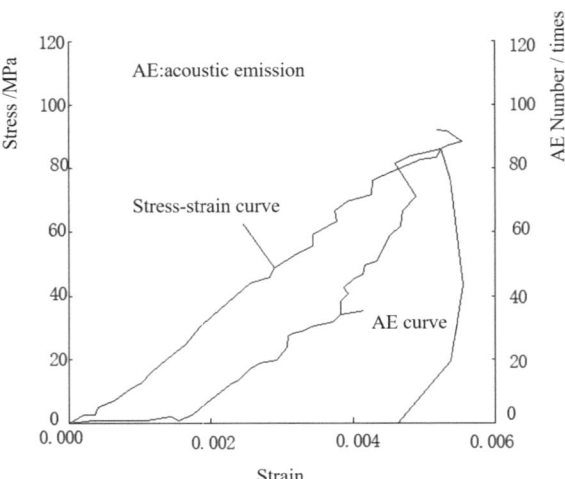

**Fig. 10.16** Typical rock acoustic emission test results

it can produce sound audible to the human ear. Most of the material deformation and fracture when the acoustic emission signal occurs, but many materials acoustic emission signal strength is very weak, the human ear cannot be directly heard, need to use sensitive electronic instruments to detect. Instruments to detect, record, analyze acoustic emission signals and the use of acoustic emission signals to deduce the source of acoustic emission technology known as acoustic emission technology, people will acoustic emission instrument image known as the material's stethoscope.

### 10.3.1.2 Fundamentals of Acoustic Emission Monitoring

Acoustic emission source (defect) in the external force induced by a kind of stress pulse wave or elastic pulse wave that is acoustic emission signal, this stress pulse wave that is acoustic emission signal is a mechanical vibration wave in the acoustic emission source where the material propagation. The so-called acoustic emission monitoring is to detect and receive the above acoustic emission signals and analyze them to obtain information about the acoustic emission source (defect).

The basic principle of acoustic emission monitoring is shown in Fig. 10.17. Elastic waves emitted from an acoustic emission source eventually propagate to reach the surface of the coal and rock body material, causing surface shifts that can be detected by acoustic emission sensors. These detectors convert the mechanical vibrations of the material into electrical signals, which are then amplified, processed, and recorded. Changes in internal stresses in solid materials, such as dislocation motion, crack initiation and extension, fracture, diffusion less phase transitions, magnetic domain wall motion, thermal expansion and contraction, and changes in applied loads. The observed acoustic emission signals are analyzed and extrapolated to understand the mechanism of acoustic emission from coal rock materials. When carrying out a specific inspection (monitoring) work, one should first know the approximate frequency range of the acoustic emission generated by the defect to be monitored under the action of external forces, and then select an optimal frequency window from this total range in order to filter out the noise interference. Generally mechanical and electrical noises have low frequencies because the lower limit of the frequency window (also known as the threshold) is first determined in acoustic emission detection. After the frequency window has been determined, the transducer and bandpass filter can be selected on that basis. Frequencies greater than or lower than those outside the frequency window are considered interfering frequencies or noise and should be eliminated from the analysis.

The main purpose of acoustic emission is:

Determine the location of the acoustic emission source;
Analyze the nature of the acoustic emission source;
Determine the time and load at which the acoustic emission occurs;
Evaluate the severity and danger of acoustic emission sources.

At present, acoustic emission technology has become a mature non-destructive monitoring method, which has been widely used in petrochemical industry, electric

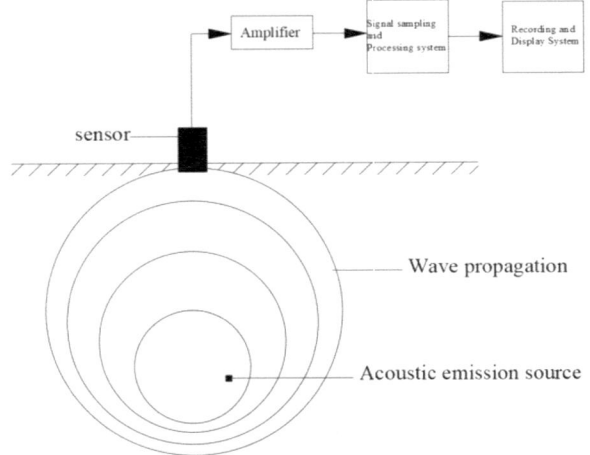

(a) Probe surface monitoring method

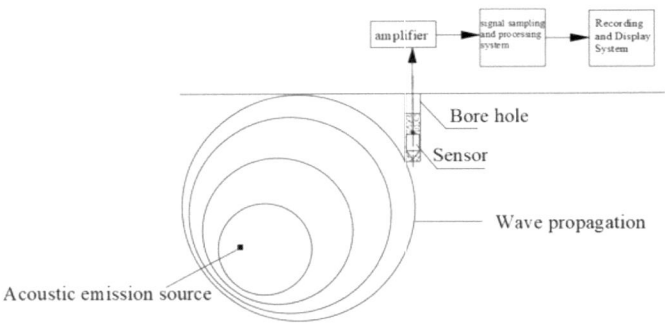

(b) Borehole mounted probe method

**Fig. 10.17** Fundamentals of acoustic emission monitoring (detection)

power industry, civil engineering and construction, geological prospecting, seismic monitoring, industrial flaw detection and so on. In recent years, it has begun to be applied in the field of mining engineering, such as mining earthquake, physical exploration, impact pressure and protrusion, etc.

### 10.3.1.3 Characteristics of Acoustic Emission Monitoring (Inspection)

Acoustic emission monitoring is also a non-destructive monitoring method, however, in many ways different from other conventional non-destructive testing methods, the advantages are mainly manifested in:

Emission is a dynamic inspection method, acoustic emission detects energy from the object under test itself, unlike ultrasonic or radiographic (e.g., ground-penetrating radar, electromagnetic radiation) methods, which are supplied by the testing instrument itself;

The acoustic emission detection method is more sensitive to linear defects, it can detect the activity of these defects under applied structural stresses, and stable defects do not produce acoustic emission signals;

During a single test, the acoustic emission test is capable of detecting and evaluating the state of active defects in the entire rock structure as a whole;

Real-time or continuous information on changes in active defects with external variables such as load, time, temperature, etc. Can be provided, and thus it is suitable for on-line real-time monitoring of engineering processes and early or near damage prediction;

Because the proximity of the monitored (inspection) parts is not required, it is suitable for monitoring (inspection) under the environment that is difficult or inaccessible by other methods, such as the real-time information in the environment of high and low temperatures, nuclear radiation, flammable, explosive, and extremely poisonous;

For complex production environments, the probe can be installed far away from inconvenient environments for monitoring.

Since acoustic emission monitoring is a dynamic detection method and detects mechanical waves, it has the following characteristics:

Acoustic emission characteristics are very sensitive to material properties and are susceptible to interference from electromechanical noise, thus, the correct interpretation of the data requires a richer database and field monitoring experience to support the analysis and application;

Acoustic emission detection generally requires appropriate loading procedures, and in most cases, ready-made loading conditions can be utilized, although special preparations are sometimes required;

Acoustic emission testing currently only gives the location, activity and intensity of the acoustic emission source, not the nature and size of the defects within the acoustic emission source, and still relies on other non-destructive testing methods for retesting. Table 10.8 lists a comparison of the characteristics of acoustic emission monitoring methods and other non-destructive (acoustic) testing methods.

**Table 10.8** Comparison of the characteristics of the acoustic emission method and other non-destructive testing methods

| Acoustic emission monitoring methods | Other conventional non-destructive testing methods |
|---|---|
| Defective growth/activity | Static presence of defects |
| Related to the acting stress or energy | Associated with the formation of defects |
| Higher sensitivity to material properties | Less sensitive to material properties |
| Poor sensitivity to geometry | High sensitivity to geometry |
| Fewer requirements for access to subjects | Higher requirements for access to examinees |
| Conducting large-scale corporate monitoring | Perform localized scans |
| Main issue: the difference between noise and data interpretation | Main issues: proximity and distinction between geometries |

#### 10.3.1.4 Acoustic Emission Sensors

(1) Acoustic emission sensors

Acoustic emission signals propagating in solid-type media contain characteristic information about the source of the acoustic emission, and to utilize this information to reflect material properties or the state of development of defects, it is necessary to receive such acoustic emission signals on solid surfaces. Acoustic emission signal is a transient random wave signal, frequency distribution in the infrasound to ultrasonic frequency range (a few Hz to tens of megahertz). This requires acoustic emission monitoring instruments with high response speed, high sensitivity, high gain, wide dynamic range, strong blocking recovery capability and frequency monitoring window can be selected and other properties. In the actual acoustic emission monitoring process, the monitored signals are often complex signals that undergo multiple reflections and transitions. The acoustic emission signal is received by the sensor and converted into an electrical signal, which, according to a specific calibration method, gives a frequency sensitivity curve, according to which different types of sensors with different frequencies and sensitivities can be selected according to the purpose of monitoring and the environment.

Sensors are made using the principle that the physical properties of certain substances (such as semiconductors, ceramics, piezoelectric crystals, strong magnets and superconductors, etc.) change in response to the external action to be measured. It utilizes a number of effects (including physical effects, chemical effects and biological effects) and physical phenomena, such as the use of materials, piezoresistive, moisture-sensitive, heat-sensitive, light-sensitive, magnetic and gas-sensitive effects, such as strain, humidity, temperature, displacement, magnetic field, gas, etc., acoustic signals are measured and transformed into electricity. The discovery and utilization of new principles and effects, and the development and application of new physical materials have led to the development of physical sensors. Therefore, understanding the various effects on which the sensors are based is necessary for their understanding, development and application. In the acoustic emission monitoring process, the piezoelectric effect is usually used.

The piezoelectric effect is reversible and it is a general term for both the positive and inverse piezoelectric effects. It is customary to refer to the positive piezoelectric effect as the piezoelectric effect. When some of the dielectric along a certain direction by the external force and deformation, in its certain two surfaces produce positive and negative charges, when the external force is removed, and then return to the uncharged state, this phenomenon is known as the positive piezoelectric effect. When the direction of polarization in the dielectric applied electric field, some dielectric in a certain direction will produce mechanical deformation or mechanical stress, when the external electric field withdrawn, the deformation or stress also disappeared, this physical phenomenon is known as the inverse piezoelectric effect. The use of the inverse piezoelectric effect can be made of ultrasonic generators, piezoelectric speakers, highly stable frequency crystal oscillator (such as per day and night error is less than $2 \times 10^{-3}$ of the quartz clock, table) and so on. The inverse piezoelectric effect can be applied to acoustic emission signal generation.

There are many types of sensors with a wide range of applications. However, in order to meet the monitoring of various parameters, in addition to the need to develop new sensitive components, increase the variety of components as well as improve their performance, but also need to use the correct composition of the sensor method, that is, the use of sensitive components, conversion components, conversion of electrical appliances in different combinations of methods, to achieve the purpose of detecting a variety of parameters. This section focuses on the piezoelectric principle of acoustic emission sensors.

(2) Sensor structure

Acoustic emission transducers (probes) generally consist of a housing, a protective film, a piezoelectric element (sheet), a damping block, a connecting wire and a high-frequency socket. Piezoelectric elements are usually berkelium lead titanate, barium titanate and lithium niobate. Sensors with different structures and properties are used according to different detection purposes and environments. Among them, the resonant high sensitivity sensor is one of the most used in acoustic emission monitoring. The structure of the single-ended resonant sensor is simple, the negative electrode side of the piezoelectric element is pasted on the base with conductive adhesive; the other side is welded with a very thin lead to connect with the core of the high-frequency socket, and the shell is grounded.

(3) Sensor coupling and mounting

Acoustic emission signals are acquired via the transmission medium, coupling medium, transducer, and measurement circuitry, and it can be seen from Fig. 10.18 that there are many factors influencing the received signals. therefore, the coupling between the surface of the transducer and the detection surface as well as the installation of the transducer and other details should be strictly required.

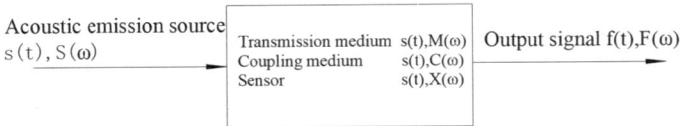

**Fig. 10.18** Acquisition of acoustic emission signals

(4) Waveguide sensors

In some cases, the acoustic emission transducer cannot be placed directly on the surface of the test object, such as high temperature, high pressure, low temperature, loose surface, deep displacement, etc., and the need to realize the acoustic energy through the waveguide (wave energy) connection, that is, through a kind of waveguide device will be received by acoustic emission signals guided to the easy to install the transducer position. Common waveguide has a metal rod or metal tube composed of waveguide components, one end of the fixed (welded or mechanically connected) in the monitoring object surface, the other end of the surface of the acoustic emission transducer placed on the surface.

(5) Sensor selection

The selection of the transducer should be based on the measured acoustic emission signal. The first is to understand the frequency range and amplitude range of the measured acoustic emission signal, including the possible existence of noise signals. Can be empirical understanding, such as welding defects in steel produced by the source of acoustic emission experimental results that the signal frequency range of 25–750 kHz, etc., but there are conditions that are best to determine the actual test, such as the first to monitor the frequency range of the normal state. Then select the relative interest in the acoustic emission signal sensitive, and the noise signal is not sensitive to the sensor for monitoring.

## *10.3.2 Acoustic Emission Monitoring System*

Acoustic emission monitoring (detection) system according to the hardware to obtain and output the data stored distinction, can be divided into two categories of parameters and waveform acoustic emission monitoring. Parameter acoustic emission monitoring mode by the hardware to obtain and output the stored data is amplitude, count and other acoustic emission waveform signal characteristics of parameters, the amount of data is small, the information relative to the waveform data is small, but the data communication and storage is easy, usually only a few thousandths of the waveform data. Waveform acoustic emission monitoring is also by the hardware to obtain and output the stored data is waveform data form, the amount of data is rich in information, but there are difficulties in data communication.

## 10.3 Creep Destabilization Signals of Coal Rock Mass and Monitoring ...

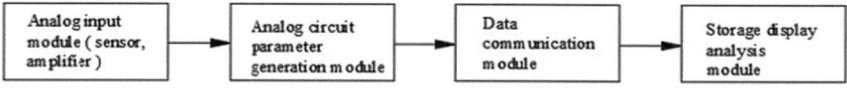

(a) Functional block diagram of typical simulation parameters acoustic emission instrument

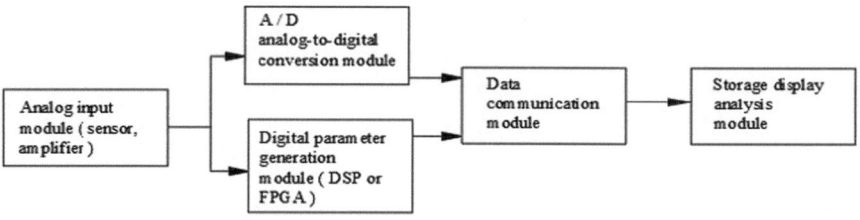

(b) Functional block diagram of typical digital parameter-waveform hybrid acoustic emission instrument

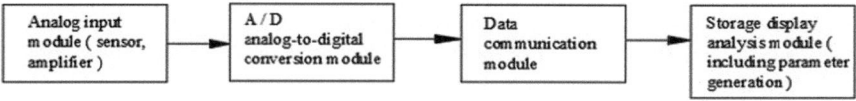

(c) Function block diagram of typical full waveform acoustic emission instrument

**Fig. 10.19** Functional block diagram of a typical acoustic emission instrument

Figure 10.19 shows a block diagram of the characteristic structure of a typical acoustic emission meter. There are three further methods of obtaining parameter data:

Hardware analog circuitry to obtain parameters, characterized by the fact that the analog acoustic emission waveform signal is not A/D analog-to-digital converted to a digital waveform signal before obtaining the parameters;
Hardware digital circuits to obtain the parameters, characterized by the fact that the analog acoustic emission waveform signals have been A/D analog-to-digital converted to digital waveform signals before obtaining the parameters;
Software analysis to generate parameters, characterized by the hardware only to obtain waveform data does not generate parameters fork called full-waveform acoustic emission monitoring.

Signal acquisition mode is divided into single-channel, multi-channel acoustic emission system software can be divided into two categories according to the type of data: parameter data-based analysis software and waveform data-based analysis software. According to the analysis can be divided into feature analysis, localization analysis and pattern recognition.

The input data of the parametric analysis software are parameters, and its characterization mainly involves the analysis of various parameter correlation diagrams, such as the distribution of amplitude, the distribution of the number of impacts in time, etc. Positioning analysis has a variety of different positioning methods, such as linear positioning, planar positioning, three-dimensional positioning, triangular positioning, and rectangular positioning area positioning.

The input data of waveform analysis software is waveform data. Its characterization is mainly a variety of waveform data in the time domain and frequency domain analysis, such as wavelet analysis spectrum analysis. Because the waveform data can produce parametric data and can be set to produce any parameter conditions (such as threshold voltage impact definition time, etc.) and even design new parameters, so the waveform analysis software can include all the functions of parametric analysis and has greater flexibility.

### 10.3.2.1 Acoustic Emission Signal Processing Methods

The current methods of acquiring and processing acoustic emission signals can be divided into two main categories. One is to express the characteristics of the acoustic emission signal with several simplified waveform characteristic parameters, and then analyze and process these waveform characteristic parameters; the other is to store and record the waveform of the acoustic emission signal and analyze the waveform spectrally. Simplified waveform characteristic parameter analysis method is a classic acoustic emission signal analysis method widely used since the 1950s, and is still widely used in acoustic emission monitoring, and almost all acoustic emission monitoring standards on the acoustic emission source of the criterion are used to simplify the waveform characteristic parameters.

(1) Waveform characterization parameters

Figure 10.20 shows the definition of simplified waveform parameters for a burst-type standard acoustic emission signal. From this model, the following parameters are obtained: wave strike (event) counts; ringing counts; energy; amplitude; duration; and rise time.

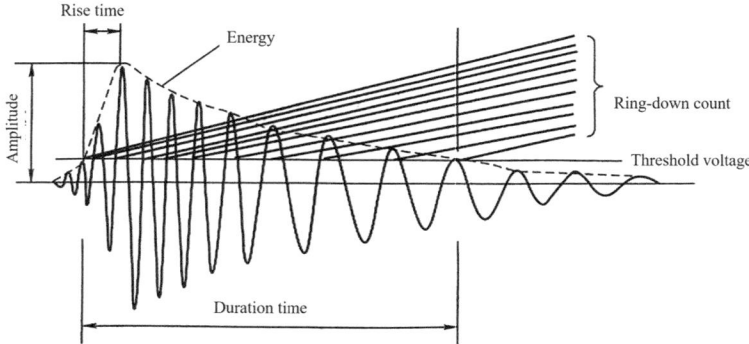

**Fig. 10.20** Simplified waveform structural parameter description of the acoustic emission signal

## 10.3 Creep Destabilization Signals of Coal Rock Mass and Monitoring ...

For continuous acoustic emission signals, only the ringing count and energy parameters of the above model can be applied. In order to characterize continuous acoustic emission signals more precisely, the following two additional parameters are introduced: average signal level and RMS voltage.

The amplitude of the acoustic emission signal is usually expressed in dBae, defining 1 μV of the transducer output as 0 dB, then the dBae amplitude of an acoustic emission signal with amplitude Vae can be calculated by the following formula.

$$dB_{ae} = 20lg(V_{ae}/1\mu V)$$

Table 10.9 lists the sensor output voltage values corresponding to commonly used integer amplitudes dBae.

For actual acoustic emission signals, the acoustic emission signal waveform is a series of waveform envelopes as shown in Fig. 10.21, due to the geometric effects of the specimen or monitored member. Therefore, for each acoustic emission channel, the series of waveform envelopes are drawn into an impact or divided into different impact signals by eliciting the acoustic emission signal impact definition time (HDT). For the waveform in Fig. 10.21, when the HDT set by the instrument is greater than the time interval T between two wave packets crossing the threshold, then the two wave packets are classified as one acoustic emission impact signal. However, if the HDT set by the instrument is less than the time interval T between the two wave packets crossing the threshold, the two wave packets are classified as two acoustic emission impact signals.

Table 10.10 lists the meanings and uses of commonly used acoustic emission signal characterization parameters. The accumulation of these parameters can be defined as a function of time or test parameters (e.g., pressure, temperature, etc.), such as total event counts, total ringer counts, and total energy counts. These parameters can also be defined as a function of time or test parameters, such as acoustic emission event count rate, acoustic emission ringer count rate, and acoustic emission signal energy rate. Correlation analysis can also be performed between any two

**Table 10.9** Sensing output voltage values corresponding to common integer amplitudes dBae

| dBae | 0 | 20 | 40 | 60 | 80 | 100 |
|---|---|---|---|---|---|---|
| Vae | 1 μV | 10 μV | 100 μV | 1 mV | 10 mV | 100 mV |

**Fig. 10.21** Characterization of the emitted impact signal

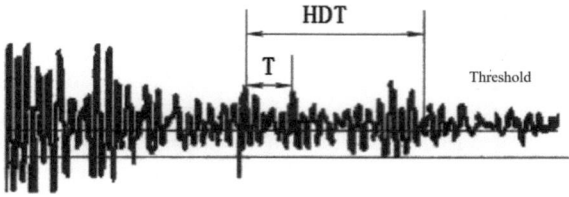

combinations of these parameters, such as acoustic emission event one amplitude distributions, acoustic emission event energy duration correlation plots, and so on.

(2) Analytical identification techniques

List display and analysis of acoustic emission signal parameters. The list display is used to chronologically arrange and directly display each acoustic emission signal parameter, including the signal arrival time, each acoustic emission signal parameter, external variables, and the coordinates of the acoustic emission source.

Table 10.11 shows a list of parametric data for crack extension acoustic emission signals collected during pressurization of a pressure vessel. The data list is directly observed during sensitivity determination of the acoustic emission system and simulated source positioning accuracy testing prior to acoustic emission detection, and is often used to display and analyze the intensity of the acoustic emission source when it has been accurately analyzed.

Single-parameter analysis of acoustic emission signals. Because the early acoustic emission instruments can only get the count, energy or amplitude and so on very few parameters, so at that time the analysis and evaluation of acoustic emission signals are usually used in the single parameter analysis method, the most commonly used single parameter analysis method for the count analysis method, the energy analysis method and amplitude analysis method.

> Counting method: the counting method is a common method for processing acoustic emission pulse signals. The counting methods currently used include acoustic emission event counting rate and ringing counting rate and their total counts, and there is also a kind of amplitude-weighted counting method, called "weighted ringing "counting method. Acoustic emission event is a single sudden impact signal generated by localized changes in the coal rock body, the acoustic emission count (ringing count) is the number of times the acoustic emission signal exceeds a certain threshold, the number of times the signal exceeds the threshold per unit of time is the counting rate, the acoustic emission count rate is dependent on the frequency of response of the transducer, the damping characteristics of the transducer, the structure of the damping characteristics and the level of the threshold.
> Energy analysis method: because of the above disadvantages of the counting method for measuring acoustic emission signals, especially for continuous acoustic emission signals, the energy of the acoustic emission signals is usually measured to analyze the continuous acoustic emission signals.
> Amplitude analysis method: the signal peak amplitude and amplitude distribution is a processing method that can reflect more information about the acoustic emission source. The signal amplitude is directly related to the intensity of the acoustic emission source generated in the material. The amplitude distribution is related to the deformation mechanism of the material.
> Experiential graph analysis method: the acoustic emission signal experiential analysis method is used to obtain the activity and development trend of the acoustic emission source by analyzing the acoustic emission signal parameters as they

**Table 10.10** Acoustic emission signal parameters

| Parameters | Hidden meaning | Features and usage |
|---|---|---|
| Hit and hit counting | Any signal that exceeds the threshold and causes a channel to acquire data is called a bump. The number of impacts measured can be categorized into total counts, count rate | Reflects the total amount and frequency of acoustic emission activity and is commonly used in acoustic emission activity evaluations |
| Event count | The first localized change in material that produces acoustic emission is called an acoustic emission event. It can be categorized into total counts, and count rates. One or several impacts in an array correspond to one event | Reflects the total amount and frequency of acoustic emission activity for the evaluation of source activity and localization concentration |
| Ringer ID | The number of oscillations of the signal crossing the threshold, which can be divided into total counts and count rates | Signal processing is simple, suitable for two types of signals, and can roughly reflect the signal strength and frequency, widely used in acoustic emission activity evaluation, but affected by the size of the threshold, see Fig. 10.22 |
| Amplitude | Maximum amplitude value of the signal waveform, usually expressed in dB ae (0 dB for 1 μV sensor output) | Directly related to the size of the event, independent of the threshold, directly determines the measurability of the event, commonly used in the identification of the type of wave source, intensity and attenuation measurements |
| Energy technology (MARSE) | The area under the signal detection envelope, which can be divided into total counts and count rates | Reflects the relative energy or intensity of an event. Less sensitive to threshold, operating frequency, and propagation characteristics, and can be used in lieu of ringing counts and also for type identification of wave sources See Fig. 10.23 |
| Span | The time interval elapsed between the first crossing of the threshold by the signal and its final descent to the threshold, expressed in μs | Very similar to ringer counting, but commonly used to identify specific wave source types and noises |
| Rising time | The time interval experienced by the signal from the first crossing of the threshold to the maximum amplitude, expressed inns | The physical significance of the noise becomes unclear due to propagation, and is sometimes used to identify electromechanical noise |

(continued)

**Table 10.10** (continued)

| Parameters | Hidden meaning | Features and usage |
|---|---|---|
| RMS voltage (RMS) | The root-mean-square value of the signal during the sampling time, expressed in V | Related to the size of the acoustic emission, easy to measure, not affected by the threshold, applicable to continuous signals, mainly used for continuous acoustic emission activity evaluation |
| Average signal level | The average value of the signal level, expressed in dB, during the acquisition time | Provides information and uses similar to RMS, and is particularly useful for continuous-type signals that require high amplitude dynamic range but not high time resolution. Also used for measurements of background noise levels |
| Arrival time | Time for an acoustic emission wave to reach the sensor, expressed in μs | The position of the wave source, the sensor spacing and the propagation speed are determined for the calculation of the position of the wave source |
| Exogenous variable | Additive variables to the test process, including time, load, displacement, temperature and fatigue cycles | Data sets that are not signal parameters, but are wave impact signal parameters, for acoustic emission activity analysis |

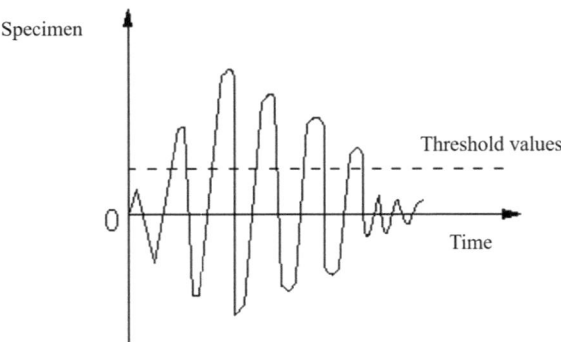

**Fig. 10.22** Schematic diagram of ringing method of counting

change from time to time or from external variables. The most common and intuitive method is graphical analysis.

Distribution analysis method: the acoustic emission signal distribution analysis method is to analyze the acoustic emission signal impact count or event count according to the signal parameter value.

## 10.3 Creep Destabilization Signals of Coal Rock Mass and Monitoring ...

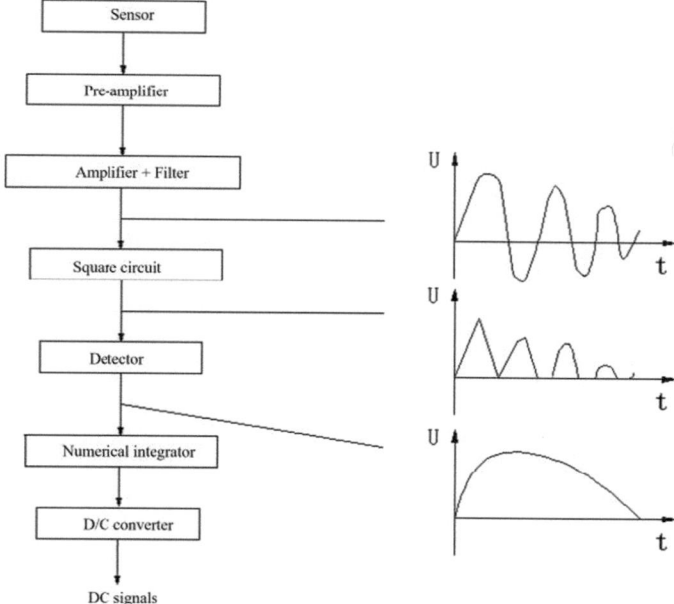

**Fig. 10.23** Energy measurement and corresponding signal waveform diagrams

**Table 10.11** Data list of characteristic parameters of acoustic emission signals

| Arrival time | Stresses | Conduit | Rising time | Counting | Energy | Duration | Amplitude |
|---|---|---|---|---|---|---|---|
| 01:18.9101730 | 36.60 | 3 | 81 | 92 | 57 | 3222 | 59 |
| 01:18.9103205 | 36.60 | 12 | 133 | 49 | 48 | 6243 | 51 |
| 01:18.9104999 | 36.60 | 4 | 69 | 62 | 86 | 6899 | 55 |
| 01:18.9112070 | 36.60 | 8 | 29 | 27 | 53 | 1947 | 51 |

Correlation analysis method: the correlation analysis method is also the most commonly used method in the analysis of acoustic emission signals. The characteristic parameters of the waveforms of any two acoustic emission signals can be analyzed by making a correlation diagram between them.

## 10.4 Acoustic Detection Technology of Rock Anomaly Defects and Loose Zones

The second order of physical quantities: the physical signals of this order of magnitude are more suitable for the monitoring of the existing or inherent properties, morphology, performance, and character of coal and rock bodies for research purposes. It is not suitable for the prediction and forecasting of the key information of the catastrophic evolution process of the coal and rock body from creep to drastic changes. Although the changes in the information signals of the second order of magnitude have a corresponding relationship with the action of force or energy, the real-time nature of the corresponding events is relatively weak and poor, that is, even if no event has occurred, there is already information, or the appearance of the information signals is relatively weak, for example, the pore and fissure changes, resistivity (electrical conductivity), temperature changes, magnetic field strength fluctuations, and the amount of the change. For example, pore and fissure changes, resistivity (electrical conductivity), temperature variations, fluctuations in magnetic field strength, and so on. The monitoring method is based on transmitting acoustic signals to the area to be measured by artificial means, and then receiving the signals on the side of the area to be measured, and using the information such as sound time, wave range, and signal change as a monitoring study.

### 10.4.1 Electromagnetic Radiation Monitoring Techniques

Through the analysis and study of relevant experimental data and literature reports, when the coal rock body is loaded with deformation and rupture, as well as when there is a high temperature or anomalous temperature region inside, the electrons and other electrically charged particles in the crystalline structure of the coal rock body will radiate electromagnetic waves outward at variable speeds, and this is the phenomenon of electromagnetic radiation. Carrying out research on rock electromagnetic radiation started from the discovery of pre-seismic electromagnetic anomalies by earthquake workers. China and the former Soviet Union are the earlier countries to carry out this research, Japan, Greece, the United States, Sweden, Germany and other countries have also carried out research in this area. It is said that during the Wenchuan earthquake, cell phone signals were interrupted due to electromagnetic signal interference during a short period of time (several minutes). Scholars of Chinese Academy of Sciences Geophysical Institute and Professor Xueqiu He of China University of Mining and Technology once combined with their respective professional research, through the laboratory research shows that the loaded rock and coal rock deformation and rupture process, resulting in temperature changes and electromagnetic radiation changes in the signal basically showed a gradual increase in the trend. However, in terms of the actual monitoring of the project, this trend of change and the degree of damage does not show a good correspondence between the relationship is not enough to meet for the coal mine quarry coal rock body disaster signal monitoring. From the earth geomagnetism can be seen, mine coal rock body, quarry perimeter rock even if there is no stress action and elastic wave energy impact, mine coal rock body, coal body also exists in the magnetic field, the same phenomenon of electromagnetic radiation. Therefore, for the mine quarry, electromagnetic radiation monitoring methods to obtain the information signal there is a great ambiguity and uncertainty, is not suitable for the application of real-time monitoring of the creep instability process of coal and rock body, much less suitable for mobile detection methods, monitoring information application difficulties, inconvenient to analyze the location of the signal release source, the characteristics and nature of the signal release source, and cannot be established in real-time correspondence of creep instability, the monitoring information can only be The monitoring information can only react to the damage phenomenon and loosening degree (depth) of the coal and rock body. At present, electromagnetic radiation, like the phenomenon of temperature change, is found to have a relationship with pressure when the rock is pressurized in a laboratory press. The electromagnetic signal and temperature signal generated by the pressure failure of the surrounding rock and coal rock in the mine stope are weak. Moreover, there are also signals when there is no pressure (failure). Therefore, the change of electromagnetic signal or temperature signal in the coal rock and rock in the mine stope is not necessarily related to the change of pressure, but also related to the geological structure such as loosening zone, fault, joint plane and dislocation. Electromagnetic signal anomalies often occur when approaching these structures, and the signal of such sites is stronger than the electromagnetic signal that the strength of coal rock mass is not damaged by pressure, that is, the radiation signal. From the perspective of monitoring instrument development technology, it is not the best choice to choose electromagnetic or temperature and pressure (shock, outburst) relationship signals as prediction information.

## *10.4.2 Ultrasound Detection Technology*

Acoustic testing technology is a new technology that has developed very rapidly in recent years, and it belongs to the non-destructive detection technology by using high-frequency acoustic waves emitted to the detected area or path medium and then monitored. The basic principle of ultrasonic detection is to use artificial methods in the monitoring area or along the path of the medium and structure to stimulate a certain frequency of elastic waves, this elastic wave with a variety of waveforms in the material and structure of the internal propagation and reception by the receiving instrument, through the analysis and study of received and recorded fluctuation signals to determine the regional scope or along the path of the medium and the structure of the mechanical properties of the defects that already exist in the interior of the understanding of them, such as coal and rock body density changes, faults, karst water and other underground structures.

At a certain point in the elastic medium, using some (man-made) reasons and cause the initial disturbance or vibration of the rock body, this disturbance or vibration will be in the form of wave propagation in the elastic medium, the formation of elastic waves. Acoustic wave is a kind of elastic wave, if regard coal rock body, concrete, rock a kind of medium as elastic body, then the propagation of acoustic wave in coal rock body medium obeys the law of elastic wave propagation, in order to achieve the monitoring becomes possible.

Coal rock bodies often contain various structural surfaces such as layers, joints and crevices, which are deformed under dynamic loads (wave energy impacts) and have a series of effects on the fluctuation (wave propagation) process in coal rock bodies, such as reflection, refraction, bypassing and scattering. In this way, the nodal interfaces of the coal rock mass play a role in consuming energy and changing the wave propagation pathway, and lead to the non-homogeneity and phase heterogeneity of the fluctuations in the coal rock mass. therefore, the structure of coal rock body affects the propagation process, direction and speed of elastic waves in the rock body, which means that the fluctuation characteristics of elastic waves in coal rock body reflect the structural characteristics of the rock body, so the elastic wave detection technology has become an effective, simple and reliable means of engineering rock body research.

Three types of elastic waves, longitudinal (P-wave), transverse (S-wave) and surface (R-wave), are generated in coal rock bodies under dynamic pressure. Their propagation can be described by wave speed, amplitude, frequency and waveform. The elastic wave tests currently used focus on longitudinal wave speed, followed by transverse wave speed, and have begun to study their amplitude characteristics one by one. From the results of field and laboratory studies, it is shown that the propagation speed of elastic waves in coal rock bodies is related to the type of coal rock bodies, elastic parameters, structural surfaces, physical–mechanical parameters, stress state, degree of weathering and water content, etc., and has the following laws.

> When the modulus of elasticity decreases (e.g., when the coal rock body is loosened), the acoustic wave velocity of the coal rock body decreases accordingly, which is consistent with the theoretical formula of wave velocity.
>
> The denser the coal rock body is, the higher the sound speed of the coal rock body is. Wave velocity formula, wave velocity and density is proportional to the density, but the density increases, the modulus of elasticity will have a substantial increase, and thus the wave velocity will be higher. The longitudinal wave velocity of sound waves of several common intact rock bodies is: 5500–6000 m/s for metamorphic rocks; 5000–5500 m/s for igneous rocks, limestone and well cemented sandstone; 1500–3000 m/s for sedimentary rocks and poorly cemented clastic rocks.
>
> The presence of structural surfaces reduces the speed of sound and causes phase anisotropy in the propagation of sound waves through the coal and rock mass. The speed of sound is low in the direction perpendicular to the structural surface and high in the direction parallel to the structural surface.
>
> A large degree of weathering of the coal rock body is associated with a low sound velocity.

## 10.4 Acoustic Detection Technology of Rock Anomaly Defects and Loose …

High acoustic wave velocity in the direction of action of compressive stress (or wave energy propagation).

A larger porosity n results in a low wave velocity. A coal rock body with high density and high uniaxial compressive strength has a high wave velocity.

The amplitude of the sound wave is also related to the characteristics of the coal rock body, when the coal rock body is more fragmented and the joints and fissures are developed, the amplitude of the sound wave is small, and vice versa, the amplitude of the sound wave is larger. The amplitude of the sound wave propagating perpendicular to the structural surface is smaller than that in the parallel direction.

(1) Common instruments and monitoring methods for ultrasonic detection

The whole process of acoustic wave detection in coal rock body is acoustic wave emission, propagation and receiving display, and its corresponding instruments are: ① transmitting transducer, ② receiving transducer, and ③ acoustic wave receiver. Acoustic transducer is the conversion device of acoustic and electric energy, commonly known as the probe. Transducer is the conversion device of acoustic and electric energy, commonly known as the probe. The transducer generally works by utilizing the principle of piezoelectric effect of piezoelectric ceramic crystals. One of the transmitting transducers is the acoustic wave instrument transmitter output with a certain power of electrical signals into acoustic signals emitted into the rock, its working principle is to use the inverse piezoelectric effect of the crystal. The receiving transducer is to convert the acoustic signal propagated in the rock body into an electrical signal, which is input to the input system of the sonde receiver, and its working principle is to utilize the piezoelectric effect of the crystal. Due to the actual measurement of the transducer and frequency band, different requirements of the mode of operation, so made with different structures and different vibration mode of the piezoelectric transducer, China's current production of piezoelectric series of transducers see Table 10.12

The main performance indicators of a transducer are probe directivity or diffusion angle, quality quotes and bandwidth. The directivity of the transducer depends on its diffusion angle, and the size of the diffusion angle depends on the ratio of the diameter of the sound wave to the wavelength of the sound wave in the medium. With a large ratio, the diffusion angle of the sound wave is small, and with a small ratio (too small a source size), the sound wave is poorly directional and disperses in all directions. Quality factor depends not only on the characteristics of the transducer itself, but also with the measured medium and fixed method, quality factor is small, the probe radiated out of the energy is small, low efficiency, but improved resolution. In addition, the bandwidth of the transducer is good and mechanical quality factor is inversely proportional.

(2) Transducer arrangement method

**Table 10.12** Piezoelectric Series Transducers

| Test condition | Object | Transducer form | | Frequency Kc/Hz | Characteristics | Main application |
|---|---|---|---|---|---|---|
| Field | Rock surface | Loudspeaker style curved | | 10–40 15–20 | Separate reception for sending and receiving | (1) Modulus (2) Rock classification |
| | Hole wall rock | Two-hole | Pressurized | 25–35 | Separate sending and receiving | |
| | | Single hole | Round tube radial type Round tube bending type Plate bending type | 30–40 15–20 15 ~ 20 | Lit.one shot, two receipts Lit.one shot, two receipts Lit.one shot, two receipts | (1) Elastic modulus (2) Loosening ring |
| Interior | Specimen | Round block axial | | 100–200 | Separate sending and receiving | (1) Elastic modulus rock specimens (2) Strength of specimen |

The main types of transducer arrangements are as follows:

Penetration method. It is to place the acoustic wave transmitting transducer and the receiving transducer on two opposite surfaces of the medium, and judge the quality of the medium according to the changes in wave speed and energy after the sound wave penetrates the medium (see Fig. 10.24). This method can be used in the case of large thickness and easy placement of transducers on both surfaces, in the classification of underground engineering surrounding rock, measurement of physical and mechanical parameters of rock mass and quality inspection of underground mass concrete components. It is often used to drill two parallel drilling holes or bury two parallel measuring tubes, and the transmitting transducer and the receiving transducer are respectively put into two drilling holes or two tubes, and the holes are filled with fluid oil or water as the coupling agent, which becomes the double-hole perforation method (referred to as the double-hole method). Penetration method is a widely used method with high sensitivity, simple and clear waveform, less interference and easy identification of various waveforms. However, this method requires high relative accuracy of transducer installation.

Reflection method. It is the transducer to the medium to transmit sound waves, fluctuations along the direction of emission propagation to the bottom surface of the medium, was reflected back and then received by the transducer, the pattern of reflected wave propagation time and display waveform to determine the defects within the medium and the nature of the material method, this method applies to the other side of the medium cannot be placed transducer situation, in the structure of the rock layer, the thickness of the concrete test and the integrity of the pile foundation testing, that is, the use of reflective This method is suitable for the case that the transducer cannot be placed on the other side of the medium.

**Fig. 10.24** Probe arrangement of the penetration method

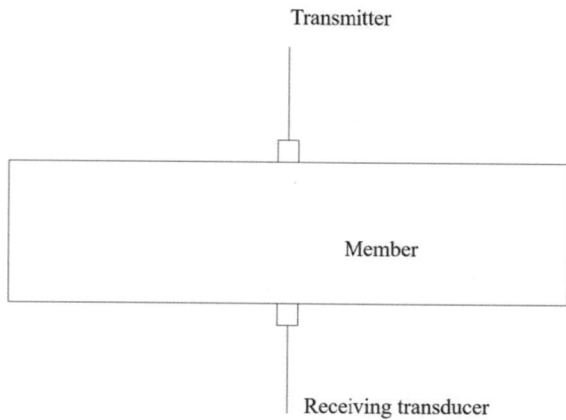

Profile method, also known as along the surface method. This method is the transmitting transducer transmits longitudinal waves through a certain angle of incidence into the medium, is converted into surface waves, through the determination of the surface wave characteristics to determine the defects of the medium and the performance of the material. If the transmitting transducer and the receiving transducer are put into a borehole at the same time, it is a single-hole logging.

### *10.4.3 Technology for Ground-Penetrating Radar Monitoring Applications*

Ground Penetrating Radar (GPR) is an electromagnetic technology that uses wireless electromagnetic wave to monitor the distribution of underground media and scan the invisible target (density difference, lithology change) or underground interface (karst water, geological layer) to determine its internal structure shape or location. The working principle is as follows: High-frequency electromagnetic waves are transmitted by the transmitting antenna in the form of wide-band pulses, reflected or transmitted by the target body, and received by the receiving antenna. When high-frequency electromagnetic waves propagate in the medium, their path, electromagnetic field intensity and waveform will change with the electrical properties and collective forms of the medium. The spatial position or structural state of the underground interface or the target can be determined. The geological radar has the characteristics of high resolution, non-destructive, simple operation, strong anti-interference ability, etc. It is suitable for various environmental conditions. Due to the high attenuation of high frequency electromagnetic wave in the medium, the application of this method is limited.

Although the bottoming radar utilizes the reflection of high-frequency electromagnetic waves to detect the target body as does the air-searching radar, the propagation of electromagnetic waves in the former is much more complicated than in the latter due to the stronger electromagnetic attenuation characteristics of the underground medium as compared with the air, coupled with the discrete nature of the geological situation. Geological radar application in the early stage, limited to the electromagnetic wave absorption is very weak ice, rock salt and other detection, with the electronic technology gradually expanded to the soil layer, coal beds and rock layers and other consumptive media, covering archaeology, mineral resources exploration, geologic hazards survey, geotechnical engineering surveys, engineering building structure monitoring and many other fields.

At present, the geological radars put into field monitoring in China are mainly pulse time domain types, and the SIR Series produced by the United States "Geophysical Measurement System Company" (GSSI) and the EKKO series produced by Canada "probe and Software Company" (SSI) can be divided into the following four parts.

> Antenna: Its function is to couple high frequency electromagnetic waves from the geo-radar transmission line to the propagation medium or from the propagation medium to the transmission line.
> Transmitter: Its function is to generate high-frequency electromagnetic waves at the desired power level.
> Receiver: Its function is to receive a weak target signal and amplify the signal to a usable level.
> Display: Its function is to display the target information to the user.

(1) Profile method of measurement

Currently, the commonly used time-domain and geological radar measurement methods include the profile method, wide-angle method, loop method, multi-antenna method, etc., with the profile method combined with the multiple coverage technique being the most widely used. The profile method is a measurement method in which the transmitting antenna (T) and receiving antenna (R) move synchronously along the measurement line with a fixed spacing. When the spacing between the transmitter and receiver antennas is zero, i.e., the transmitter and receiver antennas are combined into one, it is called a single-antenna form, and vice versa, it is called a dual-antenna form. The measurement results of the profile method can be expressed in a geo-radar time profile image, where the horizontal coordinate records the position of the antenna at the surface, and the vertical coordinate is the reflected wave bi-directional travel time, which expresses the time it takes for the radar pulse to travel from the transmitting antenna through the subsurface interface and be reflected back to the receiving antenna. This kind of record can accurately describe the measurement line to put down the subsurface each reflected wave will be due to the signal-to-noise ratio is too small for repeated measurements, and then the measurement record of the same position in the record is superimposed to enhance the ability to discriminate the deep subsurface medium.

(2) On-site monitoring techniques

Analysis of monitoring objects. The success of ground-penetrating radar monitoring is directly related to the detailed analysis of the monitoring object and the enabling environment, of which the depth of the object is a very important issue. If the depth of the object exceeds 50% of the detection range of the radar system, the geo-radar detection method should be excluded. The geometry of the object must be investigated as clearly as possible, including height, length and width. The geometry of the object determines the possible resolution of the radar system and relates to the choice of antenna center frequency. The conductivity and dielectric constant of the object must also be known, which will affect the system's ability to recognize energy reflections or scattering. For monitoring in rocky media, the unevenness of the surrounding rock should be limited to a certain range, so as not to monitor the response of the object is submerged in the surrounding rock changes in the nature of the state and cannot be recognized. Finally, there should be no extensive metallic structures or radio frequency sources in the monitoring area to avoid serious interference with the monitoring results.

Layout of the survey network. Before the monitoring work is carried out, the coordinates of the survey area should first be established in order to determine the planar position of the survey line, which usually follows the following principles:

When the direction of the distribution of the monitoring object is known, the measurement line shall be perpendicular to the direction of the long axis of the monitoring object. If the direction is unknown, it should be arranged as a square grid.

When the volume of the monitoring object is limited, only a small-scale preliminary survey with a large network is used to determine the extent of the target body, and then a detailed survey is carried out with a small grid and a large-scale network. The grid size is equal to the size of the monitoring body.

When conducting surveys of two-dimensional bodies, such as bedrock surfaces, the survey area should be perpendicular to the strike of the two-dimensional body, and the line spacing depends on the degree of change of the monitoring object along the direction of the strike.

Data processing and information interpretation methods

Data processing: The objectives of ground-penetrating radar data processing are to suppress random and regular interferences, to display reflected waves on the image profile with the maximum possible resolution, and to extract various useful parameters of the reflected waves, including amplitude, waveform, frequency, etc., to aid in the interpretation of monitoring results. Since the electromagnetic wave theory and the reflected seismic wave theory are completely similar in terms of kinematic features such as reflection, refraction, and bypassing-radar usually introduces quite mature seismic processing methods as its main processing means, such as digital filtering techniques and offset bypassing processing.

Digital Filtering Technique: In order to keep more reflected wave characteristics in geo-radar monitoring, wide bandwidth is usually utilized for recording, and thus, while various effective waves are recorded, various interfering waves are also recorded. Digital filtering technology is to use the difference of spectral characteristics to suppress the interference wave to highlight the effective wave. Commonly used digital filtering techniques are frequency domain filtering and time domain filtering.

Offset Bypass Processing: Geological radar monitors the reflection of waves from the interface of underground media. The reflection point at the interface of underground media deviating from the measuring point can be recorded as long as its normal plane passes through the measuring point. Offset diffraction processing moves each reflection point in the radar record to its original position. The radar profile with offset processing can reflect the real location of underground media.

Data interpretation method: Ground-penetrating radar data reflect the electrical distribution of the underground medium, and to convert them into the distribution of the geological body, it is necessary to combine the data from geology, drilling, and geo-radar by mail in order to obtain an overall image of the monitoring object concerned.

(4) Detection of underground mining areas

The ground-penetrating radar method is a broad-spectrum (1 MHz–1 GHz) electromagnetic technology used to determine the distribution of underground media, i.e., ground-penetrating radar is the use of high-frequency electromagnetic waves in the form of broadband short pulses, sent from the ground into the ground through the antenna T, and then returned to the ground after being reflected by the underground strata or the destination body to be received by the other antenna R. The pulse wave travels when expressed as

$$t = \frac{\sqrt{(4z^2 + x^2)}}{v} \qquad (10.15)$$

where t is the two-way travel time of electromagnetic wave; z is the depth of the reflector; v is the wave speed in the underground medium; x is the distance between the transmitting and receiving antennas.

When the wave velocity v in the lower medium is known, the depth z of the reflector can be obtained from the above formula according to the exact measured time t value. When the electromagnetic wave propagates in the medium, its electromagnetic wave intensity and waveform will change with the electrical properties and geometry of the medium through which it passes. Therefore, based on the received wave travel time (also known as two-way travel time), amplitude and waveform form data, the structure of the medium can be inferred. Radar graphics are usually recorded in the form of pulse reflected waves, the positive and negative peaks of the waveform are represented in black, white, or gray scale or color. In this way, the in-phase axis can represent the subsurface reflector with equal gray scale and equal color line.

Figure 10.25 is a schematic diagram of radar detection and waveform recording. The waveform recording is drawn against a simple geological model. On the waveform recording diagram, each measuring point records the waveform in the vertical

## 10.4 Acoustic Detection Technology of Rock Anomaly Defects and Loose ...

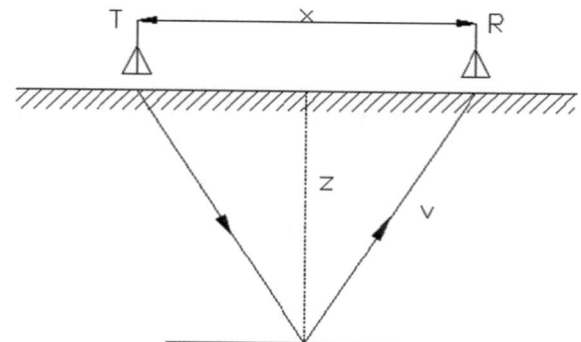

**Fig. 10.25** Radar reflection

direction of the survey line, forming the radar time profile. Because the electromagnetic wave segment emitted by GPR is usually more than $10^7$ order of magnitude, and the radar wavelength in the formation medium is generally 0.1–2 m, GPR has a higher resolution than seismic wave method when detecting shallow formation medium, and has the advantages of economy, non-destructive and fast.

The radar pattern is often recorded in the form of pulsed reflected waveforms, and Fig. 10.26 shows the model of reflected waveform recording. In the waveform recording, each measurement point is recorded in the plumb direction of the measurement line, constituting a radar profile, after post-processing of the data can be obtained from the distribution of different underground media and the location of the dielectric constant change surface and other parameters.

Electromagnetic wave propagation in the medium its path, electromagnetic field strength and waveform will change with the electrical properties and geometry of the medium through which it passes. Therefore, according to the travel time of the received wave (also known as two-way travel time), amplitude and waveform information, can be inferred from the structure of the medium. When there is an empty or collapsed area under the ground, the electromagnetic wave forms a strong reflection wave on both sides of the interface between the rock body and the empty area, and at the same time, the high-frequency pulsed electromagnetic wave propagates through the mass of the underground consumptive medium with attenuation characteristics, and by analyzing the attenuation behavior and characteristics of the high-frequency pulsed electromagnetic wave in the underground consumptive medium, the differences in attenuation and dispersion characteristics of the different lithologies and rock structures can be utilized to provide a good basis for identifying and This characteristic can be utilized to provide a basis for identifying and deducing the nature of rocks and the internal structure of the rock body.

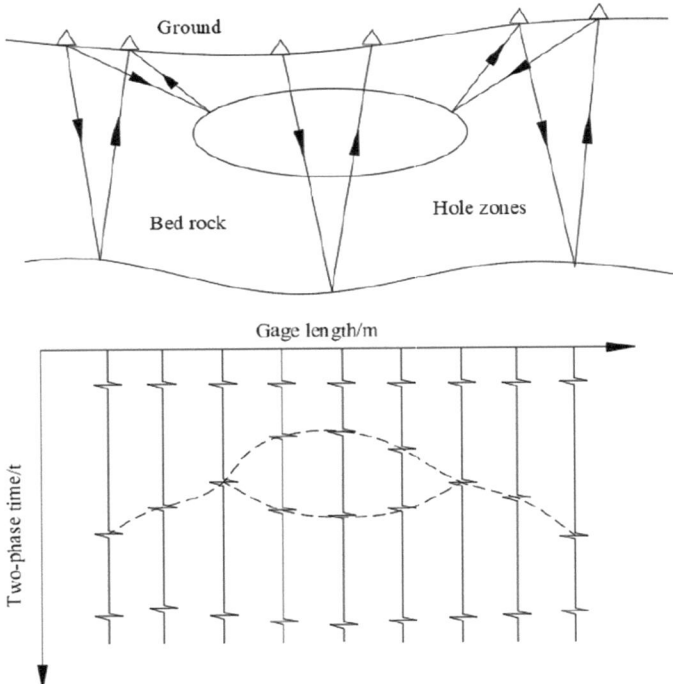

**Fig. 10.26** Schematic of radar detection airspace profile

## 10.4.4 Principles of Elastic Wave CT Technology

(1) Overview of CT technology

As early as 1917, Austrian mathematician J. Radon's famous paper "On Determination of a Function by its integral along certain manifold forms" (i.e. Randon transformation) was instrumental in the formation and development of CT technology. CT technology refers to the use of a certain ray source, according to the projection data obtained by the external detection equipment of the object, according to certain physical and mathematical relations, the use of computer special software system to invert the distribution function of some unknown physical quantity inside the object, generate two-dimensional and three-dimensional images, and reproduce the internal characteristics of the object. The so-called CT technology is to invert the internal image of the object according to the projection data, so it can also be called image reconstruction.

(2) Classification of elastic wave CT technology

According to the detection equipment and signal frequency, the elastic wave CT can be divided into acoustic wave CT and seismic wave CT, the principle of the

## 10.4 Acoustic Detection Technology of Rock Anomaly Defects and Loose ...

two imaging methods is exactly the same, the difference is that the former has the advantages of high precision, fast and simple testing, but the energy is small for short-distance transmittance, which is mainly used for concrete quality testing and rock classification; the latter reads with lower precision, but the signal energy is very large for long-range transmittance, which is generally used for the detection of large geological formations by inter-hole transmission method. The latter has a lower reading accuracy, but the signal energy is large enough for long-distance transmission, and is generally used for the detection of large geological formations using the inter-hole transmission method.

Based on the inversion physical property parameters, elastic waves can be subdivided into wave velocity CT and absorption coefficient CT, and the most commonly used imaging method is the level expansion method based on ray theory. In the level expansion method, the forward and inverse processes need to determine the ray path matrix. The key to determining the ray path matrix is how to quickly calculate the distance of each ray through each cell and trace the first wave propagation path and travel time. The simplest method is to approximate the wave propagation path as a straight line, which is characterized by simple calculation, and when the observation system is determined, the ray path matrix is also fixed. However, since the wave speeds at each point on the imaging profile are not equal, especially when the wave speed (i.e., the wave impedance at the anomaly interface) varies greatly between points, the first wave propagation path will tend to undergo a large bending propagation tendency. Therefore, fast tracing of bending rays is one of the key techniques to improve the accuracy of CT imaging and realize bending ray tomography.

Ray tracing technology

The shortest path algorithm is the basic content of computational mathematics "graph theory, the essence of which is that if certain discrete points and the distance between the points are known, the optimal calculation method and computational techniques are used to find the shortest distance from the starting point to the end point. Obviously, if the weight between the points is the required travel time, the result is the minimum travel time path. To distinguish it from the shortest path, it is called the optimal path. There are methods to calculate the shortest path for specific problems, among which the algorithm proposed by Dijkstra to generate the shortest path in non-decreasing order according to the length is particularly suitable for fast curved ray tracing, which is one of the most commonly used and effective methods for curved ray CT, and this method has the advantages of simple programming, high computational efficiency and wide range of applications.

Radiation short-wave CT

Elastic wave velocity CT based on ray theory can be subdivided into straight ray CT and bending ray CT, the only difference between the two is the different ray tracing methods. Since straight ray CT uses a straight line to approximate the wave velocity, the imaging accuracy is lower than that of bent ray CT, but the imaging speed is faster. therefore, the most commonly used method is: firstly, straight ray CT is used to image the profile quickly; then based on this result, curved ray CT is used to image the measured object with high accuracy.

Attenuation factor CT

It is now gradually recognized that, on the basis of elastic wave velocity imaging, it is also possible to reconstruct the distribution of the attenuation properties of the object under test by using the attenuation information reflected in the waveform. Under many conditions, the attenuation of elastic waves is much more sensitive to the nature of defects than the wave speed, so attenuation tomography is also an important means to fully understand the internal structure of the object. The elastic wave attenuation mechanism is an extremely complex process, and the elastic wave attenuation characteristics can be described by the attenuation coefficient (i.e., absorption coefficient) or quality factor. Elastic wave attenuation imaging can be carried out in the time domain, the most commonly used methods are amplitude attenuation method, rise time method, spectral ratio method and spectral center-of-mass offset method. The spectral center-of-mass offset method is a relatively new method, which has good prospects for application as it performs more consistently in general compared to the other methods.

### 10.4.5 Acoustic Monitoring Technology Use Case Profile

(1) Arrangement of acoustic integrated monitoring system in the mining area

The theory of active acoustic wave method and passive acoustic emission method can be used for comprehensive monitoring and early warning of regional and dynamic characteristics of mining face or roadway. Acoustic wave transmission method is used to detect the stress state of coal and rock mass in the coal seam plane, and then combined with the information monitored by acoustic emission, jointly determine the dangerous area of coal and rock impact tendency. Figure 10.27 shows the case of equipment layout of acoustic integrated monitoring system in mining area.

The system takes KJ series mine environment and safety monitoring system as the communication platform, uses acoustic wave monitor and acoustic emission monitor as the sensors to connect to KJ series sub-station, transmits the acoustic wave transmitter and receiver and acoustic emission monitor to the computer monitoring center through KJ series monitoring system, collects and analyzes and forecasts the acoustic wave and acoustic emission signals of the coal-rock dynamics hazards through monitoring of the terminals and the data processing of the related analyzing software, and realizes overall monitoring in a wide range and continuous monitoring for a specific hazardous area. Through the monitoring and related analysis software data processing in the terminal, the acoustic wave and acoustic emission signals received from the coal-rock power disaster are collected and analyzed for forecasting, which can realize the overall monitoring within a large range and the continuous monitoring for a specific dangerous area. Thus, it can provide early warning hints in

## 10.4 Acoustic Detection Technology of Rock Anomaly Defects and Loose …

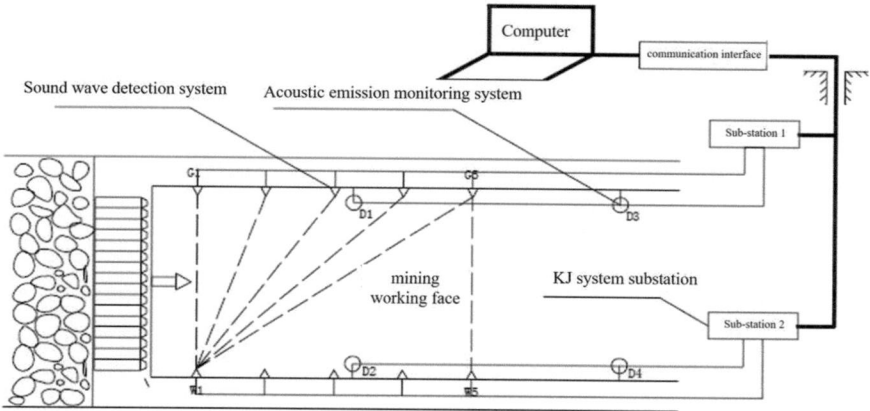

**Fig. 10.27** Schematic diagram of the integrated acoustic monitoring system in the extraction zone

combination with the site conditions and the pattern of mine pressure manifestation, and prompt the adoption of preventive and control measures when the degree of danger increases, so as to achieve the purpose of guaranteeing the safety of the mining personnel and safeguarding the continuous production of the mine.

(2) Sonic perforation method to monitor at depth the rock fragmentation zone

In environmental geological investigation, it is necessary to investigate the stability of rock strata. The former Research Institute of Methods and Technology of the Ministry of Geology and Mineral Resources (Qingzeng Wu, Jianshe Zhan), has undertaken drilling of more than 300 m deep strata, acoustic hole measurement (well), aimed at the exploration and analysis of the fault fracture zone of weak structure, and make geological and physical characteristics evaluation. Using the self-developed "one T double R" transducer and automatic acoustic monitoring instrument (i.e. ultrasonic monitoring instrument), the test results are very good. Figure 10.28 shows the test record of one of the boreholes at 90.6–166.5 m. The geological condition of this section is complicated, but the reaction to the fracture zone is very obvious. It can also be seen that in the 140.5–147.61 m section, the soft rock formation structure that cannot be divided by drilling coring can be clearly responded by acoustic logging, while in the relatively complete mixed rock formation of 91.6–124.21 m, the sound time is almost unchanged, and the sound velocity is 4140 m/s. Acoustic monitoring (inspection) fully improves the drilling coring data, and provides technical guarantee for objectively evaluating the stability of rock mass.

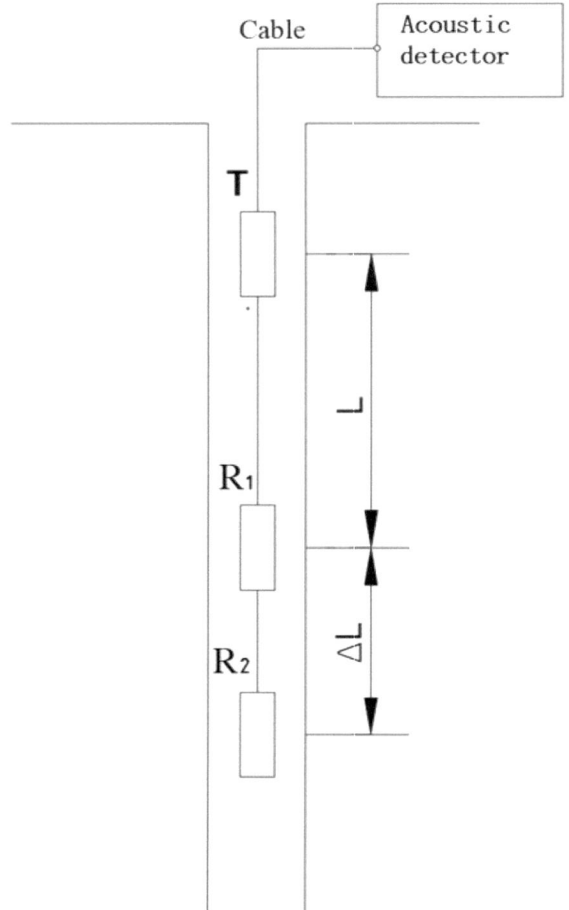

(a) Acoustic monitoring method of drilling (deep) into rock formations

**Fig. 10.28** Schematic diagram depicting the acoustic sounding method and results of the rock analysis

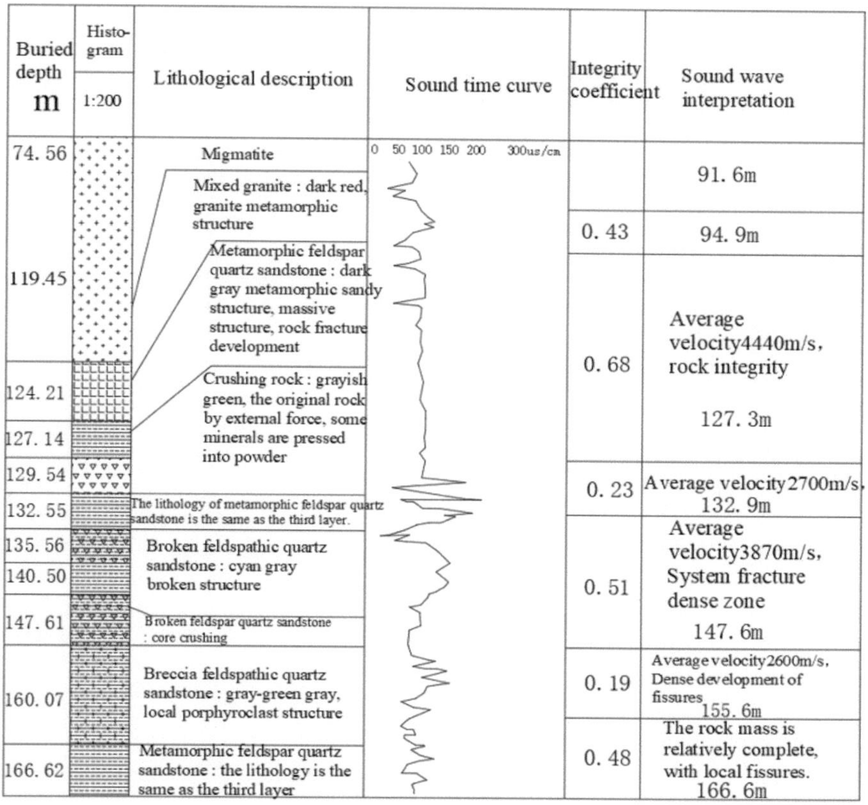

10.4 Acoustic Detection Technology of Rock Anomaly Defects and Loose …

| Buried depth m | Histogram 1:200 | Lithological description | Sound time curve 0 50 100 150 200 300us/cm | Integrity coefficient | Sound wave interpretation |
|---|---|---|---|---|---|
| 74.56 | | Migmatite | | | 91.6m |
| 119.45 | | Mixed granite : dark red, granite metamorphic structure | | 0.43 | 94.9m |
| 124.21 | | Metamorphic feldspar quartz sandstone : dark gray metamorphic sandy structure, massive structure, rock fracture development | | 0.68 | Average velocity 4440m/s, rock integrity 127.3m |
| 127.14 | | Crushing rock : grayish green, the original rock by external force, some minerals are pressed into powder | | | |
| 129.54 | | | | 0.23 | Average velocity 2700m/s, 132.9m |
| 132.55 | | The lithology of metamorphic feldspar quartz sandstone is the same as the third layer. | | | Average velocity 3870m/s, System fracture dense zone 147.6m |
| 135.56 | | Broken feldspathic quartz sandstone : cyan gray broken structure | | 0.51 | |
| 140.50 | | | | | |
| 147.61 | | Broken feldspar quartz sandstone : core crushing | | | |
| 160.07 | | Breccia feldspathic quartz sandstone : gray-green gray, local porphyroclast structure | | 0.19 | Average velocity 2600m/s, Dense development of fissures 155.6m |
| 166.62 | | Metamorphic feldspar quartz sandstone : the lithology is the same as the third layer | | 0.48 | The rock mass is relatively complete, with local fissures. 166.6m |

(b) Acoustic evaluation method of rock stability

**Fig. 10.28** (continued)

**Open Access** This chapter is licensed under the terms of the Creative Commons Attribution-NonCommercial-NoDerivatives 4.0 International License (http://creativecommons.org/licenses/by-nc-nd/4.0/), which permits any noncommercial use, sharing, distribution and reproduction in any medium or format, as long as you give appropriate credit to the original author(s) and the source, provide a link to the Creative Commons license and indicate if you modified the licensed material. You do not have permission under this license to share adapted material derived from this chapter or parts of it.

The images or other third party material in this chapter are included in the chapter's Creative Commons license, unless indicated otherwise in a credit line to the material. If material is not included in the chapter's Creative Commons license and your intended use is not permitted by statutory regulation or exceeds the permitted use, you will need to obtain permission directly from the copyright holder.

# References

1. Brady BHG, Brown ET (1990) Rock mechanics of underground mining. Coal Industry Press, Beijing
2. Borbely T, Kuso O, Jensner B (1994) Acoustics of pore media. Petroleum Industry Press, Beijing
3. Zou D, Duan W (2008) Transmission and monitoring method of elastic stress wave in coal rock. Coal Sci Technol (02):88–91
4. Zou D, Song J, Xu P et al (2007) Development and application of impact ground pressure prediction system based on stress method. Coal Sci Technol (4)
5. Zou D, Jiang F (2004) Research on the mechanism and prediction method of stored energy and impact pressure in coal rock body. J Coal Sci (2)
6. Zou D, Liu X (2002) Unified prediction and prevention technology of impact pressure and protrusion. Min Res Dev (1)
7. Chen G, Qian M (1994) Control of peripheral rocks in Chinese coal mine quarries. China University of Mining and Technology Press, Xuzhou
8. Feng S (1986) Rock mechanics of underground mining. Coal Industry Press, Beijing
9. Xu B (1992) Contact mechanics. Higher Education Press, Beijing
10. Kang H (1993) Mechanism and prevention of soft rock tunnel undercutting. Coal Industry Press, Beijing
11. He M (2010) Basic research and application of deep rock mechanics. Science Press, Beijing
12. Petukhov (1980) Impact ground pressure in coal mines. Coal Industry Press, Beijing, pp 2–11
13. Zhang S, Ren Z et al (1996) Ideas and programs of the "mesoscale earthquake prediction experimental field", an observation and application study of the Mentougou coal mine earthquake. J Seismol 18(4):529–537
14. Zhao B, Teng X (1995) Impact ground pressure and its prevention. Coal Industry Publishing, Beijing, pp 54–62
15. Song Z (1988) Practical mine pressure control. China University of Mining and Technology Press, Xuzhou
16. Qian M, Shi P, Xu J (2010) Mine pressure and rock control. China University of Mining and Technology Press, Xuzhou
17. Buhoyno (1985) Mine pressure and impact ground pressure. Coal Industry Press, Beijing, pp 6–18
18. Pan L, Zhang L, Liu X (2006) Impact ground pressure prediction and prevention of practical technology. China University of Mining and Technology Press, Xuzhou
19. Qi Q, Do L (2008) Impact ground pressure theory and technology. China University of Mining and Technology Press, Xuzhou, pp 2–8
20. Dou L, Gong S, Liu P, Ding E, Huagang, He J, Liu P (2015) Remote online early warning platform for mining shock disaster. Coal Sci Technol 43(06):48–53

21. Dou Linming, Zhao C, Yang S, Wu X (2006) Prevention and control of impact ground pressure disasters in coal mining. China University of Mining and Technology Press, Xuzhou
22. Yang Y, Wang D, Guo M, Li B (2014) Research on rock damage characterization based on triaxial compression acoustic emission test. J Rock Mech Eng 33(01):98–104
23. Zhou Y, Wu S, Xu X, Sun W, Zhang X (2013) Particle flow analysis of acoustic emission characteristics during rock fracture. J Rock Mech and Eng 32(05):951–959
24. Qihu Q (2014) Definition, mechanism, classification of rock burst and impact ground pressure and their quantitative prediction models. Geotech Mech 35(01):1–6
25. Weimert WR, Li D (1975) New techniques and ideas on ground pressure control by the U.S. Burea of Mines. Metall Saf (1):60–67
26. Ma T, Tang C, Tang L, Zhang W, Wang L (2016) Research on rock burst prediction mechanism based on microseismical monitoring technology. J Rock Mech Eng 35(03):470–483
27. Li Z-H, Do L-M, Guan X-Q et al (2009) Zonal monitoring method and application of mine earthquake precursors. J Coal Sci 34(5):614–618
28. Li N, Wang E, Ge M (2017) Microseismical monitoring technology and its application in coal mines status and prospect. J Coal 42(S1):83–96
29. Ministry of Coal Industry (1987) Provisional regulations on safe mining of coal seams under impact pressure. Coal Industry Press, Beijing
30. Zou D, Jiang F (2004) Research on the mechanism and prediction method of stored energy and impact pressure in coal rock body. J Coal Sci (2):159–163
31. Ge MC (2005) Efficient mine microseismic monitoring. Int J Coal Geol (1):44–56
32. Li N, Ge M, Wang E (2014) Two types of multiple solutions for micro seismic source location based on arrival-time-difference approach. Nat Hazards 2:829–847
33. Liu Y, Tian Y, Feng S, Zheng Z, Chi R (2013) A review of micro seismic technology and application. Adv Geophys 28(04):1801–1808
34. Pan Y, Li Z, Zhang M (2003) Distribution, type, mechanism and prevention of impact ground pressure in China. J Rock Mech Eng (11):1844–1851
35. Bieniawski AT, Dellkhaus Hqvogler UW (1969) Failure of fracture rock. Int J Roek Mech Min Sei 6:323–341
36. Zou DY, Duan W, Liu ZG, Shi H, Liu YL (2010) Coal rock body vibration signal monitor for mining: China, 200920226992.4, 2010-06-02
37. Zou DY, Duan W, Liu ZG, Shi H, Liu YL (2011) Coal rock body elastic stress wave sensing device: China, 200910019022.1, 2011-01-26
38. Zou DY, Duan W, Liu ZG, Shi H, Liu YL (2010) Coal rock body vibration signal sensor for mining: China, 200920226993.9, 2010-06-02
39. Zou D, Wu Y, Shang W (2004) Discussion on the mechanism of energy storage and impact mineral pressure in coal rock body. Mine Pressure Roof Manag (1):91–93
40. Tan Y et al (2011) Damage and control of peripheral rock in deep roadway. Coal Industry Press, Beijing, pp 281–288
41. Xie H, Peng S, He M (2006) Basic theory and engineering practice of deep mining. Science Press, Beijing
42. Brady BHG, Brown ET (2005) Rock mechanics for underground mining
43. Hirata, Kameoka Y, Hirano T (2007) Safety management based on detection of possible rock Bursts by AE monitoring during tunnel excavation. Rock Mech Rock Eng (6):563–576
44. Vogel M, Andrast HP (2000) Alp transit-safety in construction as a challenge health and safety aspects in very deep tunnel construction. Tunneling Undergr Space Technol 15(4):481–484
45. Urbancic TI, Trifu C-I (2000) Recent advances in seismic monitoring technology at Canadian mines. J Appl Geophys (4):225–237
46. Xie H (2019) Progress of research on deep rock mechanics and mining theory. J Coal 44(05):1283–1305
47. Hu S, Wang E, Li Z, Shen R, Liu J (2014) Dynamic nonlinear characteristics of electromagnetic radiation from loaded coal body. J China Univ Min Technol (03):380–387
48. Liu J (2015) Characterization of electromagnetic radiation signals in coal-rock power disasters. Chin J Saf Sci 25(12):105–110

# References

49. Cai M (2000) Principles and techniques of geo stress measurements. Science Press, Beijing
50. Tan Y, Liu C (1999) Prediction and control of tunnel rock stability. China University of Mining and Technology Press, Xuzhou
51. Cai M, Brown ET (2017) Challenges in the extraction and utilization of deep mineral resources. Engineering 3(04):9–12
52. Kang HP, Lin J (2007) Geomechanical tests and their applications in rock anchorage design. In: Proceedings of the 11th congress of international society for rock mechanics, Lisbon, Portugal, pp, 303–305
53. Yuan C, Wang W, Feng T, Yu W, Wu H, Peng W (2017) Research on the principle of roadway perimeter rock control based on plastic zone extension. J Min Saf Eng 34(06):1051–1059
54. Hu C (2017) Research on key technology of deep roadway perimeter rock control. J China Univ Min Technol 46(05):970–978
55. Yu X, Zheng Y, Liu H, Fang Z (1983) Stability analysis of underground engineering surrounding rocks. Coal Industry Press, Beijing
56. Jiang JQ (1994) Pressure control of peripheral rock support in mining field. Coal Industry Press, Beijing
57. Zhu W, He M (1995) Stability of surrounding rock and dynamic construction mechanics of rock mass under complex conditions, Science Press, Beijing
58. Shen M, Chen J (eds) (2006) Rock mechanics. Tongji University Press, Shanghai
59. He M (2005) Conceptual system and engineering evaluation indexes of deep. J Rock Mech Eng (16):2854–2858
60. Cao M, Sun H, Dai L, Sun D, Wang B, Miao F (2018) Simulation study on the dynamic effect of coal and gas protrusion. J China Univ Min Technol 47(01):113–120+154
61. Malan DF (1999) Time-dependent behavior of deep level tabular excavations in hard rock. Rock Mech Rock Eng 2:123–155
62. Liu ZG (2011) Research and application of impact ground pressure monitoring equipment based on acoustic emission principle. Shandong University of Science and Technology, Qingdao
63. Wang G, Wu M, Wang H, Huang Q, Zhong Y (2015) Sensitivity analysis of factors affecting coal and gas protrusion based on energy balance model. J Rock Mech Eng 34(02):238–248
64. Wang H, Zhang Q, Yuan L, Xue J, Li Q, Zhang B (2015) Coal and gas protrusion simulation system based on CSIRO model with experimental application. J Rock Mech Eng 34(11):2301–2308
65. Fu H, Li H, Lu W, Xu Y, Wang Y (2016) An improved limit learning machine coal and gas protrusion prediction model. J Sens Technol 29(01):69–74
66. Pan Y, Zhang Y, Wang Z (2008) Application of mutation theory in dynamic instability of rock system. Science Press, Beijing
67. Song Z (2019) Discussion on the current situation of the development of mining engineering discipline in China and its deep-rooted development problems. Tunneling Undergr Eng Disaster Prev Control 1(02):7–12
68. Song Z, Hao J, Shi Y, Tang J, Liu J (2019) Connotation and development of practical mine pressure control theory. J Shandong Univ Sci Technol (Nat Sci Ed) 38(01):1–15
69. Wen Z, Jing S, Song Z, Jiang Y, Tang J, Zhao R, Xiao Q, Zhang T, Wang H, Zhao H, Sun G, Zhang T, Kong C (2019) Research on the spatial structure model of mining field and related dynamic disaster control. Coal Sci Technol 47(01):52–61
70. Wu Q (2014) Research progress, problems and prospects of mine water prevention and control and resource utilization in China J Coal 39(05):795–805
71. Zhang W, Zhang G, Li W, Hua X (2013) Fisher's discriminant analysis model for the risk of water breakout in coal bed floor. J Coal 38(10):1831–1836
72. Gao Y (1991) Research on rock body strength theory and deformation damage law of quarry floor. China University of Mining and Technology, Beijing
73. Guo W, Chen S, Chang X, Zhu X (2012) Research on deformation and evolution law of overburden rock body in deep mining. Coal Industry Press, Beijing

74. Wen G, Li J, Zou Y, Lv G (2011) A preliminary study on the applicable conditions for acoustic emission monitoring of coal-rock dynamic hazards in mines. Coal J 36(02):278–282
75. Zhao K, Yan D, Zhong C, Zhi X, Wang X, Xiong X (2012) Comprehensive analysis method and experimental validation of ground stress by acoustic emission measurement. J Geotech Eng 34(08):1403–1411
76. He M (1996) Theory and practice of soft rock engineering mechanics. China University of Mining and Technology Press, Xuzhou
77. Tan Y, Wu S, Yin Z, Ning J (2007) Mine pressure and rock control. Coal Industry Press, Beijing
78. Fan KK, Zhai DY (2004) Analysis of destabilization and non-symmetric control mechanism of weak structural damage in roadway enclosure. Coal Industry Press, Beijing
79. Yang YJ, Zhou G, Shou Y, Li et al (2010) Geostress measurement and supporting technology of comprehensive tunneling along the void. Coal Industry Press, Beijing
80. Lin W (2008) Non-destructive testing technology of civil engineering quality. China Electric Power Press, Beijing
81. Zhao K, Yuan H (2009) Mine ground pressure monitoring. Chemical Industry Press, Beijing
82. Li S (2010) Discussion on the control theory of the surrounding rock in mine back-mining roadway. J Coal 35(11):1842–1853
83. Miao F, Sun D, Hu C (2013) Formation mechanism of shock wave in coal and gas herniation. J Coal Sci 38(03):367–372
84. Shao T-B, Ji S-S (2015) Earthquake-induced mechanisms in subduction zones: a review of research progress. Geol Rev 61(02):245–268
85. Zhang S, Xuan X, Zhou J, Liu G, Jiang X, Teng J (2017) Characterization of shale acoustic emission and energy distribution based on Brazilian splitting test. J Coal 42(S2):346–353
86. Wang SY (2012) Numerical simulation of the failure mechanism of circular tunnels in transversely isotropic rock masses. Undergr Space Technol 32:231–244
87. Jiang Y, Xian X, Yang C (2008) Prediction of destabilization zone of creep fracture in roadway rock. J Geotech Eng (06):906–910
88. Li Y, Pan Y, Zhang M (2006) Study on the time effect of fracturing process in deep rock zones. China J Geol Hazards Prev (04):119–122
89. She S, Lin P (2014) Some progress and challenges of rock engineering in China. J Rock Mech Eng 33(03):433–457
90. Hou C (2017) Research on key technology of deep roadway perimeter rock control. J China Univ Min Technol 46(05):970–978
91. Zhu W, Niu L, Li S, Li S (2019) Research on rock creep-impact test-current status and outlook. J Min Rock Control Eng 1(02):77–87
92. Moosavi M, Grayeli R (2006) A model for cable bolt-rock mass interaction: integration with discontinuous deformation analysis (DDA) algorithm. Int J Rock Mech Min Sci 43:661–670
93. Wang W, Yu C, Yu W, Wu H, Peng W, Peng G, Liu X, Dong E (2016) Research on stability control method of surrounding rock in deep large deformation roadway. J Coal 41(12):2921–2931
94. Wang Y, Ren W (2007) Creep damage characteristics and optimal support time of weak surrounding rock. China Railway Sci (01):50–55
95. Sellers EJ, Klerck P (2000) Modeling of the effect of discontinuities on the extent of the cracking zone surrounding deep tunnels. Tunneling Undergr Space Technol 15(4):463–469
96. Zhang H, Song W, Fu J (2014) Critical parameters and stability analysis of roof instability in large-span open area. J Min Saf Eng 31(01):66–71
97. Zou D, Zhang G, Xin C, Zhang M (2009) Characterization and prediction method of large-area roof pressure dynamics. Coal Sci Technol 37(06):53–56
98. Zhong X (1994) Mutation mode of roof deformation and destabilization in longwall face. J Xiangtan Min Coll 1994(02):1–6
99. Lin HL (2016) Research on the stability of shallow buried large-scale mining area based on thin plate theory. Coal Min 21(03):28–30+105

# References

100. Han H, Wang X, Xu J, Wu Y, Ji Y (2018) Study on the motion characteristics and "re-stabilization" conditions after the destabilization of overburden key layer structure. J Min Saf Eng 35(04):734–741
101. Liu C, Qin M, Peng Y (2014) Study on time-course characteristics of air shock wave disaster induced by large-area roof instability. China Prod Saf Sci Technol 10(05):49–55
102. Zhong X (1996) Elastic stability analysis of hard roof slabs in quarries. Coal, 1996(04):15–17+44
103. Liu R, Li W, Li J (2012) Classification of hazardous sources of destabilization in large area of coal mine. Coal Mine Saf 43(05):143–146
104. Wu L, Wang J (1994) Study on the deformation mode of rock formation controlled by continuous large-area mining pallets. J Coal 1994(03):233–242
105. Song X-X, Lian Q-W, Xing P-W, Fu Y-P, Li Z-J (2009) Study on the air impact hazard of large-scale roof collapse in an air-mining area. Coal Sci Technol 37(04):1–4+81
106. Xie X-B, Deng R-N, Dong X-J, Yan Z-Z (2018) Stability of the mining zone cluster system based on mutation and rheology theory. Geotechnics 39(06):1963–1972
107. HUANG Qingxiang, XIA Xiaogang. Study on "four zones" delineation of mining rock formation and surface movement[J]. Journal of Mining and Safety Engineering,2016,33(03):393–397.
108. Wang S, Wang D, Cao K, Wang S, Pi Z (2014) Three-dimensional distribution law of void ratio in the air-mining zone and overlying rock layer. J Cent South Univ (Nat Sci Ed) 45(03):833–839
109. Wang C (2011) Comprehensive evaluation model of impact ground pressure risk based on unconfirmed measurement theory and application research. China University of Mining and Technology
110. Liu J, Zhai M, Guo X, Jiang F, Sun G Zhang Z (2014) Theory and application of joint monitoring of impact ground pressure by vibration field and stress field. J Coal 39(02):353–363
111. Jiang Y, Zhao Y (2015) Research status of impact ground pressure in coal mines in China: mechanism, early warning and control. J Rock Mech Eng 34(11):2188–2204
112. Qi Q, Ou YZ, Zhao S, Li H, Li X, Zhang N (2014). Research on the types and prevention methods of impact ground pressure mines in China. Coal Sci Technol 42(10):1–5
113. Fujii Y, Ishijima Y, Deguchi G (1997) Prediction of coal face rock bursts and micro seismicity in deep longwall. Coal Min Rock Mech Min Sci 1997(1)
114. Pan Y (2016) Research on the integration of composite dynamic hazards of coal and gas protrusion and impact ground pressure. J Coal 41(01):105–112
115. Zhang HW, Zhu F, Han J, Huo BJ, Rong H, Tang GS (2016) Geodynamic conditions and monitoring and prediction methods of impact ground pressure. J Coal 41(03):545–551
116. Pan J (2016) Shock initiation mechanism of impact ground pressure and its application. Research Academy of Coal Science
117. Lv J, Jiang Y, Zhao Y, Zhu J, Gao F (2013) Research and application of impact ground pressure hierarchical monitoring and its early warning method. J Coal 38(07):1161–1167
118. Zhao Z, Ma N, Guo X, Zhao X, Xia Y, Ma Z (2016) Conjecture on the occurrence mechanism of butterfly impact ground pressure in coal bed roadway. J Coal 41(11):2689–2697
119. Impact ground pressure theory and technology. China University of Mining and Technology Press, Qi Qingxin, 2008
120. Manchao HE, Jiong WANG, Xiaoming SUN, Xiaojie YANG (2014) Mechanical properties of negative Poisson's ratio effect anchor cable and its application in impact ground pressure prevention and control. J Coal 39(02):214–221
121. Yan J, Zhang X, Zhang Z (2013) Discussion on the geological control mechanism of coal and gas protrusion. J Coal 38(07):1174–1178
122. Frid V (2013) Electromagnetic radiation method for rock and gas outburst forecast 1997(2):97–104. Xu J, Liu D, Peng S, Zhou W, Cheng M (2013) Simulation experimental study of coal and gas protrusion under different protrusion caliber conditions. J Coal Sci 38(01):9–14

123. Wang G, Wu M, Chen W, Chen J, Du W (2015) Analysis of coal and gas protrusion energy conditions and protrusion intensity influencing factors. Geotechnics 36(10):2974–2982
124. Li X, Lin B (2010) Current status and analysis of the mechanism of coal and gas outburst. Coalfield Geol Explor 38(01):7–13
125. Ou J (2012) Experimental study on simulation of coal and gas protrusion evolution process. China University of Mining and Technology
126. Fu JH, Cheng YP (2007) Current situation and countermeasures for prevention and control of coal and gas outbursts in Chinese coal mines. J Min Saf Eng 03:253–259
127. Sun Y, Tan C, Sun W, Wang R, Wu S, Wang X, Chen Q (2008) Current status and research direction of coal gas protrusion. J Geomech (02):117–134
128. Tang J, Yang L, Wang Y, Lu J (2014) Deep coal and gas protrusion test under geo stress and gas pressure. Geotechnics 35(10):2769–2774
129. Liu YW (2011) Research on the law, mechanism and kinetic modeling of coal particle gas discharge. Henan University of Science and Technology
130. Ou J, Wang E, Ma G, Wang C, Song D, Chen P, Li N (2012) Evolution of coal body rupture during coal-gas protrusion. J Coal Sci 37(06):978–983
131. Xie X, Feng T, Wang Y, Huang S (2010) Dynamic energy balance during coal-gas protrusion. J Coal Sci 35(07):1120–1124
132. Liu J (2014) Experimental study on the evolution process and mechanism of coal and gas pressure outflow dynamics. China University of Mining and Technology
133. Wang Q, Ju J (2014) Research on the law of mine pressure manifestation in 450 m super-long comprehensive mining face. Coal Sci Technol 42(03):125–128
134. Liang Y, Li B, Yuan Y, Zou Q, Jia L (2017) Key layer movement type of large-height general mining quarry and its influence on ore pressure in working face. J Coal 42(06):1380–1391
135. Hu G, Jin Z (2006) Observation of mine pressure in large-height synthesized mining face and its manifestation law. J Taiyuan Univ Technol (02):127–130
136. Yang D, Chen Z, Sun J, Wang L, Gao Q (2016) Mechanical modeling of cracked plate for roof collapse in large mining height longwall working face. J Southeast Univ (Nat Sci Ed) 46(S1):210–216
137. Kang H, Xu G, Wang B, Wu C, Jiang P, Pan J, Ren H, Zhang Y, Pang Y (2019) Development of coal mining and rock control technology in China 40a and prospect. J Min Rock Control Eng 1(02):7–39
138. Zhou H, Dong Y, Cai H (2007) Measurement of mine pressure in large-height lengthening synthesized mining face. Coal Technol (10):50–52
139. Zhang LH, Li NN (2017) Research on the law of mine pressure manifestation in 8 m large height synthesized mining face. Coal Sci Technol 45(11):21–26+44
140. Yang D, Zhang L, Chai M, Li B, Bai Y (2016) Research on the breakage law of roof plate in comprehensive mining of extra-thick coal seam based on fracture mechanics. Geotechnics 37(07):2033–2039
141. Zhang ZZ (2013) Energy evolution mechanism during rock deformation and destruction. China University of Mining and Technology
142. Zhang X (2013) Energy zoning of coal mining rock body and analysis of power disaster prevention and control. J Anhui Univ Technol (Nat Sci Ed) 33(03):24–29
143. Xue D, Zhou H, Zhong J, Huang Y (2014) Research on energy release and catastrophic mechanism of mining rock body. J Rock Mech Eng 33(S2):3865–3872
144. Guo C (2017) Research on the relationship between coal rock energy characteristics and impact ground pressure strength. Liaoning University of Engineering and Technology
145. Rong H, Zhang H, Liang B, Han J, Wang Y (2017) Instability mechanism of coal rock dynamic system. J Coal 42(07):1663–1671
146. Xie H, Gao F, Ju Y (2015) Research and exploration of deep rock mechanics. J Rock Mech Eng 34(11):2161–2178
147. Shi H, Ma N, Xu H (2019) Discussion on the mechanism of coal and gas protrusion based on energy theory. China Saf Prod Sci Technol 15(01):88–92

# References

148. Sun H-F, Yang Y-M, Ju Y, Zhang Q-G, Peng R-D (2014) Numerical analysis of deformation damage and energy release of coal rock under excavation unloading condition. J Coal 39(02):258–272
149. Song D (2012) Research on the evolution process and energy dissipation characteristics of impact ground pressure. China University of Mining and Technology
150. Xiao F, Liu G, Shen Z, Zhang F, Wang Y (2016) Energy conversion law and acoustic emission change characteristics of coal samples under cyclic loading. J Rock Mech Eng 35(10):1954–1964
151. Luo J (2016) Research on energy source and energy dissipation mechanism of coal and gas protrusion. Chongqing University
152. Li N, Wang E, Ge M (2017) Micro seismic monitoring technology and its application in coal mines status and prospect. J Coal 42(S1):83–96
153. He X, Do L, Mou Z, Gong S, Cao A, He J (2014) Theory and technology of continuous monitoring and early warning of coal-rock impact power disaster. J Coal 39(08):1485–1491
154. Xiao X, Ding X, Pan Y, Wang X, Xu J, Luo H, Zhao X (2015) Acoustic emission and charge induction test of granular coal rock rupture process. J Coal 40(08):1796–1804
155. Xu WQ (2012) Research on monitoring technology and application of surrounding rock stress in mining space. China University of Mining and Technology
156. Wang E, Xu W, He X, Shen R, Kong X (2017) Development and application of coal rock body stress dynamic monitoring system. J Rock Mech Eng 36(S2):3935–3942
157. Lv J, Jiang Y, Zhao Y, Zhu J, Gao F (2013) Research and application of impact ground pressure hierarchical monitoring and its early warning method. J Coal 38(07):1161–1167
158. Li C, Sun X, Gao T, Xie B, Xu X (2015) Vibration damage test and micro seismic signal characteristics of coal rock body. J Coal 40(08):1834–1844
159. Cao JT, Lai XP, Cui F, Shan PF, Yang YR (2016) Research on multi-parameter prediction method for structural dynamic instability of complex coal rock body. J Xi'an Univ Sci Technol 36(03):301–307
160. He X, Wang A, Do L, Song D, Zu Z, Li Z (2018) Dynamic monitoring technology of micro seismic region in prominent hazardous coal seam. J of Coal 43(11):3122–3129
161. Zhang Z, Wang Y, Zhao C, Deng Z, Wang C (2011) Application of integrated micro seismic and ground sound monitoring in the prevention and control of impact ground pressure. Coal Sci Technol 39(01):44–47
162. Yang W (2014) Research on the correlation response mechanism and characterization of electromagnetic and microseismical signals of coal rock deformation and rupture. China University of Mining and Technology, Beijing
163. Li Y, Yuan A (2018) Research on acoustic emission characteristic parameters and damage precursor information of coal rock bodies with different lithologies. Coal Eng 50(12):86–89
164. Xia Y, Feng M, Li H (2018) Research on geophysical monitoring method of impact ground pressure. Coal Sci Technol 46(12):54–60
165. Pan Y, Luo H, Zhao Y (2013) Application of charge sensing monitoring technology in mine power disaster. Coal Sci Technol 41(09):29–33+78
166. Deng Z, Qi Q, Zhao S, Ouyang Z, Kong L, Li S (2016) Application of self-seismic microseismical monitoring technology in coal mine power disaster early warning. Coal Sci Technol 44(07):92–96

MIX
Papier aus verantwortungsvollen Quellen
Paper from responsible sources
FSC® C105338

If you have any concerns about our products,
you can contact us on
**ProductSafety@springernature.com**

In case Publisher is established outside the EU,
the EU authorized representative is:
**Springer Nature Customer Service Center GmbH
Europaplatz 3, 69115 Heidelberg, Germany**

Printed by Libri Plureos GmbH
in Hamburg, Germany